Springer Tracts on Transportation and Traffic 1

Series Editor

Roger P. Roess
Polytechnic Institute of New York University
Metrotech Center 6
11201 New York
USA
E-mail: rroess@poly.edu

For further volumes:
http://www.springer.com/series/11059

Roger P. Roess and Gene Sansone

The Wheels That Drove New York

A History of the New York City Transit System

Springer

Authors
Roger P. Roess
Professor of Transportation Engineering
Polytechnic Institute of New York University
New York
USA

Gene Sansone
Chief Mechanical Officer
MTA New York City Transit (Retired)
Adjunct Professor
Polytechnic Institute of New York University
New York
USA

ISSN 2194-8119 e-ISSN 2194-8127
ISBN 978-3-642-30483-5 e-ISBN 978-3-642-30484-2
DOI 10.1007/978-3-642-30484-2
Springer Heidelberg New York Dordrecht London

Library of Congress Control Number: 2012939522

Printed on acid-free paper

Springer is part of Springer Science+Business Media (www.springer.com)

Preface

I was born in New York in 1947, and despite focusing most of my career on traffic engineering, I have always been fascinated by the world of public transportation. When I first started teaching in 1970, my first assignment was to organize a two-course sequence in public transportation planning. At the time, there were no such courses in the U.S. Working with another young faculty member, Martin Huss, we spent a semester learning everything we could on the subject. Our search included a fascinating two-week training program hosted by the Chicago Transit Authority, and the frenetic collection of every possible piece of written material on the subject. As part of the CTA course, we got to drive elevated trains on the famous Chicago Loop, and buses on a training track. After two years, my colleague continued on his own, while I shifted to my then primary interest, traffic engineering. Along the way, I got hooked on the subject of public transportation.

Five years ago, I got the chance to put together another course on the history of the New York City transit system, and began assembling this manuscript. About 18 months ago, I was lucky enough to talk my coauthor, Gene Sansone, into joining the project. From his years in the New York City Transit Authority, Gene knows more about subway cars, their development, and their history than anyone I have ever met. He has provided the invaluable insights as an insider with a great understanding of both the technical issues faced by the NYCTA over the years, and the politics that accompanied them. He has made this labor of love even more fascinating than I could have imagined.

My father always thought that he had lived during the greatest era of technological change in history. He was, in many ways, right. His godfather helped him build an early crystal radio, and he was amazed that it could pick up a single station. By the time he died, he lived to see the era of television: first black and white, then color, and then the plethora of offerings provided by cable services. He was born a little after Kitty Hawk, but remembered Lindberg's flight and the loss of Amelia Earhart. He lived to see Neil Armstrong walk on the face of the moon. He grew up on the lower East Side, and remembered horse-drawn fire engines, cleaning up horse manure at night (it was great fertilizer), and the first car on his street. He drove until he was 86.

Technology is moving, however, at an ever-accelerating rate, and I suspect that people of my age will see even more stunning advances and changes than my father did.

The society we live in has changed rapidly as well. I spent a lot of time on public transportation in my youth, and was allowed to travel alone or with friends when I was only 11-years old. My grandmother took me on a ride on the Queensborough Bridge trolley in 1956 because she thought I ought to ride one before they disappeared into history. She also took me on the Third Avenue El to visit her friends while it still ran in upper Manhattan. Riding the El was a real thrill – there is something about being close enough to people's windows to shake hands with them. Every year, she and my mother took me to see the Christmas Show at Radio City at 7:00 AM. In the 1950s, you could get into the first morning show for 25 cents! We had to take two buses and three trains to get there, and had to leave at 5 AM, but we saw a stage show, a new-release movie, newsreels, and cartoons.

I saw my first baseball game in 1960. It was a World's Series game. My friends and I left Flushing at 5:00 AM, taking a bus and two trains to Yankee Stadium. We rode on IRT Low-V's as we exited the tunnel in the Bronx to a stunning vista of green grass on a sunny autumn day. We stood in line for three hours to purchase bleacher tickets – still sold on game day at the time, which cost $2.00 because it was a World's Series game. We sat in the bleachers for many hours as we waited for the game to start. We talked to Yankee pitchers warming up in the old bullpen, located right next to the bleachers. Ryne Duren provided hot dogs, which he threw from the bullpen, and Bobby Shantz talked to us about the strategy of pitching when you didn't have a fastball anymore. The Yankees beat the Pirates that afternoon behind Whitey Ford (always my favorite Yankee pitcher) 10–0. They won two other games by scores of 12–0 and 16–3, setting every World Series offensive record in the books. They still lost the series 4–3 when Bill Mazeroski hit the only walk-off home run in series history at that time. We survived the postgame crush, but it was 8:00 PM by the time we returned home. Miraculously, no one was worried about us – after all, what could possibly happen to you at a ball park or on the subway!

When I attended Polytechnic Institute of Brooklyn as an undergraduate, I commuted each day from Flushing, taking a bus to Main Street, hopping the No. 7 train to Grand Central, then the Lexington Avenue line downtown to Brooklyn. The mix of old and new rolling stock was amazing, even in 1964. I was stuck in the Great Blackout of 1965, fortunately at a downtown station. I walked uptown and over the 59th Street Bridge with total strangers. By the time I arrived at Queens Plaza, I knew more about them than I did about members of my own family. I caught a bus that took me to another bus that took me to another bus – anyway, I got home at about 4 AM. My mother greeted me at the door with "I kept your dinner warm," not really my first concern at the time. It was during finals week, so with the trains up and running again, I trudged back to Brooklyn a few hours later. On my arrival, they were just posting notices that Con Ed had asked everyone to close – and trudged back home again – hoping that the trains would continue running. This constituted my worst 24 hour commuting experience ever; nothing else has ever come close, but we all survived with no extraordinary effort.

During the 1970s and early 1980s, like many New Yorkers, I avoided the subways as much as possible. The graffiti-covered trains of the era offended me – I had always thought that subway cars were things of beauty, although some clearly more than others. For years, I drove to and from Brooklyn until the traffic and time reliability got to be silly. I use the LIRR these days, and take the subway up to the Institute on days that are too cold or too wet to walk.

The history of transit in New York has strong elements of urban planning and development. Without the transit system, New York could not and would not exist in its current form. The transit system played a major role in the city's dynamic history. This manuscript focuses on the history of the transit system itself, but also on its critical role in building NYC, and its impact on NYC residents and visitors.

Both Mr. Sansone and I are engineers, not historians. In writing this manuscript, we have made extensive use of secondary sources and on-line resources. Historians would rely on primary sources which may take years to search out. Nevertheless, we have been careful in trying to insure that the information presented is accurate, and we have consulted many primary sources when we could gain access to them. The existence of modern electronic archives has made this a lot easier than it was even five years ago when this effort began.

The story of New York City's transit system, and its impact on the City and its residents, is a compelling one which demonstrates the critical links amongst technology, politics, and society. We hope you enjoy the ride!

Acknowledgements

This book could not have been written without the assistance of many people who provided the authors with their expert knowledge and access to materials not readily found in the public domain.

A very special "thank you" goes out to Carey Strumm and Desiree Alden, archivists for the New York Transit Museum. They were invaluable in identifying and helping the authors gain access to rare photographs and other materials related to the transit system. Thanks also to the museum's executive director, Gabrielle Shubert, who met with the authors, reviewed the manuscript, and provided excellent advice on how to proceed.

Special thanks as well to Jana Richman, Director of Libraries at Polytechnic Institute of NYU, who provided expert guidance and assistance in navigating NYU's electronic archives of New York newspapers and other periodicals from the mid-1800s and beyond.

Thanks to Eli Bail and Fred Eisinger of the Motor Bus Society, who provided critical information and photos concerning New York City's bus fleets and services, and to William A. Luke, author of several bus histories published by Iconographics, who provided excellent suggestions for this effort. Thanks also to Anthony Russo, Bus Historian, for excellent suggestions.

Many individuals associated with MTA New York City Transit provided a wide variety of support and assistance, including Brian Cohen, NYCT; Ronald Cuomo, Assistant Chief Officer, Training, NYCT; Paul J. Fleuranges, Senior Director, Corporate and Internal Communications, MTA; Wayne Galante, Assistant Chief Officer, NYC Buses/LI Bus/MTA Bus, System Safety, FTA Compliance and Probationary Employee Programs; Arthur Murphy, NYCT; Charles Seaton, Deputy Director, Media Relations-NYC Transit Operations; and Kenneth Teu, NYCT. All are greatly appreciated.

Contents

Chapter 1
Introduction to the NYC Transit System

This chapter presents a general overview of the New York City transit system as it exists today.

Fig. 1.1 Map of the Brooklyn-Manhattan Transit Corp. (BMT) lines in 1924. The map does not show existing Interborough Rapid Transit (IRT) lines in service at the time.

The sheer size of New York City's massive public transportation system is staggering. The system includes the NYC subways (including elevated and surface lines), the Staten Island Rapid Transit, and a huge bus system. Some fundamental statistics provide a glimpse of the enormity of the package:

<u>New York City Subways and Elevateds (2010)</u>
656 revenue track-miles
468 stations
186 nonrevenue track-miles
6200 railcars
5.2 million daily ridership (weekdays); 5.3 million (2011)
5.4 million daily ridership (weekends); 5.4 million (2011)

R.P. Roess and G. Sansone: The Wheels That Drove New York, STTT 1, pp. 1–6.
springerlink.com

Staten Island Rapid Transit (2010)
14.0 route-miles
22 stations
63 railcars

New York City Bus (2010)
191 local routes
27 express routes
12,499 stops
4400+ buses
2.2 million daily ridership (weekdays); 2.1 million (2011)
2.4 million daily ridership (weekends); 2.2 million (2011)

This does not even count the miles of commuter rail lines that pass through the city. Both the Long Island Railroad and Metro North lines make stops and service passengers within New York, as well as those from eastern and northern suburban counties (Nassau, Suffolk, Westchester, etc.).

The subway alone is the largest in the world (in terms of physical size): bigger than London, Paris, Tokyo, and Seoul. If the tracks were laid end-to-end, they would stretch from New York to Chicago. The total annual ridership (rail plus bus) is over two *billion* per year. Subway ridership in 2011 was at its highest level since 1950.

1.1 The Transit System and the City

New York City as we know it could not exist without the support of the transit system. Over four million people per day enter the Manhattan business district (anything below 59th Street). Over 90% do so on the transit system. All of the highway traffic and the resulting massive delays are caused by less than 10% of the trips made.

A single lane of freeway traffic can carry approximately 2000 vehicles per hour, with an average car occupancy of about 1.5 people per car – a capacity of about 3000 people per hour, per lane. On a typical weekday, a single track of the E/F corridor through Queens carries up to 68,000 people per hour. Replacing one track of this corridor by highways would require construction of close to 29 fully-dedicated freeway lanes! The density of development in New York City, and particularly in Manhattan, cannot be supported without high-capacity public transportation systems, including heavy rail.

Thus, the daily operation of New York City and its economy are integrally tied to its transit system. The transit system is the city's bloodstream, carrying the essential ingredient, that is, people, to the vast number of places where their labor is critically needed. It also provides NYC residents with unprecedented access to jobs, culture, recreation, education, sports, and other opportunities. Where else can you go to a ballgame in the afternoon and a Broadway show at night without driving a car; or go to work, stop at a mall on the way home, and take a college course in the evening without driving a car? The transit system provides New

York's citizens with unprecedented mobility and accessibility, despite the constant crush that leaves the city and its transportation systems congested at virtually any hour of the day or night.

If the daily operation of the city is tied so heavily to its transit system, the transit system is also integral to its history. In the transportation field, the normal course of events is to predict future transportation demands and provide safe and efficient systems to accommodate them. In other words, professionals react to demands as they are expressed, and try to provide facilities to meet them.

This, however, is not New York's history. Its transit system was developed, not in response to an anticipated demand, but to allow the growth of the city to occur, and to direct its course. The need to grow was primarily an economic issue, and transit systems were built to open new lands to development, to be an enabler of a growing economy and its attendant social needs. No study of New York City's history, therefore, is complete without including the development and impact of its public transportation systems.

1.2 New York City Transit and Its People

Many of the famous residents of New York played a major role in the development of its transit resources. Because of the pivotal role transit played in the economic development of New York, many of those involved in its growth were also involved in planning, financing, and building its transit system.

William Marcy (Boss) Tweed, of Tammany Hall fame, was an early opponent of heavy rail systems and used the franchise system to skim millions of dollars of graft from early transit operators. Alfred Beach, founder and initial publisher of *Scientific American,* was an early developer of transit technologies, and built the first subway in New York City by stealth of night, with many New Yorkers unaware of what he was doing. Jay Gould, one of the so-called "robber barons" of Wall Street, played a pivotal role in the development of the city's elevated rail lines. August Belmont Jr. (of Belmont Park fame) financed the first subway system, and William Barclay Parsons (of Parsons Brinckerhoff) directed its construction. In the course of its development, the transit system would be touched by other famous New Yorkers, creating new celebrities because of their association with transit.

This book, then, is essentially a study of those unique interactions between the city, its people, and its transit system. The focus is on the critical role played by the transit system in defining the physical character of the city, and many of its social and economic characteristics.

1.3 A Transit Timeline for New York City

This book treats the history of various types of public transit service and how each affected the growth of New York City and its suburban environs. New York City has been historically served by six relatively definable transit systems, three of which remain in operation today. The six systems are:

Omnibuses. The earliest form of public transit in NYC, these were horse-drawn carriages carrying up to 12 passengers traveling along a fixed route, picking up and discharging passengers. The first such service was initiated on Broadway by Adam Brower in 1827. The last omnibus operated on Fifth Avcnuc until 1905.

Streetcars. The next evolutionary step was the introduction of horse-drawn carriages on rails, usually within the right-of-way of a city street. The NY and Harlem Railway began the first horse-streetcar operation in 1832. The last NYC streetcar operated along Bleeker Street until 1917.

Trolleys. Technically, trolleys emerged from horse streetcars as electrification was introduced. The first electrified line was along Jamaica Avenue in Queens. It began electrified operation in 1887. The last trolley in NYC provided service over the 59th Street Bridge until 1957.

Elevated Rapid Transit (the "Els"). Charles Harvey inaugurated the first elevated rapid transit service (the 9th Avenue El) in 1867. The El was powered by steam-driven cables, and went bankrupt a few years later. When it reopened in 1871, it was powered by steam engines. The 6th, 3rd , and 2nd Avenue Els soon followed. The Els were electrified in Brooklyn starting in 1898, and in Manhattan in 1902. The last section of the 3rd Avenue El in Manhattan was discontinued in 1955. Of course, many elevated sections of the "subway" system still exist in Brooklyn, Queens, and the Bronx.

Subways. The first subway (an experimental demonstration) in NYC opened in 1870. It was powered by pneumatic pressure (atmospheric pressure), and was built by Alfred Ely Beach under the cover of night, with few realizing it was being built, and without legal authorization to build it. It covered several blocks. The pneumatic pressure system, however, was never developed for a multistop, multi-train operation on a long system. The first operational subway, the Interborough Rapid Transit (IRT) was built in 1904. The Dual contracts were signed in 1919, and construction of extensions to the IRT system and for the Brooklyn Rapid Transit System (BRT) started shortly thereafter. The Independent System (IND) opened its 8th Avenue line in 1932, and its 6th Avenue line in 1840. Subways, of course, form the core of NYC's public transit system to this day, and several expansion projects are underway at this very moment (2nd Avenue subway, extension of the No. 7 line).

Buses. The first motor bus in NYC replaced battery-powered trolley service on Madison Avenue in 1919. Eventually, all surface transit in NYC would be provided by buses.

There are obvious overlaps in the time periods during which these various systems served New York City. Omnibuses and horse-drawn streetcars have a substantially overlapping history, and both shared the streets of NYC during most of their periods of influence. Steam-powered Els predate the first trolley, as the latter had to wait for electric traction technology to develop. The Els dominated the scene for about 30 years, before coexisting and coordinating with new subways. Buses were introduced almost as soon as they were invented.

Figure 1.2 shows a timeline for these systems that clarifies their overlapping histories. Because these systems overlapped in time, and were primarily started as private enterprises, they were often competitors. This made coordination of the various services difficult, and often resulted in chaotic and inefficient services. “Public” transit only became publicly owned and operated when it became clear that it would be difficult for a private entrepreneur to make money in the industry. That was indeed clear by the mid-1910s, although full public ownership and operation did not come about until 1940.

Omnibuses, horse streetcars, trolleys, and buses fall in the general category of “surface transit” services. They take place (and took place) on the surface, primarily using public streets and roadways. Elevated trains and subways form what is commonly referred to as “rapid transit,” primarily due to the higher average speeds at which they operate.

Timeline	Horse-Drawn Omnibus	Horse Streetcar	Electrified Trolley	Elevated Rapid Transit	Subways	Motor Bus
1800						
1820						
1827	Adam Brower introduces service on Broadway.					
1832	Stage Carriage Act	NY & Harlem Railway begins operation				
1840						
1854		First Brooklyn Streetcar				
1860						
1867				9th Ave El tested and opened for service		
1870					Beach Pneumatic Subway open	
1871				9th Ave El goes steam		
1878				6th and 3rd Ave Els open		
1880				2nd Ave El open		
1885				First Bkln El open		
1887			Jamaica Ave Line Electified			
1898				BRT begins to electrify		
1900						
1902/1904				Manhattan Els begin to electrify	IRT open	
1905	Last omnibus on 5th Ave discontinued.					
1913					Dual contracts signed; BRT/IRT expansion begins	
1917/1919		Last streetcar on Bleeker street ends				First motor bus replaces battery streetcar on Madison Ave
1920						
1932					8th Ave Subway (IND) open	
1939			Last Queens Trolley			
1940					6th Ave Subway (IND) open	
1955				Last El in Manhattan		
1951/1952			Last Brooklyn Trolley		Rockaway Line acquired from LIRR	
1957			Last Trolley over the 59th St Bridge			
1960						
1972					63rd St tunnel completed	
1980						
1988					Archer St opens	
1989					Service through 63rd St tunnel initiated	
2000						
????					*2nd Ave Subway*	
????					*No. 7 line extention*	
				↓	↓	↓

Fig. 1.2 A Transit Timeline

1.4 Closing Comments

Modern New Yorkers take the ability to travel extensively within the city with relative ease for granted. This was not always so. In fact, the inability to travel beyond relatively short distances limited the growth of New York for almost two centuries. This book tells the story of how the transit system became the ultimate enabler, allowing New York City to grow during the latter 1800s and 1900s, and how it continues to provide unprecedented mobility to New York City residents and visitors.

Figure Sources

1.1: *Courtesy of* Wikimedia Commons.

Chapter 2
An Overview of the History of New York: Discovery through the Civil War

It is impossible to understand the history and importance of the New York City transit system without an appreciation for how the city became a place and a people whose destiny literally depended on its invention and implementation. By 1850, New York had reached a point where its future growth was severely limited. Its people and its enterprises were squeezed into a small area in what is now downtown Manhattan, suffering with a disease-ridden environment, and intense competition for both jobs and housing which bred ethnic controversies that frequently escalated into violence. The city was ready to explode. Either a means to expand its limits would be found, or the city would collapse in ways that no one wanted to contemplate.

This chapter provides an overview of the general history of New York as a city from its inception to the onset of the Civil War. As the Civil War was ending, New York was on its way to doing away with the barriers to physical expansion. From that point, the history of the city itself converges with the history of the public transit system. In no small way, the city's subsequent history was and is enabled by its extraordinary public transportation facilities.

2.1 Introduction

The history of New York City begins with the arrival of Henry Hudson in what is now Upper New York Harbor on September 2, 1609. His arrival came more than 100 years after the official "discovery" of the New World by Christopher Columbus and Americus Vespucius (aka, Amerigus Vespucci). Neither man set foot in North America, although both explored the coast of South America during several voyages. Columbus died believing that he had found Japan, and that a route to India was nearby. Vespucius did eventually come to believe that he had found a "new world."

Both, and countless others over the next century, were searching for a trading route to the East by sailing west. They were searching for access to the lucrative spice trade with India. The search was necessitated by the failed crusade of Pope Pius II against the Turks. The war had functionally blocked most of the traditional trade routes to India. Spices were lucrative because they were necessary for the curing and preservation of meat and other food products.

R.P. Roess and G. Sansone: The Wheels That Drove New York, STTT 1, pp. 7–34.
springerlink.com

In 1524, Giovanni da Verrazano, on a mission for the French crown, was the first European to enter New York's harbor and to sail what is now the Hudson River. Bad weather, however, prevented him from exhaustively exploring the river and its surrounding lands.

2.2 Henry Hudson's Voyage

Hudson was an Englishman who had undertaken two previous journeys searching for a westerly route to the East at the behest of the British Crown. His approach was to seek a crossing very close to the Arctic Circle, which would shorten the distance traveled. He also believed that the heat of summer would melt major portions of the polar cap. In 1607, he landed on what is now Jan Mayen Island, which eventually became part of Norway. On both the 1607 voyage and the second one in 1608, ice forced Hudson to return to England in failure.

Little is known about Hudson's early life, and a portrait often used to depict him is widely thought to be someone else (unknown). After failing to get British support for a 3rd voyage, Hudson was commissioned by the Dutch East India Company to make another attempt at finding a pathway to the East through the Arctic region in 1609.

His ship for the third voyage was the *Half Moon*. He considered the ship to be substandard. Too small and too heavy, it rode high in the water and was difficult to control. He wound up with the *Half Moon* because he had exceeded his budget in crewing and provisioning the ship. The Dutch East India Company did not have a great deal of faith in Hudson, and were not willing to exceed their budget for the voyage.

Fig. 2.1 A replica of the Half Moon constructed for the 1909 Hudson–Fulton Celebration on the Hudson River

When ice problems arose once again, he decided to go south to seek a southwestern passage through North America.

On September 2, 1609, Hudson and the *Half Moon* entered the mouth of the river now named after him and viewed what was one of the greatest natural harbors in the known world at the time. He traveled upriver to what is now Albany before determining that the river would not provide passage to the East, and returned to the harbor. The great harbor became the initial impetus to build a settlement in New York. In addition, the area contained an abundance of beavers and other mammals that were attractive for the fur trade. Eventually, explorers would dream of a harbor that could safely protect 1000 sailing vessels within its natural boundaries.

The initial exploration of New York by Hudson began a unique history for the city and its role within the nation. Unlike many explorers, commissioned by governments, kings, and queens, Hudson was commissioned by what was at the time the largest corporation in the world, the Dutch East India Company.

The search for religious freedom accompanied most of the initial settlers in the New World, for example, Pilgrims in New England, Catholics in Maryland, and Quakers in Pennsylvania. The first settlers in New York came to make money. Thus, from its very first days, the history of New York is tied to commerce and relatively free exercise of capitalism. New York was only the capital of the United States for a few short years, but its role as the nation's financial capital has never been challenged. This lack of a religious foundation is evident from the fact that the first church in New York was built 17 years after the first settlers arrived. It also resulted in a settlement that from its very earliest years represented considerable ethnic and religious diversity, unknown in the other colonies.

2.3 The Dutch Era in New York

For a little more than a decade, the Dutch government attempted to attract settlers to its newly-claimed colony, though without much success. The first small contingent of settlers actually arrived in New Amsterdam in 1614, led by Hendrik Christiansen, who actually started trading fur with native Indians at a merchandising site consisting of several wooden huts on the southern tip of Manhattan. Christiansen's activities were halted when he was killed in a dispute with an Indian trader.

The first significant group of settlers arrived in the spring of 1624, sponsored by the Dutch West India Co., a newly-formed corporate entity spawned by the larger Dutch East India Co. These settlers were, by and large, not Dutch. Most were Walloons, French-speaking Protestant refugees. They began a settlement on the southern tip of Manhattan, and began to establish a fur trade.

In 1626, the directors of the Dutch West India Co. appointed one of their own – Peter Minuit – to be Director-General of the entire region of New Netherlands, which included New Amsterdam and other lands in/on what is now New Jersey and Long Island. He followed two relatively unknown predecessors in this role – Cornelius Jacobsen Mey (1620 to 1625) and Willem Verhulst (1625 to 1626). He reached New Amsterdam in April of 1626, bringing with him new settlers and cattle.

Fig. 2.2 Portrait of Peter Minuit

Later, in 1626, eight African slaves captured from Spanish and Portuguese ships were brought to New Amsterdam. While enslaved, the men were permitted to be baptized, marry, own land, sue or bear witness in courts, and were permitted to work on their own after their assigned tasks were completed.

The new settlers began to build houses and small farms in lower Manhattan. Soon after, a delegation of native Indians called on Minuit to ask whether he intended to take their land by force, or pay for it. This resulted in the famous purchase of the entire island of Manhattan for a reported 24 dollars. In reality, a meeting with the Lenape Indians and their Chief, Sagisgura, was established to set a purchase price. The Chief had a written agreement from one of Minuit's short-lived predecessors to purchase the island for 1200 Dutch guilders. Minuit didn't have this amount available, and the Indians thought that bank notes were useless in any event. The final agreement was reached when the Indians accepted some Spanish gold coins, glass pearls, and bales of cloth. Historical accounts of what was actually traded are inexact. Thus, the generally-accepted fact that the value of these goods was approximately 60 guilders (the equivalent of $24), is not viewed as certain, and some historians believe a much higher value was exchanged.

Under Minuit, the amount of furs that were exported was multiplied by five. Additional groups of settlers continued to arrive, and the area surrounding New Amsterdam continued to develop.

With the arrival of new settlers, however, new problems began to arise. Some of the larger landowners challenged the order that all fur trade had to be under the auspices of the Dutch West India Co. This rebellion led to significant losses for the company in New Amsterdam, resulting in hard times for all. Support for Minuit quickly dissipated. In 1633, the company directors removed Minuit from his post.

Minuit was followed by Wouter Van Twiller as Director-General between 1633 and 1638. His tenure did not significantly reduce the difficulties of the new settlements. Part of the reason was that Dutch emigration to the settlements was still sparse. Most prospective Dutch emigrants considered the fur trade to be a socially low-class business, and tended to migrate to other islands in the New World. New Amsterdam fast became a very seedy place. At its worst, it was noted that there was one bar for every 20 citizens within New Amsterdam – an interesting number, considering that none of the women or children used the bars. The streets of the town became dirty and muddy, and New Amsterdam's somewhat seedy reputation provided new disincentives to prospective colonists.

Van Twiller did, however, acquire additional land from the Indians. The deal included large areas of what is now Brooklyn, as well as what are now known as Wards, Randalls, Roosevelt, and Governors Islands. Van Twiller used some of this land to raise tobacco for private interests.

Van Twiller was followed by Willem Kieft from 1638–1647. If Van Twiller's reign was not particularly noteworthy, Kieft's was an unmitigated disaster. In September of 1639, Kieft levied a tax on Indians living in New Netherlands (on both sides of the river) for the purpose of providing "protection" from rival Indian tribes. Relationships with the Indians rapidly deteriorated, and by 1641, sporadic attacks by Indians on Dutch settlers in outlying areas took place. Kieft ordered an attack on the Lenape Indians in an outlying area known as Pavonia. He supposedly ordered his soldiers to spare women and children, but they ignored this order. On February 25, 1643, Dutch soldiers attacked and killed 80 men, women, and children. Several Indian tribes grouped for a retaliation that devastated virtually all of the Dutch holdings in the area surrounding Pavonia. Fortunate settlers escaped to New Amsterdam, but others were killed or captured. The year 1643 became known as the "Year of the Blood." Hostilities ended with a cease-fire agreement in August of 1645.

In 1647, the last, and most successful, Director-General arrived in New Amsterdam, namely, Peter Stuyvesant. As the result of a war wound suffered in April 1644, his right leg was amputated and replaced with his now-famous wooden peg. He was the son of a minister, and was sent to New Amsterdam with a clear mandate to clean up the deteriorating town and its rag-tag residents, to expand the settlements, and to return the company to profitability.

Fig. 2.3 Portrait of Peter Stuyvesant

He immediately banned drinking on Sundays, street fights, and fornication with Indians. He also mandated church attendance on Sunday (without restriction on *which* church). He built streets, row housing, piers, windmills, schools, and a defensive wall to protect the settlement. He also introduced slave trading to the settlement as another means of generating profits.

By 1654, the population of greater New Netherlands swelled to 4000. To provide labor for the expanded fur trade and other enterprises, Stuyvesant allowed emigrants from all parts of Europe to enter, creating a diverse population in which the Dutch were a distinct minority. In 1654, the first Jews arrived, fleeing the Spanish Inquisition. Stuyvesant petitioned the Dutch West India Co. to deny them access. He was sternly rebuked by the company, which sent a remarkable order in support of diversity.

It is during this period that some of the modern structure of New York City began to take shape:

- One of the first streets built was the widening of an Indian trail from the tip of Manhattan into the then northern frontier (only about three miles) to allow for easier transport for fur traders and to allow access to outlying farms to be established. The route of the trail is now Broadway in lower Manhattan.
- In 1646, the Dutch West India Company established the village of Breuckelen (now Brooklyn) on the western tip of Long Island. It was actually the first recognized municipality in the region, followed by the predecessors of Kingston (Wiltwyck) and New York (New Amsterdam) that were officially recognized later.
- In 1639, Jonas Bronck became the first European to settle on the mainland north of Manhattan. Visitors to this remote area were said to be traveling to "the Broncks." This historical oddity explains why modern New Yorker's refer to Manhattan, Brooklyn, Queens, and Staten Island – and to the Bronx.
- The defensive wall built by Stuyvesant is now Wall Street, the universally-recognized financial capital of the world.
- A canal built by Stuyvesant is now Broad Street.
- A farm owned by Stuyvesant was referred to as "the Bouwerie" a Dutch word for farm. It was near the current location of the Bowery.

Despite the relative success of Stuyvesant's reign, his strict and unyielding ways caused constant friction with the residents of New Netherlands. At the same time, England was developing an interest in New Netherlands, as it was a Dutch holding in the midst of English colonies to the north and south. In 1664, King Charles gave his brother (James II of England) land that included New Netherlands. On August 27, 1664, four English ships bearing 450 men seized the Dutch colony. Stuyvesant attempted to rally his forces to resist the take-over, but a large contingent of prominent local merchants (including his own son) essentially forced him to surrender. The merchants of New Amsterdam literally did not care who ruled them, and they were only too happy to continue working with the English, who were clearly a more powerful international force than the Dutch at the time. On September 6, 1664, Stuyvesant signed a treaty with the English at his Bouwerie house. Except for a brief period in 1673–1674, when the Dutch regained control of New Netherlands, the Dutch era in New York had ended.

Stuyvesant was offered safe passage back to Holland by the English. He chose, however, to live out his life on his farm. To its north lay the woods and swamps of the village of Haarlem. A pear tree which arrived from Holland in 1647 remained at the corner of Thirteenth Street and Third Avenue until 1847, bearing fruit until its last year. The Bouwerie house burned down in 1777. Stuyvesant also built an executive mansion called Whitehall (near current Whitehall St.). He died in August of 1672, and was buried at St Mark's Church-in-the-Bowery in Manhattan.

Fig. 2.4 A print depicting Stuyvesant surrendering New Amsterdam to the British in 1664

2.4 The English Era in New York

The English assault on New Netherlands in 1664 was led by Richard Nicholls. After his victory, he was appointed as the first Governor of the colony, which was now renamed New York, for the Duke of York. The area of Breuckelen was renamed Kings, an area east of Kings was named Queens, and Staten Island was named Richmond (supposedly for one of the King's illegitimate sons). It is interesting to note that the current boroughs of New York City retain their original Dutch names. However, the names of the counties that coincide with the boroughs use their English designations.

By all accounts, Nicholls was an effective Governor, and was well-liked by his constituents. After what was literally a disruption of only several days for the change from Dutch to English rule, most residents returned to work, and life and commerce in New York continued to develop.

Trade with the West Indies grew quickly, and with it new needs and new industries began to take shape in New York. Trade required ships, rope, sails, and other equipment that began to be manufactured in New York. By the end of the century (1700), the population of New York had doubled to about 8000 people. Approximately 100 ships per year were landing in New York on trade missions.

By 1740, the population had reached 12,000, of which 2000 were slaves, many of whom had been freed. Slavery was becoming a major issue in New York, as well as other nearby colonies. While New York did not have a huge number of slaves, and included a significant number of freed slaves, it was a major player in slave-trading due to the prominence of the port. Most of the slaves entering the colonies entered through the Port of New York, and banks were lending money for the purchase of slaves. A great deal of money was being made as a result of the slave trade.

The early 1700s saw gradually-escalating violence in the way slaves were treated, and in periodic rebellions. Pennsylvania imposed heavy tariffs on the import of slaves, and thus removed itself from the slave trade. Massachusetts became the first colony to absolutely prohibit the importation of slaves. New York, not wishing to abandon the profits of the slave trade, placed legal restrictions on the education of slaves, and conditions on owners who freed their slaves. Newly-freed slaves were prohibited from owning any property.

In New York, matters came to a head in 1741, the year of the so-called "New York Slave Insurrection." History is uncertain as to how much of an organized "insurrection" was present. What is clear, however, is that this period resulted in numerous trials and executions that rivaled the Salem Witch Trials for their lack of judicial process and their brutality. Torture and hideous forms of execution were revived. In little over a year, 16 Blacks and 4 whites were hung, 13 Blacks were burned at the stake, and 17 Blacks were deported.

The lead-up to these events was an extremely severe winter, leaving both Black and white poor struggling to obtain food and to survive. Rumors of a slave rebellion, and actual rebellions in the West Indies fueled significant fears among the population. When a series of mysterious fires at homes of prominent New Yorkers occurred, it was concluded that angry slaves were responsible.

Three key individuals played a pivotal role in this period, two of whom were eventually executed as a result:

- John Hughson owned a disreputable bar that was frequented by freed slaves and Irish immigrants.
- An indentured servant, Mary Burton, testified that Hughson received stolen goods from slaves, and reported on overheard discussions of a planned campaign of insurrection. She also implicated Hughson's wife and daughter in the plot.
- The third "conspirator" was John Ury, a schoolteacher who was accused of being an underground Catholic priest. He was implicated (again by Mary Burton) as one who actively instigated the slave revolt.

All semblance of fair trials vanished. Mary Burton was the star witness at most of the trials, and the defense was perfunctory at best. Many witnesses testified after having been themselves tortured and charged. Plea deals were prominently used throughout the trials in return for damning testimony against others.

The trials ran their course and petered out, partly due to a growing recognition that there were serious issues with the way justice was being dispensed. Mary Burton began making ever-wilder claims against some prominent whites, and her veracity was, for the first time, questioned. She was given a 100 pound reward for her assistance in "unearthing the insurrection," after which she left New York and disappeared into history.

2.4.1 New York Approaches the Revolution

In 1756, war engulfed Europe. Prussia, Great Britain (with the colonies), Hanover, and Ireland were pitted against Austria, France, Russia, Sweden, and Saxony. The war is frequently called the *Seven Years War*, and is occasionally referred to as a world war. Its fundamental causes were the lines of European succession, and the colonial struggle between Great Britain and France. The *French and Indian War* was the principal colonial theater in this war.

The war in Europe brought good business to New York. Between 1758 and 1763, over 30,000 British troops were garrisoned in New York. They needed food, ships, supplies, weapons, and sex. Their presence helped New York reach new levels of commercial success. Business thrived, and virtually every New Yorker could make a good living.

However, if the war was good for business, its end in 1763 was a disaster. With the departure of British troops, a depression hit the New York economy. To make matters worse, the British began to impose new taxes to recoup the costs of the war. The widespread dissatisfaction of New York's citizens, along with those in the other colonies, began to put the colonies on the road to revolution.

2.4.2 New York and the Revolution

In the Spring of 1765, Britain imposed the Stamp Act on the colonies. This act taxed over 40 business transactions that were heretofore legal, and free. The tax was so broad-based that virtually everyone was negatively affected in some way, causing almost unanimous opposition. On November 1, 1765, an angry mob gathered on the New York Commons. They marched to Bowling Green, where the mob staged a mock hanging of the King in effigy. Because of this, and similar uprisings in other colonies, the British repealed the Stamp Act later that year.

In 1773, a seventeen-year old illegitimate child arrived in New York from the West Indies. It was Alexander Hamilton, who played a critical role in the development of the nation, and who is truly the father of American banking and capitalism. This role would be dramatically played out in the course of succeeding decades. In 1774, however, Hamilton published a pamphlet openly urging rebellion against the crown.

The Revolutionary War started on April 23, 1775, with the beginning of hostilities at Lexington and Concord. While some critical battles in the war occurred in New York, the city played a relatively minor role during the course of the war, at least for the colonists. For the most part, New York City was a British garrison, and was under British control for most of the conflict. The start of the war caused a massive evacuation of New York. By January of 1776, almost 80% of the population of New York had left.

It was clear from the beginning of the war that the British wanted to control New York. Its location was strategically critical because of the great port, and because the British could effectively split the colonies into north and south by controlling New York. To counter this, in the Spring of 1776, George Washington rushed his forces to Manhattan.

On June 29, over 100 British ships entered New York harbor. In the weeks that followed, 500 British ships landed, dispatching tens of thousands of troops, and establishing positions on Staten Island. On July 9, 1776, the Declaration of Independence was read on the New York City Commons. The lead statue of King Charles II was dismantled, and its lead melted in order to make musket balls.

Washington moved his troops to the Brooklyn side of the river, correctly anticipating that the British would first attack there. Washington also believed he had a better chance of successfully defending Brooklyn than Manhattan. On August 26, 20,000 British troops began the attack on Washington's army. Within two days, Washington lost about 2000 men. In desperation, Washington moved his remaining troops back to Manhattan by boat under cover of darkness during a seven-hour period. The operation was led by a young patriot who would later earn infamy, Aaron Burr. Many were lost during the dangerous nighttime crossings. The unprotected boats were open targets for British artillery, and without light, a number of other mishaps occurred. Nevertheless, Washington moved about 10,000 troops, and retreated to northern Manhattan. By November of 1776, Washington decided that he could not defend New York, and left. His strategy was to save the troops, not the place. For the remainder of the war, the British were in control of New York.

During the initial battles, fires, possibly set by fleeing revolutionaries, destroyed over 100 buildings in New York, which was most of the western half of the city. During the remaining British occupancy, little was done to protect the infrastructure of the city, and further destruction occurred. Until 1783, New York served as the base of British military operations. Most of the people that stayed in New York were British loyalists.

In 1778, the French agreed to help out the colonies. Around the same time, the Dutch offered to provide funds to assist the colonists. The war unofficially ended on October 19, 1781 with the surrender of the British army at Yorktown. Over three years passed, however, before the war officially ended, and all treaties were signed. Washington returned to New York in triumph on November 25, 1783, as the last British troops departed. New York was an American city at last.

In the months that followed Washington's return, over 7000 loyalists left the city to return to England. Exiled citizens began to return, only to find a city in ruins and in need of massive rebuilding.

2.5 New York – The First Capital of a New Nation

The Revolutionary War officially ceased with the ratification of the Treaty of Paris by the Continental Congress on January 14, 1784. The path to full nationhood was a difficult one, with issues of states' rights vs. legitimate federal interests, trade and money policies, preservation of individual rights, slavery, and others under serious debate.

During this period, Alexander Hamilton opened a law firm in New York, organized the Bank of New York, and preached emancipation of slaves. On January 11, 1785, the continental congress relocated to New York, making it the temporary capital of United States.

The drafting of the Constitution was a long and painful process. However, after many revisions, and the inclusion of the Bill of Rights, Virginia became the ninth colony to ratify the Constitution by a slim vote of 89 to 75 on June 25, 1788. As nine colonies were required to ratify the Constitution, this action formally created the United States of America. New York finally ratified the Constitution, with the Bill of Rights, on July 26, 1788, by an even slimmer vote of 30 to 27.

On January 7, 1789, Presidential electors were chosen in the 11 states that had ratified the Constitution, except in New York, where most members were still in transit. Electors cast their ballots on February 4, 1789. The ballots were counted on April 6, 1789. George Washington was unanimously elected as President, and John Adams was elected as Vice-President. On April 30, 1789, George Washington took the oath of office on the balcony of the Federal Hall in Manhattan at the age of 57.

Fig. 2.5 An oil painting depicting George Washington taking the oath of office as the nation's first President at the Federal Hall, New York, on April 30, 1789

It was in the earliest years of the nation that New York's role as a financial capital was determined. From the beginning, there was a great divide over what the nation's focus and structure should be, and the leading voices in the debate were Alexander Hamilton and Thomas Jefferson (along with James Madison).

Hamilton strongly believed that the nation's future should be based upon business and commerce. He recognized the importance of industry to the nation's economy, and viewed it as the path to growth and stability. Jefferson envisioned a nation founded on rural agriculture and farming. Hamilton favored a strong central government. In fact, in the earliest days of the Constitutional Convention, Hamilton had actually proposed a system in which the President and Senators were elected for life, and in which state governments would be abolished. Jefferson favored strong state governments and limited federal power. As a result of this long-running controversy, Hamilton eventually helped form the Federalist Party (with Washington and Adams), and Jefferson and Madison, the Republican Party, which came to be known as the Democratic-Republican Party.

Both Hamilton and Jefferson played key roles during Washington's administration, namely, Hamilton as Secretary of the Treasury, Jefferson as Secretary of State and the Speaker of the House of Representatives. Hamilton had particularly strong ties to Washington, having served as his chief-of-staff during the Revolutionary War, and having played a key role in the defeat of the British at Yorktown.

In the Spring of 1790, a major dispute arose over the dispensation of the enormous war debts that had been incurred by the colonies during the Revolutionary War. Hamilton proposed that the federal government assume all of these debts (New York had the highest debt of all the colonies), arguing that if left to each state, many would fail financially under the burden of repayment. Jefferson strongly opposed this, believing that the southern states would be strongly disadvantaged by federalization of debt, and argued that the constitution prohibited such an action. Congress deadlocked over this issue.

On June 20, 1790, Hamilton and Jefferson reached an historic compromise that would have tremendous impact on New York, and on the nation. Jefferson agreed to let Hamilton federalize the war debt of the states. Hamilton agreed to let the capitol move to an area along the Potomac River – the current Washington D.C. – which was, at the time, an undeveloped swamp. Hamilton then had the treasury issue $80 million in government bonds to pay off the war debt of all of the states.

In agreeing to give up the capitol, Hamilton argued that it was more important to control *what* the federal government did than to control *where* it did it. Relieved of its role as the nation's capitol, New York was free to develop its role as the financial center of the nation with full abandon.

On May 17, 1792, 22 brokers and businessmen met to determine a more systematic means of conducting business. This meeting led to the signing of the "Buttonwood Agreement," which formed a system of trading that would eventually become the New York Stock Exchange.

As the century closed, New York State adopted the "Act for Gradual Emancipation of Negro and Other Slaves." While certainly a step in the right direction, the act did not lead to a complete abolishment of slavery in New York.

New York was off and running. As the new century approached, New York was poised to become a world center of commerce and business. By 1800, its population approached 80,000, and a century of unbridled growth was about to begin.

2.6 New York – 1800–1825

2.6.1 Hamilton and Burr

In 1804, Alexander Hamilton was killed in a famous duel with his long-time political rival, Aaron Burr. In effect, New York lost two of its most famous and influential residents – Hamilton to death, and Burr to infamy that drove him from the region. The story of the Hamilton–Burr feud and eventual duel is indeed a sad one. It might be argued that the only good thing to result was public recognition of the barbarous nature of the "gentlemanly duel" and its eventual criminalization throughout the U.S., although they would continue to occur for many years.

Hamilton and Burr began as cordial acquaintances. Both were active in the New York Bar, and although they were adversaries on several issues, they both reportedly liked and respected each other, often dining with each other after long work days. This changed when Burr defeated General Philip Schuyler, Hamilton's father-in-law, for a seat in the U.S. Senate. Hamilton never forgave him, and the two became bitter rivals.

The election of 1800 was the nation's fourth presidential election, and one of its most historic. It was the first time that candidates actively campaigned, and the first in which each presidential candidate ran with a "running mate" for vice-president. The Federalists ran John Adams for a second term, with Charles Pinckney. Adams, however, had become quite unpopular during his term, primarily due to the passage of the Alien and Sedition Acts, which were viewed as a direct attack on the freedom of speech and of the press. The Democratic-Republicans ran Thomas Jefferson, with Aaron Burr as his running mate. Burr was rewarded with a place on the ticket because he had been instrumental in gathering New York support for Jefferson's candidacy through the Tammany Society.

Hamilton played a unique role in the election. Although of the same party, Hamilton intensely disliked John Adams. During and after the French Revolution, there was great turmoil over whether the U.S. should assist the French (who had assisted us) or remain neutral. This was made more difficult, as the English threatened to declare war on us if we supported the French. Adams wanted neutrality; Jefferson wanted to openly support the French Revolution. As a result of not supporting France, the French threatened hostile actions against the U.S. The Continental Army, which had been disbanded after the Revolution, was called back into service. George Washington, not well at the time, was named General, but he insisted that Hamilton be named second-in-command. After Washington died, Hamilton wanted to take lands to the west by force, and advocated invading Mexico. Adams, having successfully negotiated peace with France, disbanded the army to prevent Hamilton from following through wish this plan.

As the election of 1800 approached, Hamilton wrote a pamphlet, "The Public Conduct and Character of John Adams," that was highly critical of Adams and attacked his integrity. Hamilton, however, did not intend the pamphlet to be publicly distributed. Aaron Burr proceeded to publish and widely distribute the pamphlet without Hamilton's permission. This severely damaged Adams' prospects for re-election. At the same time, Hamilton was bitterly opposed to Jefferson. The election resulted in the Jefferson/Burr ticket being chosen.

At the time, the Electoral College only permitted a vote for President. Each elector had two votes. They could not vote for the same person twice, and only one could be a candidate from their home state. The candidate with the most votes was President, and the candidate with the second-most votes was the Vice-President. To avoid a tie, one of the Jefferson/Burr electors was to submit a blank ballot on the second vote. The arrangement, for reasons that were never clear, didn't work, and the electoral ballot resulted in a tie between Jefferson and Burr.

Under the Constitution, the election was thrown into the House of Representatives, dominated by the Federalists, who now had to choose between two Democratic-Republicans, that is, Jefferson and Burr.

Over the following days, 35 votes were taken, each resulting in a tie between Jefferson and Burr. Some Federalists believed that if they kept voting a tie, that they could maintain effective control of the country. Some historians believe that an armed conflict might have erupted if this strategy had continued, and that Jefferson himself might have supported it. Behind the scenes, John Adams suggested several compromises that Jefferson could make to win support. At the same time, Hamilton lobbied heavily for electing Jefferson over Burr. On the 36th ballot, Thomas Jefferson was elected, avoiding a constitutional crisis for the young nation. Through the entire crisis, Aaron Burr never left New York City.

The election had many repercussions: (1) the growing feud between Hamilton and Burr was intensified, (2) the 12th Amendment was passed, changing the voting procedures in the Electoral College, and allowing separate votes for president and vice-president, (3) the Alien and Sedition Acts were repealed and allowed to lapse, and (4) Jefferson, now quite distrustful of Burr, virtually ignored his presence as V.P., and dropped him from his ticket when he ran for reelection in 1804.

Knowing that he would be dropped from the ticket in 1804, Aaron Burr ran for Governor of New York. Burr lost the election and was furious at Hamilton for having written several vitriolic letters against him. He was most angry, however, at remarks Hamilton reportedly made at a political dinner at which he was quoted as saying that he could deliver a "still more despicable opinion" of Burr. Burr demanded that Hamilton recant the statement publicly; Hamilton refused, claiming to be unable to even recall the statement. Burr challenged Hamilton to a duel under the existing, but antiquated, rules of dueling. Interestingly, both Burr and Hamilton had been involved in previous duels, and Hamilton's eldest son had been killed in a duel in 1802. Dueling had been outlawed in both New York and New Jersey. However, Hamilton and Burr were not citizens of New Jersey, so they met on July 11, 1804 near Weehawken, NJ.

Accounts of the actual duel vary based upon accounts from the seconds who were the only witnesses historians are certain were present. One claimed that Hamilton never fired. The other claimed that there was a three to four second gap between the first shot and the second shot. In any event, Hamilton's shot missed Burr, while Burr's hit Hamilton in the abdomen. Second-hand accounts state that Hamilton deliberately missed Burr (called "throwing away one's fire"). However, Hamilton did not invoke the specific dueling rules for doing so, which could have been matched by Burr doing the same. Hamilton was brought to the home of a friend in Manhattan where he died after several days at the age of 47.

As a result of the duel, Burr was indicted for murder and other crimes in both New York and New Jersey, but was not tried in either jurisdiction. He subsequently returned to Washington to complete his term as Vice-President, where he earned praise for his handling of the impeachment trial of Samuel Chase, a Supreme Court justice. Burr was later accused of treason for allegedly being part of a conspiracy to form a new nation in the west, with the objective of controlling the agricultural economy of the U.S. He was tried and acquitted on this charge, but his influence in U.S. politics effectively ended.

2.6.2 Early Days

The 1800s comprised a century of explosive growth in New York, spurred by the Industrial Revolution, and the success of Hamilton's vision of a nation built around commerce and business. The population of the city grew from just under 80,000 in 1800 to almost 3.5 million in 1900. This growth, however, did not come easily, and did not come without significant problems.

During the first decade of the century, inventors and entrepreneurs flooded into the city. In 1807, Robert Fulton launched the first successful steamship in New York Harbor, freeing ships forever from the dependence on the wind for propulsion. Cornelius Vanderbilt launched a ferry service between Brooklyn and Manhattan in 1809, and established a ship-building company which became one of the cornerstones of his economic empire.

Fig. 2.6 Fulton's steamboat, the Clermont, at the Hudson–Fulton Celebration in 1909

In 1809, Deidrik Knickerbocker published the *History of New York*, which rapidly became a best-seller. Deidrik is the Dutch word for "father." The history, however, was historical fiction. The book detailed the actions of the fictional Knickerbocker family over several generations coinciding with the history of the city. Because of the transient nature of its population, there was very little real history of New York available. For generations, prominent New Yorkers would cite their relationship to the fictional Knickerbocker family. "Father Knickerbocker" was, in fact, Washington Irving, the first U.S. citizen known to make his living exclusively through writing.

2.6.3 Preparing the City for Growth

The existence of a great natural harbor and the entrepreneurial spirit of its citizens were not alone sufficient to propel New York to status as a world center of commerce. Four key developments greatly advanced the journey.

2.6.3.1 The Commissioners' Plan of 1811

In 1800, New York City was still primarily confined to a small area on the southern tip of Manhattan. The landscape to the north was hilly and heavily forested, making construction difficult. In March of 1807, a three-man committee was appointed to develop a plan and layout for the future streets of Manhattan between 14th Street and Washington Heights. The committee consisted of Gouveneur Morris, a lawyer, John Rutherfurd, and Simeon DeWitt, a surveyor. DeWitt was a relative of DeWitt Clinton (by marriage), a state senator at the time, and a future governor of New York State. In April of the same year, the State legislature gave the commission exclusive rights to develop the plan.

In 1811, the commission returned with a recommendation of staggering dimensions. Rather than attempting to build roads and buildings around and over the natural terrain, the commission recommended completely leveling over 11,000 acres to build a rectangular grid pattern of streets, creating relatively uniform blocks. The plan created 12 north-south avenues spaced approximately 800 ft apart. Crosstown streets were laid out to produce 20 per mile (average spacing about 260 ft). Over 2000 long, narrow parcels were created by the plan for development. This plan was implemented over the decades that followed, and spurred an orderly pattern of development as the city exploded northward. It should be noted that this original plan did not include an accommodation for what would become Central Park.

There were exceptions to the plan: Broadway would remain a diagonal street due to its historic significance, and the irregular street pattern of Greenwich Village would remain untouched due to the political oppositions of its residents.

Over the years, the plan has been much discussed. Some criticized its uniformity and uninteresting layout. The leveling of the terrain was a frequent target of critics over the years, as it removed any local uniqueness. Others have praised the plan as visionary, as it laid out an attractive pattern for easy development, and provided for wide avenues and right-angled corners which were thought to be both efficient and attractive for potential builders. Despite its mixed reception, the plan set a course for development, removing the geographic barriers for northward expansion of the city.

Fig. 2.7 Commissioner's Plan for Streets and Avenues in Manhattan, approved in 1811

2.6.3.2 DeWitt Clinton and the Erie Canal

DeWitt Clinton began his career as secretary to his uncle, George Clinton, who was a popular and influential Governor of New York, after receiving a law degree from what is now Columbia University. He soon became an active member of the Anti-Federalist Party, although, in later years, he would be accused of sympathizing with the Federalists. He was a member of the NYS legislature from 1797 to 1802, at which time he was elected to the U.S. Senate. He resigned from the Senate a year later, reportedly due to his unhappiness with living conditions of the new capital in Washington D.C. On his return, he became Mayor of New York City (an appointed office at the time), and served several terms (1803–1807; 1808–1810; 1811–1815). In 1812, he ran for President, losing a convoluted electoral battle to James Madison. He was elected Governor of New York in 1817, and served until his death on February 11, 1828.

Clinton was a true visionary who recognized the almost limitless opportunities available to New York City. He realized that two key things had to happen if New York was to become the predominant world port of the U.S. Access to the Midwest was needed to become part of the growing industry and trade in the Great Lakes region, and New York had to expand northward to accommodate a significantly larger population.

Fig. 2.8 Portrait of DeWitt Clinton

By the strict rules of nature, New Orleans should have been the nation's greatest port. It had access to virtually all of the nation's midsection via the Mississippi River system, and would have been the natural conduit for raw materials and goods coming from and going to the region. In 1811, DeWitt Clinton proposed that a 363 mile canal be built connecting the Great Lakes with the Hudson River. Since most existing canals of the era were in the order of 5–10 miles long, this recommendation was viewed by many as pure fantasy. It was variously referred to as "Clinton's Ditch" and "Clinton's Folly." Thomas Jefferson's initial response to a request for federal funding was hardly sympathetic: "It's a splendid project, and may be executed a century hence." Despite a lack of federal support, in April 1817, the NYS legislature passed a bill guaranteeing funding of the project. On July 4, 1817, New York's new Governor, DeWitt Clinton, presided over the groundbreaking for the Erie Canal in Utica, and construction commenced simultaneously to the east and west.

Without federal support (indeed, Jefferson pronounced the plan "madness"), Clinton lobbied long and hard to obtain financing for the canal. He created a business plan in which public funds were used to entice private investment. Clinton eventually convinced the major banks that they would make money by investing in the project. New York State was then able to issue bonds to build the canal.

The canal was finished in seven years, and had several operational legs running before construction was complete. The project finished on time, within budget, and opened three years ahead of schedule. From very early in its operation, it was making money, and investors were satisfied with the results of their investment. The project included 83 locks and one water bridge.

With the opening of the Erie Canal, shipments that once took 11 weeks to reach New York from the Midwest, now took one week. The cost of shipping was reduced significantly. The trade route for all of the goods and materials to and from the Midwest now ran through the Port of New York.

2.6.3.3 Initiation of Scheduled Packet Service

In 1818, the Black Ball Line inaugurated regular packet service between New York and Liverpool. Before this time, all ships made their trips on an irregular basis depending upon when they had sufficient goods booked to make the trip profitable. The Black Ball Line, owned by a group of Quaker merchants, namely, Isaac and William Wright, and Francis and Jeremiah Thompson, began running ships at regularly scheduled times. The service attracted a great deal of freight, as shippers could now count on when ships would leave and when their goods would arrive, and several other shippers inaugurated similar services over the next decade. This seemingly small change resulted in a revolution in shipping services, and served to further strengthen New York's place as the nation's leading port.

2.6.3.4 The Cotton Trade

The last major achievement was less spectacular, but was indicative of the financial influence of New York. Using control of shipping lines and banks, New York's financiers virtually forced most of the rapidly-growing cotton trade to flow through

New York. Southern cotton growers would have to send their raw materials to Europe through the New York port. The returning goods flowed through New York as well. By the 1820s, New York had a major piece of the cotton market, even though they were not directly involved in production or consumption of cotton. New York was making a fortune via financial manipulation of trade markets.

In the period between 1800 and 1825, New York had effectively defeated the chaos of nature, setting the stage for a period of explosive growth and development. In doing so, however, it set the stage for the chaos of man that would accompany that growth, and that would shape the future of the city in ways no one at that time could have imagined.

2.7 New York, the Civil War, and the Industrial Revolution

In 1825, the population of Manhattan had reached 170,000, virtually all confined to the southern tip of the island. Only two miles from lower Manhattan, the terrain was dominated by farms and forests. The city had no regular police force, few public schools, no fire department, and no public transportation. By 1900, the population had exploded to 3,437,202, and the city had left its confines of lower Manhattan to engulf the entire island and the beginnings of the spread to the "outer boroughs." The city was poorly situated to accommodate the explosive growth in population that would soon engulf it. By 1860, the population was four times what it was in 1800; by 1890, it was eight times what it was in 1800.

2.7.1 The Flood of Immigrants Begins

The Erie Canal, the great harbor, regular packet service, rapid growth of small manufacturing, and money made through the cotton and slave markets made New York a magnate for immigrants. The first waves of immigrants were primarily Irish, fleeing English control of their homeland and the poverty of their economy. The flow of Irish immigrants would become a tidal wave in 1845, when the great potato famine in Ireland forced millions from their homeland. By 1865, there were more Irish in New York than in Dublin. Significant German immigration also occurred during this period. On one record day, June 2, 1836, an astonishing 15,825 immigrants from Europe flowed into New York.

Small manufacturing businesses sprung up throughout lower Manhattan. Originally, small factories located themselves within tenement neighborhoods. By the 1830s, however, they began to centralize, creating the first purely business districts in the city.

Immigrants flooded into the tenements of the "Five Towns" district, better known as the present "lower east side" of Manhattan. Conditions in the tenements were unregulated, and the housing conditions for immigrants were generally horrible. By 1840, the density of these tenements was 300 people per acre. Disease became a principal concern for residents, as cholera and other illnesses become major risks to life. The tenements became the first NYC slums.

Fig. 2.9 Five Towns (i.e., Five Points) in 1827. Anthony Street veers off on the right; Cross Street on the left; Orange Street goes from left to right in the foreground

While some Irish immigrants were skilled workers, most were unskilled. They began to compete for and take the lowest paying jobs which had been dominated by Blacks for many years. The living conditions of the tenements, and the intense competition for unskilled jobs and housing created racial, economic, and social tensions on a scale that not been previously experienced. Profiteers capitalized on these tensions, displacing Blacks because the Irish would work for less. When scattered strikes occurred, the same profiteers would hire Blacks as strike-breakers. These tensions would periodically explode, and would play a major role in the development of the city for over a century.

2.7.2 *Birth of Print Media*

The 1840s gave birth to an explosion of print media. Numerous daily "penny papers" were published, and print journalism became a major medium for communication and information. In earlier times, the city was small enough that most people knew what was going on by word-of-mouth. By the 1840s, the city had become too large and too complex for most to understand, and the market for newspapers was born.

The first major daily was the *New York Herald*, published by James Gordon Bennett. Most papers were tabloids, focusing on the sensational: scandal, disease, murder, rape, and fires. By 1850, New York boasted 20 daily and 40 weekly newspapers, most located in a small area that became known as Printhouse Row (near the present City Hall). Morse also opened a telegraph office there.

In 1841, Walt Whitman arrived in Manhattan, looking for work at one of the newspapers. He found work at the *Aurora*, one of the "penny papers" in circulation, and began to write his revolutionary poetry, discarding the historic rhyme and rhythm of European poets for a freer, more open style, and exuding the vibrancy of urban life, including sexuality. His work did not find an immediate audience, and he achieved most of his acceptance late in life and posthumously.

He self-published the original edition of his famed poetry collection, *Leaves of Grass*, on July 5, 1855. He would spend most of his life adding to it, and revising it, releasing several versions over the years, including a "deathbed" version he was working on when he died.

Another significant event in 1841 was the opening of the American Museum at Broadway and Printhouse Row. The museum was a sideshow of oddities, created and managed by Phinius T. Barnum (better known as "P.T."). The show included oddities such as "General" Tom Thumb, Chang and Eng (Siamese twins), and others. It also included a "lecture hall" in which patriotic displays and shows were presented. It became a popular distraction for New Yorkers. In the 27 years that it survived, the museum sold 42 million tickets. The museum was thought to be a reflection of the complex relationships between the city and its people.

Fig. 2.10 P.T. Barnum's American Museum at Printhouse Row in 1858

2.7.3 *The Immigrants and the Rise of Tammany Hall*

A movement to improve the "slums" was started by Protestant missionaries, and spurred the creation of the Children's Aid Society. While providing for improved medical care, the movement included Protestant instruction, and was viewed with much suspicion by immigrants. Street violence became a significant problem and a common occurrence.

It is during this turbulent period, between 1830 and 1860, that immigrants become a major political force in New York City. The Tammany Society had been formed in the 1786 as a fraternal organization, adopting its name from Tamanend, a leader of the Lenape Indians. The society adopted many Indian words and customs, and referred to its hall as a "wigwam" and its leader as the "Grand Sachem."

By 1798, the Society's activities had become substantially politicized. One of its early leaders was Aaron Burr, under whom the Society became the center of support for Jeffersonian Republicans in New York. Many historians believe that John Adams would have won the Presidential election of 1800 were it not for the support of the Tammany Society in New York. After 1839, the Society became the NYC center of the Democratic Party. Eventually, the organization would take on the same name as its headquarters, that is, Tammany Hall.

The politicians of Tammany Hall were the first to realize the immense power of immigrants. They helped immigrants find jobs and housing, and aided them in becoming citizens, after which they could vote in large numbers for candidates supported by the Hall. Their first major success occurred with the election of Fernando Wood as mayor in 1854, based largely upon a huge Irish and German vote. Immigrants were mobilized with promises of jobs and other benefits. The upper class was outraged. They lobbied for and obtained a change in the NYC charter that limited the power of the mayor. This led to a nasty confrontation between city and state police on June 17, 1857. Though the state prevailed in the confrontation, the power of the mayor was soon restored.

Fig. 2.11 Fernando Wood, the first Tammany Hall mayor, circa 1860

In 1858, Tammany Hall saw a new leader, that is, William M. (Boss) Tweed. Tweed became synonymous with rampant corruption during his reign. Tweed had been elected to the U.S. House of Representatives in 1852, the NYC Board of Advisors in 1856, and the NYS Senate in 1867. Tweed became a close friend of Jay Gould, who made him a director of the Erie Railroad; Tweed repaid the favor by arranging legislation favorable to the railroad's interests.

Fig. 2.12 Portrait of William Marcy "Boss" Tweed

In 1858, Tweed was appointed to the New York County Board of Supervisors. Together with a number of cronies on the Board, contractors were forced to turn over 15% of their city contracts to Tweed and friends.

In 1870, Tweed successfully secured the passage of a new charter for NYC, placing full control of the city in the hands of the Mayor, the comptroller, and commissioners of parks and public works. Using control of these bodies, Tweed stepped up his activities to plunder the city. While the total amount stolen was never precisely established, estimates range between $25 million and $200 million. Proving that there is no honor among thieves, Tweed was indicted and convicted of fraud in 1871 after an associate, disgruntled with his take, gave the *New York Times* evidence that money was being stolen.

2.7.4 The Financial Collapse of 1857

In August of 1857, a collapse of the financial markets on Wall Street caused a substantial panic. Over 25,000 workers, mostly poor unskilled laborers, were thrown out of work, while another 100,000 were placed in great difficulty. The Fall of 1857 followed with many banks failing, and many businesses closing, creating even more unemployment and economic deprivation.

2.7.5 Central Park

In the midst of the economic and social turmoil of the era, a defining public works project of unimaginable scale was born: Central Park. The Commissioners' Plan of 1811 did not include Central Park, or any other park, for that matter. With the rapid expansion of New York's population, the need for open space became obvious. William Cullen Bryant and Andrew Jackson Downing began to publicize the need for a major park in New York in 1844. In 1858, the New York State legislature designated a 700 acre area from 59th Street to 106th Street for creation of the park, at a cost of $5,000,000 for the land alone. The state appointed a Central Park Commission, which selected a plan and design created by writer Frederick Law Olmstead and architect Calvert Vaux.

The vision for the park was that it would be a great equalizer, a magnet for rich and poor alike, providing a place of respite and reflection for both. The mixing of ethnicities and economic classes would, it was hoped, bring about understanding, reducing the ever-climbing tensions between and among the diverse cultural and economic populations of New York.

From the very beginning, the dream was sabotaged. Before work could begin, 1600 residents within the park's designated area had to be removed, using the power of eminent domain. Those displaced were primarily Irish and Black. Among the displaced was Seneca Village, a Black settlement of about 270 people that included a school and three churches. Residents of the settlement were scattered throughout Manhattan, and their cohesive community was effectively destroyed.

Further undermining its lofty goal, construction jobs for the park went mainly to the Irish. Because of historic tensions between the Irish and Blacks, Blacks were specifically barred from working on the park.

While the construction of Central Park would continue until 1873, the first section was opened to the public in the summer of 1858. From the very first day, it became clear that Central Park would be a refuge for the rich. Few working class people could walk from their homes to the park, and few could afford the cost of public conveyances to get to it. A series of regulations on the use of the park, instituted by Olmstead, reduced its desirability to the masses. Group picnics and walking on the grass were prohibited. Games, such as baseball, were similarly prohibited. A park police force was created to enforce these rules and to insure the gentility prevailed.

2.7.6 The Civil War Approaches

As the election of 1860 approached, the park was put on hold as tensions between the north and south strained to the breaking point over the issue of slavery.

Simmering beneath the surface since the drafting of the Constitution, the issue of slavery had moved to the forefront. Talk of secession was already rampant in some southern states. New York played a unique role in both the election of 1860 and the Civil War which followed shortly thereafter.

On February 25, 1860, a relatively unknown Illinois politician arrived in Manhattan to gain support for his Presidential bid. On Sunday, Abraham Lincoln took a ferry to Brooklyn to hear a heated sermon against slavery. He was scheduled to give a major address at Cooper Union Hall on Monday. He worried about how well he would go over with an urban crowd, given his rural roots. Deep in thought, he walked around lower Manhattan, stopping to buy a black top hat that would become his trademark. The weather was poor, heavy rains prevailed, yet, 1500 people arrived to hear his speech.

The speech, which made a clear case for restricting the growth of slavery, was extremely well received. Many papers published the speech in full, and it soon received national distribution and attention. Lincoln would later say that "the Cooper Institute made me President."

The rapid rise of Lincoln, however, alarmed the south greatly, and talk of secession increased. Many in New York were similarly alarmed. New York was still making substantial money from its various trade enterprises with the south, and southern states owed over $40 million to New York financial institutions. On January 21, 1861, Mayor Fernando Woods actually introduced a motion to have New York secede from the union and declare New York a "free city." On April 13, 1861, news of the attack on Fort Sumter, the initial battle of the Civil War, reached New York. On April 20, 1861, over 200,000 New Yorkers took to the street in support of the Union. Secession plans were never implemented, and New York's support of the Union was never questioned thereafter.

Fig. 2.13 Portrait of Abraham Lincoln

As with most war efforts, New York initially gained with the start of hostilities, as the manufacturing sector shifted into high gear to produce materials for the war. However, high inflation set in, and rents soared. As a result, the poor of New York began to suffer greatly. As long-standing racial animosities increased with inflation-related problems, three things brought them to a boiling point:

- On January 1, 1863, Abraham Lincoln issued the Emancipation Proclamation. In New York, white immigrants feared that freed slaves would storm New York and take most of the available unskilled jobs.
- In March of 1863, Congress passed the Conscription Act. Many immigrant workers feared that they would be forced to fight a war to free slaves who would take their jobs. The Act allowed any person to buy out of the need to serve for $300, or by providing a replacement. Virtually all of the middle class and wealthy bought out of their responsibility; the poor were placed in a lottery for selection in the draft.
- In Spring of 1863, General Robert E. Lee invaded northward through Virginia. As the Battle of Gettysburg approached, New Yorkers feared that the city could be overrun by the confederate army.

On July 11, 1863, New York's draft lottery commenced. The Conscription Act called for draft lotteries in states that did not meet their enlistment quotas. Despite legal challenges, the lottery took place, and by 6:00 in the evening, 1200 names had been drawn. Unrest over the draft simmered over the next few days, and exploded for three days, that is, July 14–17, 1863.

It is not clear how the so-called Draft Riots got started. It was not thought to be a planned event, but one that snow-balled out of control in a city virtually unprotected by police or militia, most of whom were engaged at Gettysburg.

What is clear is that on the evening of July 14, small gangs of primarily Irish workers began to move toward the draft office. By the time they reached the office, the groups had merged into a single mob of about 15,000. Before long, the crowd turned ugly, burning down the draft office and cutting telegraph wires. The crowd turned on the few militiamen who had tried to protect the office, killing one.

After burning down the draft office, the crowd dispersed into smaller mobs that continued to rage throughout Manhattan for the next three days. The mobs wreaked havoc on two very different groups, namely, the wealthy and powerful on one side, and Blacks on the other. Homes of the wealthy, politicians, and financiers were burned to the ground. Newspapers were attacked and burned. An orphanage housing over 200 Black children was surrounded and threatened. The children were led out by a back door just before the building was burned down. They had to be placed on a barge to protect them from the ever-widening barrage of violence. No one is sure how many were killed, though official figures placed the number at 119 (18 Blacks, 85 rioters, and 16 soldiers). However, many more were simply thrown into the river and left to drown.

After three days, the Union army began to arrive from Gettysburg, and order was restored. Ironically, most of the militiamen and police who restored order were Irish.

Proving that there is no situation bad enough not to be manipulated for political gain, Boss Tweed arranged for Tammany Hall to provide loans to poor workingmen to buy out of their service.

As war drew to a close, Blacks were leaving New York in large numbers, and a new Board of Public Health began to initiate the first new housing laws since 1817. A professional Fire Department was established, and the size of the Police Department was greatly expanded.

On April 15, 1865, New Yorkers were stunned to learn of the assassination of President Abraham Lincoln. When his funeral train reached New York, a huge parade was held to honor him. Despite initially being banned, Blacks were permitted to march in this final New York tribute to the fallen President.

2.8 New York Beyond the Civil War

There is, of course, a great deal of New York history beyond the Civil War. Increasing industrialization continued, and New York's role as the financial center of the nation was solidified.

The period before and during the First World War brought new waves of immigrants, this time from Eastern Europe, to the city. After the war, the city embarked on over a decade of unparalleled economic and cultural growth, known as the Roaring Twenties. Broadway became the center of a new popular opera form, and Harlem gave rise to some of the greatest jazz musicians in the world. In 1928, New York's Catholic, reformist governor, Al Smith, ran for President, making Tammany Hall a national player for the first time. He lost in a landslide to Herbert Hoover in a campaign filled with virulent anti-Catholic, and anti-New York fervor. Al Smith was also a strong supporter of the repeal of Prohibition, a major political issue at the time.

The Roaring '20s came to an abrupt end with the collapse of the stock market in late 1929. As the fallout from the crash spread, the Great Depression took hold of the city and the nation. From the desolation of the depression, however, came some of most critical accomplishments in restoring the city. Tammany Hall's last prewar Mayor, the infamous Jimmy Walker, was forced out of office when investigations uncovered massive graft, much of it lining the pockets of the Mayor himself and his Tammany Hall cronies. He was replaced by one of the most dynamic mayors the city has ever known, that is, Fiorello LaGuardia.

LaGuardia served for 12 years, and spent a good deal of his time dismantling as much of the Tammany machine as possible. Virtually every appointed official with Tammany ties was fired and replaced, with one critical exception.

LaGuardia formed a close bond with New York's former governor, then President Franklin Delano Roosevelt, and with another man who would become the principal architect of its infrastructure, Robert Moses, initially a protégé of Tammany Hall's Al Smith. When LaGuardia first took office, unemployment was at 50% in the city, and the city was in debt to virtually every bank. He was forced to cut many services in a city that was clearly in dire need of more services.

FDR's New Deal brought about a new era for the funding of public works. Amassing funds from new federal programs, Robert Moses, acting as LaGuardia's Commissioner of Parks, began the systematic rebuilding of every city park, including Central Park. The park system was reclaimed in less than a year after decades of neglect, and tens of thousands of workers were put to work. Moses proceeded to pick up the Triborough Bridge project, which had been started and abandoned for years. Finally, he took on the challenge of creating one of the world's greatest beaches on the south shore of Long Island – Jones Beach – and a system of highways and bridges to allow all New Yorkers to reach it. Moses, who had started his career in the service of Al Smith, was the most powerful figure in the state for over 50 years. Building an insulated system of commissionerships and public authority directorships, he was virtually beyond the control of the political system, and insulated from the people. Nevertheless, he made the city accessible to automobiles, which he recognized as the new populist form of transportation. Along the way, he built and improved parks, pools, and the Niagara Power project. Historians will long argue over whether the balance of his works were good or bad (or both), and anguish over the sometimes-strong-armed tactics he applied to implement his plans. He was the architect of the 1939 World's Fair, and his career ended with his direction of the 1964 World's Fair, which was seen as less than successful, at least financially.

Ahead lay the challenges of World War II, the great prosperity of the postwar period, the baby-boomers, new waves of immigrants, the ever-expanding role of the automobile in American life, and numerous other challenges.

At the end of the Civil War, however, much of the city's essential character, however, was certain. It was, and would remain, the financial capital of the nation, and perhaps, of the world. It was, and would remain, one of the world's largest and most densely populated cities, with an ethnic, national, and religious diversity unknown anywhere else, with all of its benefits and problems. The era when most people could walk to work was over, and relief from the awful density of lower

Manhattan required that new areas be made accessible for development. The unification of the five boroughs into a single consolidated New York City didn't occur until 1898. However, by then, the need to access these "outer" boroughs for development was clear.

From this point, the history of the city and the history of its transportation systems substantially merge. The growth and prosperity of New York would forevermore require an infrastructure of public and private transportation services. Transit, already begun and taking baby steps, was, by 1860, ready to play a major role in the city's future.

References

1. Burns, R.: A Documentary Film, Public Broadcasting System Home Video, New York (2004)
2. Mushabac, J., Wigan, A.: A Short and Remarkable History of New York City. Fordham University Press, New York (1999)
3. Bernstein, I.: The New York City Draft Riots and Their Significance for American Society and Politics in the Age of the Civil War. Oxford University Press, New York (1990)

Figure Sources

2.1: *Courtesy of* Library of Congress, Prints & Photographs Division, Detroit Publishing Company Collection, LC-D418-22636.
2.2: *Courtesy of* Wikimedia Commons.
2.3: *Courtesy of* Wikimedia Commons.
2.4: *Courtesy of* Library of Congress, Prints & Photographs Division, LC-DIG-pga-001466]. Harris CX, Etching 1908, "The Surrender of Nieuw Amsterdam"
2.5: *Courtesy of* Wikimedia Commons.
2.6: *Courtesy of* Library of Congress, Prints & Photographs Division, HC White & Co. Collection, LC-USZ62-131469.
2.7: *Courtesy of* English Wikimedia.
2.8: *Courtesy of* Library of Congress, Prints & Photographs Division, Childs & Inman Collection, LC-USZ62-50394.
2.9: *Courtesy of Wikimedia Commons,* George Caitlin, *The Five Points*, oil 1927.
2.10: *Courtesy of* Wikimedia Commons.
2.11: *Courtesy of* Library of Congress, Prints & Photographs Division, Brandy-Handy Photographic Collection, LC BH82 5229B.
2.12: *Courtesy of* Library of Congress, Prints & Photographs Division, LC-USZ62-22467.
2.13: *Courtesy of* Library of Congress, Prints and Photographs Division, Jackson, C, October 1, 1858, Library of Congress,USLCZ62-16377.

Chapter 3
Enabling Technologies

For the modern transit system to develop, a number of critical enabling technologies were needed. By the beginning of the 19th century, some basic inventions had already been developed, but others were not yet on the immediate horizon.

The modern bus followed the invention and growing numbers of automobiles. Rail rapid transit, the high-capacity centerpiece of any major city's public transportation system, and most certainly, of New York's public transportation system, required four fundamental technologies to be developed:

Tunneling
The Steam Engine
Electricity
The Electric Traction Motor

There are others, of course. Modern signaling and train control systems were of great value, but rail transit systems were originally based upon relatively simple systems, which have become more sophisticated and reliable over the years.

This chapter attempts to provide a very basic and introductory description of the four critical technologies noted. Without them, the NYC transit system as we know it could not have been developed.

3.1 Tunneling

It is virtually impossible to date the first tunnel ever constructed. There is enough physical evidence and historical record to support the fact that tunnels were built throughout antiquity by several different civilizations for a wide variety of purposes.

Over 4000 years ago, during the period 2180–2160 B.C., it is believed that the Babylonians built a 3000 ft long, 600 ft wide tunnel under the Euphrates River to connect the Royal Palace with the Temple of Jupiter on opposite sides of the river. The methodology was novel: During the dry season, the flow of water in the Euphrates was greatly diminished. The Babylonians literally built a new waterbed to divert the diminished flow, while slaves dug a trench into the now-dry riverbed, constructing brick walls, floors, and ceilings to enclose it. When the structure was completed, slaves back-filled over the tunnel roof, and the diverted path of the river was dammed, forcing the river back to its original path, over the now-constructed tunnel. There is no historic record to suggest that any other

R.P. Roess and G. Sansone: The Wheels That Drove New York, STTT 1, pp. 35–52.
springerlink.com

underwater tunnel through soft earth was constructed until the middle 1800s, when the Thames Tunnel was built by Mark Brunel.

Most tunnels in antiquity were built for reasons other than for human conveyance. Egyptian tombs commonly included tunnels as early as 1250 B.C. The tomb of Ramses II included several corridors penetrating sandstone cliffs to a depth of several hundred feet. Babylonian irrigation tunnels in the Indus Valley existed as far back as 2500 B.C.

Tunnels were also commonly used in warfare. Short tunnels were often carved under enemy walls to either transport troops, or (more commonly) to undermine the integrity of the wall and cause it to collapse. Even the Old Testament in the Bible, namely, Joshua 6:20, refers to tunneling under a protective city wall, setting a fire under it, and causing it to collapse.

The first road tunnel is thought to have been built in Campania, Italy, sometime between the assassination of Julius Caesar and the coronation of Augustus. This tunnel still exists, and is approximately dated as 36 B.C. Medieval legend has it that the tunnel was built by a sorcerer.

The most common reason for the building of early tunnels, however, was mining. Tunnels into the earth were (and are) needed to obtain iron ore, gold, silver, and coal.

Early tunneling techniques utilized hand tools, including picks and shovels, copper hand drills, and hollow-reed drills. A common technique was to cut grooves along the rocks' surface, with a series of strategically-placed drill holes. The holes were filled with porous wood and filled with water. The wood would expand, forcing the rock apart along predetermined fracture lines. The Romans added a technique called "fire quenching" in which the rock surface was heated by fire, and suddenly quenched with cold water, causing the rock to fracture. While a very effective procedure, it was extremely dangerous to the slaves who had to carry it out.

Except for the early Babylonian example, virtually all tunnels in antiquity were built through material strong enough to be self-supporting until interior supports and structures could be built.

3.1.1 Mark Brunel's Tunneling Shield and the Thames Tunnel

Mark Isambard Brunel was born in France on April 25, 1769. Displaying a talent for drawing and mathematics, Brunel showed an early interest in engineering. After serving in the French Navy, Brunel developed strong royalist sympathies, and was forced to flee to the United States in 1793 as the French Revolution took hold. He practiced as an architect and civil engineer in New York, and actually served as the city's chief engineer for some time. He submitted a plan for the new capital in Washington D.C., but economic and other considerations led to L'Enfant's plan being selected and implemented instead.

He returned to England in 1799, where he became a well-known inventor. He invented a machine to mass-produce ship pulleys, and machines for the bending and cutting of timber. During the War of 1812, his factory supplied the English army with boots, using novel mass production methods. With the end of the war in

1815, he was left with 80,000 pairs of surplus boots which the army refused to buy. His lumber factory had burned down in 1814, which left him essentially bankrupt. He was imprisoned in 1821 for indebtedness, but his friends obtained his release within several months.

At the beginning of the 20th century, in 1801 and 1807, attempts were made to build a tunnel under the Thames River between Rotherhithe and Wapping. Both attempts failed when initial excavations hit quicksand. In 1818, Brunel patented a design for a tunneling shield, a new device that would make tunneling through water-bearing strata possible and relatively safe.

Fig. 3.1 Brunel's Tunneling Shield. Side view shows one bay of worker cells, which were actually three deep

Despite the fact that his original patent described a circular shield, he began work using a rectangular shield that was more practical for provision of isolated working spaces. The shield contained individual cells in which single laborers would excavate material. Iron plates at the top and bottom supported the earth as the shield moved forward. Behind the shield, a shelf allowed workers to construct a brick structure to permanently support the earth. The shield was forced forward by two screw jacks positioned against the constructed brick wall behind the shield. By turning the screw jacks, the shield was forced forward in increments of 114 mm, the length of one brick. As the shield moved forward, workers would add another later of brick to the tunnel structure.

In 1825, construction of the Thames Tunnel began again, using a shield that included 12 separate frames, or 36 cells, each holding one laborer. After 2 ½ years, the tunnel was half finished when a series of deadly accidents caused the work to be shut down. The shield itself was bricked up and work suspended in July of 1827. Work was stopped for 10 years, and was restarted in March of 1837. The brick wall was removed, and a new shield of similar, but improved, design was installed. Despite additional accidents, some involving deaths, the tunnel was completed in 1842. It was officially opened to pedestrian traffic on March 25, 1843. Within 24 hours, more than 50,000 Londoners had passed through it; within 15 weeks, one million people had walked through it.

Despite its successful completion and dramatic usage, it remained a pedestrian tunnel for 23 years, with a toll of one penny per usage. In 1865, the tunnel was purchased by the East London Railway. It eventually became part of the London Underground, and is still in use today.

Fig. 3.2 Thames Tunnel used by pedestrians in 1843

Fig. 3.3 Thames Tunnel used by steam trains in 1870

3.1.2 Barlow and Greathead's Tower Tunnel

Peter William Barlow obtained a contract to build a second tunnel under the Thames, which became known as the "Tower Tunnel." He engaged James Henry Greathead as his chief engineer. While they did not invent the tunneling shield, Barlow and Greathead were credited with making improvements that enabled modern tunneling shields to develop. He replaced the rectangular shield with a cylindrical one, which was smaller and lighter than Brunel's, and contained fewer moving parts. He also replaced bricks as a tunnel lining with prefabricated cast iron sections. Construction was started in 1869 and completed in 1870. Because the size of the tunnel was far smaller than Brunel's, the tunnel was only able to accommodate pedestrian traffic, which it did through 1894. At that time, the Tower Bridge, built nearby, was opened, and the tunnel was connected to the water main system.

With exception of automated devices replacing human laborers, the Barlow/Greathead shield design is basically still in use in modern tunneling.

3.1.3 Modern Tunneling Shields

Modern tunneling shields can be used to evacuate tunnels with a circular cross-section through virtually every type of rock, soil, or sand generally encountered. A Tunnel Boring Machine (TBM) typically consists of one or two shields with trailing support mechanisms. The front of the shield is a rotary cutting wheel. Behind it is a chamber into which soil (sometimes mixed with slurry) is evacuated, as well as systems for removal of the soil. Hydraulic jacks are located behind the shield, braced against the completed portion of the tunnel, to propel the shield forward. Inside the completed portion of the tunnel, other supporting systems are located, for example, slurry pipelines (where needed), dirt removal systems, rails for transport of precast tunnel segments, and control rooms.

Modern tunnels are usually supported by precast reinforced concrete sections that may be jacked into place. If the material is hard enough to be self-supporting for a period of time, forms may be poured and reinforced concrete structures built in place.

Cutting wheels rotate at 1–10 rpm, depending on the geological conditions that are present. Rock chips or soil will usually fall onto a conveyer belt system that carries the material out of the tunnel. If slurry is used, material will be pumped back to the surface using a pipe system.

Fig. 3.4 A modern tunnel boring machine (TBM) used to dig rock tunnels at Yucca Mountain

3.1.4 Other Tunneling Systems

There are, of course, other tunneling methodologies still in use. Use of blasting in hard rock areas is still relatively common. Where tunnels are to be close to the surface, cut-and-cover techniques are still used, though only where the disruption to surface activity will not create significant problems.

The early subways in New York were built using cut-and-cover technology. In this method, an open trench is built within which the tunnel structure is built. Eventually, it is covered over and the surface restored. If a "top-down" system is used, the ceiling of the trench is the first thing to be built, and the surface can more quickly be restored. In such a case, temporary supports for tunnel walls must be in place if the material is not strong enough to be self-supporting. New tunneling technologies are available to construct close-to-surface tunnels without disrupting the surface, but these are expensive and time-consuming.

In the building of the Bay Area Rapid Transit (BART) system tunnel under San Francisco Bay, a unique tunneling methodology was adopted which was later used for constructing the 63rd Street Subway Tunnel in New York. Because the bay bed was mostly soft mud, and since the tunnel had to pass through a major earthquake fault, the tunnel was designed to literally "float" on the mud, with key tunnel joints specially-designed to allow for significant movement without rupture. A trench on the bottom of the bay was dredged, into which fully preconstructed tunnel sections were "sunk" into place, with sections being connected by divers. Sandy backfill was then used to cover the tunnel. In New York, traversing an earthquake fault was not a problem, but the sinking of fully preconstructed tunnel sections into a dredged trench both significantly reduced construction cost and shortened the construction process. Unfortunately, speed was not of the essence in this case. The tunnel was completed in 1972. The first train did not run through it until 1989. The tunnel has two levels, one for the NYC Subway, and another for the Long Island Railroad. The railroad level awaits completion of the Grand Central LIRR terminal, and, at this writing, has never been used.

3.2 The Steam Engine

Modern rail transit would not have been possible without the invention of the steam engine, and the steam locomotive. While trains powered by electricity and electric traction motors became the system of choice in the early 1900s, the first transit systems featured trains hauled by steam locomotives.

3.2.1 The First Steam Devices – Hero of Alexandria

The earliest records of steam-powered devices were invented by Hero of Alexandria (often referred to as Heron) during the first century A.D. His devices were interesting, but not considered useful or practical. They nevertheless mark the recorded beginning of man's fascination with steam, and its ability to provide mechanical energy.

Heron's first device was a toy. Water was heated to create steam inside of a closed vessel. Using tubes, the steam enters a ball with opposing exit valves. The escaping steam creates an "equal and opposite" force, causing the ball to spin.

Fig. 3.5 Hero of Alexandria's revolving steam ball

Hero of Alexandria also invented a steam device to open a temple door. The priest would light a fire, thus heating water in an enclosed sphere. Steam pressure would force water through a tube into an evacuation bucket, which would move downward under the weight of the water. The bucket was attached by a pulley and rope system that would open the temple door. The device was considered impractical, as manual labor was needed to reset it after each opening. The priest could just as easily open the door by hand.

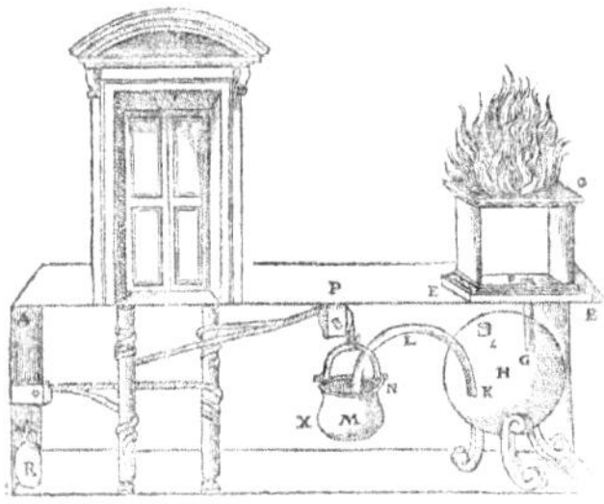

Fig. 3.6 Hero of Alexandria's steam-powered temple door-opener

3.2.2 Thomas Savery's Steam "Fire Engine"

In 1680, Denis Papin, a French physicist, invented a "steam digester" to soften bones. It was, in effect, the world's first pressure cooker, and the basic technology is still in use today. Because the initial device was likely to explode under steam pressure, he later invented a steam-release valve to guard against such explosions. Papin observed the pressure valve moving up and down in a regular pattern, and conceived the idea of a piston and cylinder engine. While he created early drawings, he never built such an engine.

In 1698, Thomas Savery, an English military engineer and inventor, greatly enhanced Papin's design and patented the first steam engine. It was originally designed to pump water out of mines. Hot steam was pumped into a chamber, which was subsequently cooled. The cooled chamber created a negative pressure that sucked water into the chamber.

Fig. 3.7 Thomas Savery's Steam Engine in 1702

Savery was the first inventor to use the term "horse-power." He characterized the power of his engine as the equivalent of about 10 horses. His engine was impractical, however, for its intended use. Mines were too deep, and boilers could not withstand the necessary pressure to function properly.

3.2.3 Thomas Newcomen and the "Atmospheric Engine"

In 1712, Thomas Newcomen introduced his atmospheric steam engine, built around a piston that was moved upward by steam pressure, and downward by atmospheric pressure.

Fig. 3.8 Thomas Newcomen's Atmospheric Steam Engine in 1705

Newcomen added a piston to his engine in the chamber. The negative pressure created when the chamber is cooled sucks the piston upward. Normal atmospheric pressure brings the piston back down. The piston is connected to a lever, which continuously rises and falls with the piston's motion.

On the other side of the lever, a long rod is attached to a pump deep inside the mine shaft. The "pump" consists of a piston that allows water to pass in only one direction. The pump, therefore, pushes the water upward through a number of check valves, to the top of the shaft.

Initially, the engine worked so slowly that valves were opened and closed manually. Eventually, this was replaced by lengths of rope that could open and close the valves based upon the movement of the engine itself. Newcomen's engine was the first deemed useful in industrial applications, principally in removing water from mine shafts.

3.2.4 James Watt and the Condensing Steam Engine

In the 1760s, Humphrey Gainsborough produced a model steam engine in which the heating and cooling cycles were separated, with the cooling cycle now referred to as "condensation." He showed it to a member of the Lunar Society. When James Watt (also a member of the Lunar Society) patented his first steam engine based upon the condensation principle, Gainsborough believed that Watt was using his design. He was never able, however, to prove this.

After studying steam engines in great detail over a period several years, Watt determined that there was great inefficiency in the process of heating and recooling the same chamber on a continuous basis.

Watt studied the operation of Newcomen's engine in great detail. Given the fuel-intensive need to reheat the cylinder chamber after each cycle, and the inefficiency of alternatively heating and cooling the chamber, he introduced two new concepts to the design:

- Thermal insulation of the cylinder chamber (or vessel): for efficiency, the vessel needed to be as hot as the steam itself.
- The condenser: The vessel where steam was to be condensed should be separated from the cylinder, and it should be kept as cold as the outside temperature.

Steam from the boiler forces the piston upward to its highest position in the cylinder. Using the differential in pressure between the steam in the cylinder and the air in the condenser, steam is forced into the condensing unit, where it is cooled. This creates a negative pressure in the cylinder, forcing the piston downward. Eventually, Watt made additional modifications to the steam engine, making it both more efficient and safer. He used the natural expansion of steam to reduce the amount of steam needed to propel the piston. He also invented the fly-ball governor, a device which automatically regulated the introduction of steam into the cylinder.

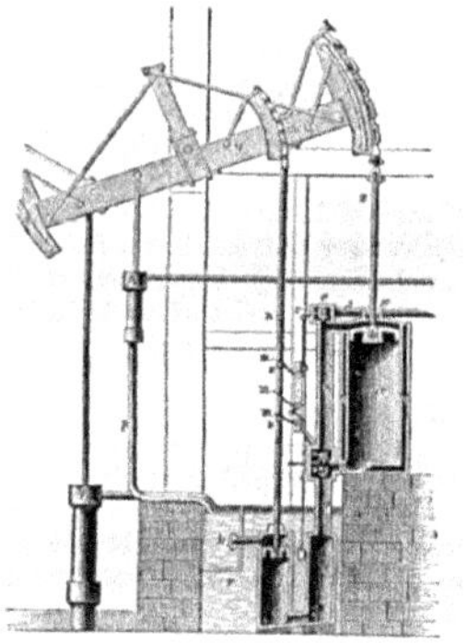

Fig. 3.9 James Watt's Steam Engine in 1774

Watt formed a partnership with Matthew Boulton, who provided financial support for his efforts. Watt and Boulton set up a factory to produce steam engines, which became very successful and made Watt a wealthy man.

While previous patents existed, Watt eventually developed a design in which the up-and-down movement of the piston could be converted into rotary power – the step that made the steam engine adaptable to power vehicular movement.

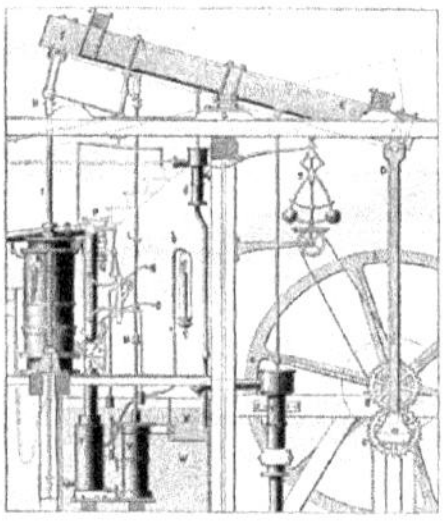

Fig. 3.10 Watt's steam engine in 1781, converting piston movement to rotary movement using the "sun and planet gear," seen at the center of the large wheel at the right

The mechanism was called a "sun and planet gear." As the drive rod is moved downward by the piston, it forces the upper wheel to rotate. The lower wheel must also turn, and it is connected to a larger wheel, which serves as a drive wheel. It can be connected to belts or chains to cause machine parts to rotate, or, by further gears, to a wheel propelling a vehicle. It is thought that the mechanism was actually invented by one of Watt's employees, that is, James Moulton, though the patent was in Watt's name.

Another important modification was the development of a reciprocating engine, in which steam propelled the piston in *both* directions (up and down), thereby providing a much smoother cycle, particularly for creating rotary motion. The result was called a "reciprocating" engine.

The addition of a slide valve allows steam to be directed to either side of the piston, depending upon its position in the cycle. Thus, when the piston is positioned to the left, steam is introduced on that side to push it forward. When the piston is on the right, the valve switches, and steam is introduced on the right side of the piston to push it to the left. The system worked best, however, using high-pressure steam, as opposed to the low pressures at which virtually all of Watt's engines operated.

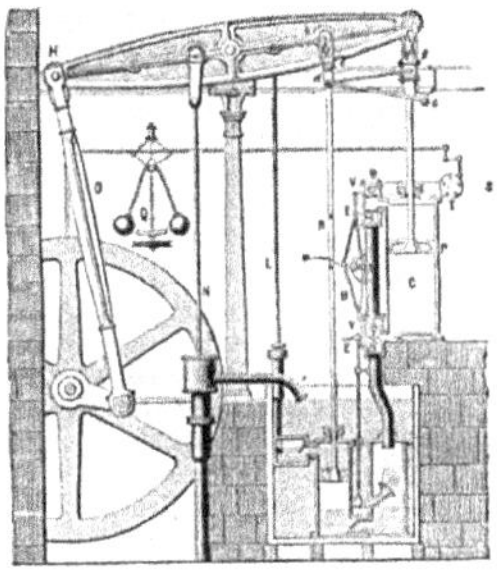

Fig. 3.11 James Watt's Reciprocating or Dual-Action Piston Engine in 1784

3.2.5 Steam Locomotives

One of the advantages of Watt's steam engine was that it worked using low-pressure steam. Power could be increased by building larger cylinders and pistons. It could also be increased by increasing the steam pressure within the cylinder. This, however, was a safety issue, as the construction of a vessel strong enough to house high-pressure steam was a problem, as was a foolproof system of preventing the pressure from getting too high. There had been several spectacular and deadly explosions of Savery engines, and although Watt patented designs for high-pressure engines, he never built one.

While a number of inventors conceived of using steam engines to power locomotives, they all recognized that sufficient power to do so could only be provided by high-pressure steam engines.

The use of railed vehicles greatly predates the steam engine. Railed vehicles had been used in mines for over a century. They were pulled, however, by human- or horse-power. The use of rails greatly reduced wheel friction, and made pulling the vehicles much easier, which (conversely) allowed human- or horse-power to pull heavier loads. The rails also eliminated the problem of steering the vehicle along an appropriate path.

The first successful steam engine was built by Richard Trevithick in 1804. It was reportedly developed due to a bet with a mine-owner in South Wales. The challenge was to create an engine that could haul 10 tons of material along a 10-mile rail system in the mines. It ran with limited success on the mine-owner's 10-mile narrow-gauge railway for a number of years. Safety of the high-pressure steam locomotives, however, became a problem, and several explosions limited the commercial success of Trevithick's engine.

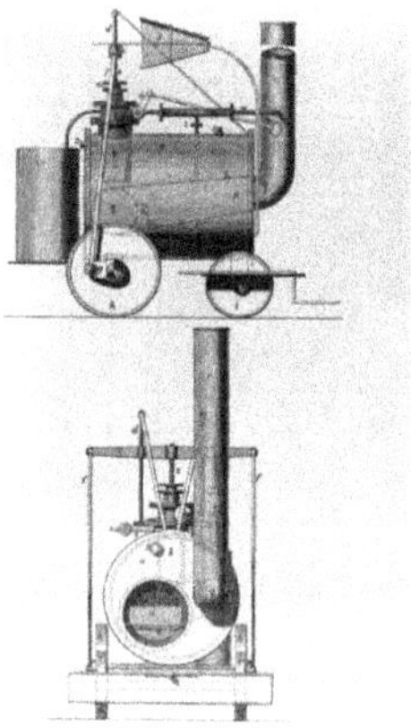

Fig. 3.12 Trevithick's Steam Locomotive,1804

No significant progress on steam locomotives was made until 1829. In England, a competition was held to design a locomotive to operate along a short railroad built between two trading cities. George Stephenson's "Rocket" won the competition and was pressed into service. The great innovation of this engine was the use of a tubular boiler, and pistons that operated directly on two drive wheels in the front of the engine.

Fig. 3.13 George Stephenson's Rocket

At the same time, the U.S. had started to develop steam locomotives, with the first appearing in 1829, that is, the Baltimore and Ohio Railroad's *Tom Thumb*. However, by 1830, in both England and the U.S., the age of steam-powered railroads was off and running, with improved designs occurring on a virtually continuous basis.

In 1804, Trevithick's steam locomotive could haul approximately 15 tons at a top speed of 5 mph. By the mid-1800s, sophisticated steam locomotives of the type illustrated in Figure 3.14, could approach speeds of 100 mph and haul significantly greater loads.

Fig. 3.14 Baltimore & Ohio's Royal Limited in 1898

With the development of steam railroads, urban public transit had found its initial power source.

3.3 Electricity

The history of electricity can fill volumes. The ancients obviously had some knowledge of electricity and electrical current. The famous "Baghdad Battery," discovered in 1936, is thought to be Parthian in origin, which would date the device between 250 B.C. and 220 A.D. There is some dispute over the dating of the device, and some scientists believe that it is more recent. In any event, man's knowledge of electricity apparently began thousands of years ago.

The years between 1700 and 1850 produced many inventors and scientists who studied and documented the behavior of electricity and charged particles. These included Luigi Galvani, Alessandro Volta, Michael Faraday, Andre-Marie Ampere, and Georg Ohm. The common measures of electricity, volts, amps, and ohms, are named after some of these scientists.

Ben Franklin conducted his famous experiment proving that lightening was essentially an electrical current at the beginning of this period, and literally sparked the interest of others in electrical characteristics and properties. He did not conduct the experiment as often pictured, that is, him standing on the ground holding a kite in a thunderstorm, as he would have been electrocuted, which he apparently understood.

The important steps in the development of practical electricity were those invented by Thomas Edison, one of the most prolific inventors ever known.

Fig. 3.15 Thomas Edison seated beside his phonograph

One of Edison's greatest innovations was his development of a "mass production" facility specifically devoted to developing new technology. His was the first industrial research laboratory, established in Menlo Park, New Jersey. Because he directed and owned the laboratory, he held title to all of its patents, even though he was assisted by many consultants, engineers, and scientists working for him.

His most original invention was the phonograph, the first device ever to successfully reproduce sound. This was one of his few totally "new" patents, as much of what he did was significantly improve upon the work of others, making previous inventions commercially viable.

The light bulb was not originally invented by Edison. Several other inventors had developed light bulbs previously, but they were all impractical due to extremely short life, high production expense, and high current needs. Edison invented the first practical incandescent light bulb. By 1879, he had developed a new concept using a high-resistance carbon filament in a very high vacuum. The first successful test bulb lasted 13.5 h, a significant, but not yet commercially viable, service life. By 1880, he submitted a patent for a bulb that would last over 1200 h. The patent was contested, but was eventually awarded in 1889, based upon his use of the high-resistance carbon filament that was demonstrably different from previous patents.

He put William Joseph Hammer, a consulting electrical engineer, in charge of Edison Light Works. In its first year of operation, the plant turned out 50,000 incandescent bulbs. This was truly Edison's genius, namely, the ability to take an invention from the laboratory to the factory in a very short time.

In order to capitalize on the invention of the light bulb, Edison needed to develop a system for delivery of electricity to businesses and residents. In 1878, Edison formed the Edison Electric Company in New York, with backing from several famous financiers, including J. P. Morgan and members of the Vanderbilt family. He patented a distribution system in 1880, and on September 4, 1882, switched on the Pearl Street generating system. It provided 110 volt direct current (DC) to 59 customers in lower Manhattan.

The principal difficulty with Edison's distribution of DC was that it could only be delivered to customers that were relatively close to the generating station. His main competition came from George Westinghouse, who was backing the use of alternating current (AC), a form developed by Nikola Tesla. AC current could be stepped up to very high voltages using transformers, and could be sent over thinner wires and much longer distances. AC current would be stepped down to a standard voltage for delivery to individual users. Edison went to "war" with Westinghouse by publicly touting the great danger of high-voltage transmissions. He even staged public "electrocutions" of dogs and cats using AC current. In one case, he electrocuted an elephant that had been condemned to death after killing several trainers. Despite his best efforts, however, the AC system was eventually adopted for mass distribution.

In 2005, there were still 1600 Manhattan customers of DC power when the service was finally discontinued by Con Edison. Edison, however, had, by the mid-1880s developed a power source that would eventually replace steam in urban rail transit systems.

3.4 The Electric Traction Motor

Electric traction motors, and their application to the powering of railroad vehicles occurred over a period of 50–60 years, and involved a number of inventors.

In 1835, Thomas Davenport exhibited a small railway operated by a miniature electric motor. In 1838, Robert Davidson, a Scotsman, developed an electric locomotive that attained a speed of 4 mph. In 1847, Farmer operated an electric car carrying two people in Dover, New Hampshire, and on April 29, 1851, a small road car built by Page (of the Smithsonian Institute) attained a speed of 19 mph. All of these vehicles, however, were powered by electric storage batteries with limited range and very high recharge times.

However, it was the development of the dynamo by Pacinette, Seimens, Farmer, and others that led to a practical electric traction motor. The principle of the dynamo is that when a wire is moved in the force field of a magnet, an electric current can be induced, that is, caused to flow in the wire. When wound on a drum armature, electric current is produced when the drum is made to revolve. A traction motor is a dynamo in reverse: it revolves when current is applied to its windings.

In 1879, Werner von Siemens built a short electric railway of 600 yd for the Berlin Exhibition of that year. Current was drawn from a third rail between the tracks, which also acted as the return for the dynamo. The locomotive carried up to 30 people at a speed of about 4 mph.

The man known as the "Father of Electric Traction," however, was Frank Julian Sprague. Sprague joined Edison's laboratory in 1883 after resigning from his Navy commission. While he only worked for Edison for a little over a year, he is credited with introducing mathematical analysis to make invention more efficient, and less of a "trial and error" process. He also improved Edison's system of mains and feeders for the distribution of DC power from a central generating station.

In 1884, he formed his own company, the Sprague Electric Railway and Motor Company. In just two years, he had introduced two important inventions. The first was a constant-speed, nonsparking traction motor with fixed brushes. It maintained constant revolutions per minute under different loads. The second was a method to return power to the main supply systems of equipment driven by electric motors.

Sprague also invented a system for collecting electricity from overhead wires. His "trolley pole" used a spring-loaded wheel to travel along the electrical wire. In 1887 and 1888, he installed the first electrified trolley system in the world in Richmond, Virginia. Richmond contained grades as steep as 10%, and the system provided evidence that his new technology could be applied in other cities. By 1889, 110 other electric railways employing Sprague's equipment had been planned or were under construction in the U.S. and abroad.

In 1890, Edison, who manufactured most of Sprague's equipment, bought him out, and Sprague turned his efforts to applying traction motors to the electric elevator.

Fig. 3.16 An old postcard shows Sprague's Richmond trolley system on Broad Street, the center of Richmond's theatrical district

In 1892, Sprague formed the Sprague Electric Elevator Company. In conjunction with Charles Pratt, he developed floor control, automatic elevators, acceleration control of car safeties and other advances. His elevators ran faster and with bigger loads than existing hydraulic or steam elevators, and within a year, over 500 had been installed. Like his railway business before, once he proved the utility of his elevator, he sold his company to the Otis Elevator Company.

Sprague's traction motor design led directly to the development of electric railways. This enabled modern rail transit to develop in cities. Likewise, his elevator designs allowed for the age of skyscrapers and the vertical development of cities, particularly, New York. He went on to develop a system for multiple-unit train operation, which allowed each car of a train to be electrically powered and operated in tandem.

3.5 Technologies Converge

By the middle of the 1800s, need and technology converged to allow elevated railways to appear in New York City, primarily powered by steam locomotives. Steam power, however, was generally considered impractical in tunnel operation, as the heat of the engine and the inability to dissipate steam in an enclosed space created serious environmental problems that were recognized even at that time. Steam operation of elevateds in New York also caused significant environmental disruption, given their closeness to buildings alongside the routes.

As the century turned, however, the existence of electric traction motors and their application to rail vehicles made the planning and construction of a true subway system possible. The city's first subway, the Interborough Rapid Transit system (IRT), would be operating by the end of 1904.

References

1. Sandstrom, G.: The History of Tunneling. Barrie and Rockliff, London (1963)
2. Bobrick, B.: Labyrinths of Steel: A History of the World's Subways. Newsweek Books, New York (1981)
3. Black, A.: The Story of Tunnels. Whittlesey House, McGraw-Hill, New York
4. Vogel, R.: Tunnel Engineering: A Museum Treatment, U.S. National Museum Bulletin 240. Smithsonian Institution, Washington, DC (1964)
5. Thurston, R.: A History of the Growth of the Steam Engine. D. Appelton & Co., New York (1878)
6. Trevithick, F.: Life of Richard Trevithick. E&N Spon (1872)

Figure Sources

3.1: *Courtesy of* Wikimedia Commons.

3.2: *Courtesy of* Library of Congress, Prints & Photographs Division, Washington DC, LC-USZ62-120899.

3.3: *Originally published in* Illustrated London News, January 8, 1870.

3.4: *Courtesy of* Wikimedia Commons.

3.5: *Originally published in* Knight's American Mechanical Dictionary, 1876.

3.6: *Courtesy of* Library of Congress, Prints & Photographs Division, Washington DC, LC-USZ62-110278.

3.7: *Originally published in* R.H. Thurston, "A History of the Growth of Steam Engines," D. Appelton & Co., New York, NY 1878, Figure 13.

3.8: *Originally published in* R.H. Thurston, "A History of the Growth of Steam Engines," D. Appelton & Co., New York, NY 1878, Figure 19.

3.9: *Originally published in* R.H. Thurston, "A History of the Growth of Steam Engines," D. Appelton & Co., New York, NY 1878, Figure 26.

3.10: *Originally published in* R.H. Thurston, "A History of the Growth of Steam Engines," D. Appelton & Co., New York, NY 1878, Figure 27.

3.11: *Originally published in* R.H. Thurston, "A History of the Growth of Steam Engines," D. Appelton & Co., New York, NY 1878, Figure 31.
3.12: *Originally published in* F. Trevithick, "Life of Richard Trevithick," E & N Spon, 1872, pg. 127, *courtesy of* Library of Congress, Prints & Photographs Division, Washington D.C., LC-USZ62-110377.
3.13: *Originally published in* R.H. Thurston, "A History of the Growth of Steam Engines," D. Appelton & Co., New York, NY 1878, Figure 55.
3.14: *Courtesy of* Wikimedia Commons.
3.15: *Courtesy of* Library of Congress, Prints & Photographs Division, Washington D.C., LC-USZ62-38662.
3.16: *Courtesy of* Wikimedia Commons; Lewis Kaufmann & Sons, "Theatrical District, Broad Street, Richmond, Virginia," from Wikimedia Commons.

Chapter 4
The Beginnings of Public Transportation in New York: Omnibuses and Street Railways

In 1825, virtually all of New York's 160,000+ residents lived on lower Manhattan, beneath what is now Canal Street. The length and breadth of the city could be walked in under 30 minutes. The vast majority of the population traveled on foot, while some traveled on horseback. The rapid growth of the city over the rest of the century, however, would soon require some form of public transportation to allow people to live further away from their workplaces. In the early- to mid-1800s, if the city was to grow, it would have to grow northward, to the unpopulated portions of Manhattan, and eastward, to Brooklyn. The age of skyscrapers was in the future, so vertical growth was not a serious option.

4.1 Early Transit Forms: Stagecoaches and Hackney Carriages

It is virtually impossible to date the first use of a horse-drawn vehicle for public conveyance. Crude stagecoaches were known to exist in Europe during the 1500s. Both in Europe and the U.S., stagecoaches were primarily used to haul mail over long distances. In the U.S., mail service to the west coast via stagecoach predated the famous Pony Express by three years.

In the U.S., the initial way west was the wagon train, followed by the stagecoach. The initial government contracts for mail service via stagecoach required the trip be completed in 25 days or less. To support the mail service, stagecoach depots with overnight accommodations were developed along several routes. The addition of passenger service was a very short leap. The American stagecoach has been memorialized in thousands of movie and T.V. westerns. The trip was rough, dirty, and dangerous. The threat of being "held up" was very real, as the stagecoaches usually carried gold and/or money for banks as well as mail and passengers.

Even after railroads replaced stagecoaches as a viable means of cross-country travel, stagecoaches continued to thrive, primarily to carry passengers from rail stations to outlying communities not on the mainline. Such services were in existence until the turn of the century.

R.P. Roess and G. Sansone: The Wheels That Drove New York, STTT 1, pp. 53–61.
springerlink.com

Stagecoaches, however, were clearly used for intercity travel as opposed to travel within a developed area.

In England, the use of horse-drawn vehicles on a for-hire basis, called "hackney carriages," began in the 17th century. These services effectively became the first taxicabs. In 1834, they were modernized as hansom cabs that also operated on a for-hire basis.

4.2 The Introduction of Omnibus Services

The concept of a horse-drawn service operating along a fixed route with designated stops, and a fixed or graduated fare, is generally accredited to the French. In the mid-17th century, French inventor Blaise Pascal convinced the Governor of Poitou to build seven coaches to provide passenger service along a fixed route in Paris. The service began on March 18, 1662, and charged a small fare. After some initial success, the service failed and was terminated after only a few weeks.

At one time, it was thought that banker Jacques LaFitte had run an omnibus service in Paris starting in 1819. There is virtually no evidence that such a service existed, and it is more likely that LaFitte, whose brother ran a stagecoach business in Paris in the early 1920s, financed the building of coaches for this intercity enterprise.

The first successful service is usually credited to Stanislas Baudry, who initiated a regular service in Nantes, France, to connect the center of town to public bath-houses that he owned. He is the first to refer to the service vehicles as "omnibuses," meaning (roughly), "carriage for all." It is thought that the idea for the term came from a store along Baudry's original route. The store was named Omnes Omnibus – "Omnes for all." Baudry quickly noted that many of his passengers would frequent intermediate shops, and began to make regular pick-up and drop-offs between the hotel and the baths. He eventually abandoned the baths, but continued the service. By 1827, Baudry had a competitor, entrepreneur Edmee Fouquet who introduced another service in Nantes under the company name of "Dames Blanche" (White Lady). In 1828, both Baudry and Fouquet received permission to begin running regular services in Paris on over 18 designated routes.

The service quickly expanded to many cities, including both London and New York. An Englishman named Shillibeer introduced a regular service in Yorkshire. Despite ridicule and the formal opposition of hackney coachmen, the service caught on and became quite popular.

In 1827, Abram Brower, a New York stagecoach operator and stable owner, initiated an omnibus service along Broadway. Initially, he operated one vehicle as far as Bleeker Street. His first coach was built by Wade & Leverich, and was a modification of the stage coaches in use at the time. It had open sides, and was divided into two sections, each with two facing seats accommodating three people. In each section, three people would be facing forward, and three would be facing backward. His first coach was named "Accommodation."

Two years later, Brower added a second coach, also built by Wade & Leverich. The second coach, called the "Sociable," had a single compartment, with the seats running front to back. A single door was located at the back of the coach, with a set of steel steps for access and egress.

Fig. 4.1 A depiction of Shillibeer's Omnibus, introduced into service in London, 1827

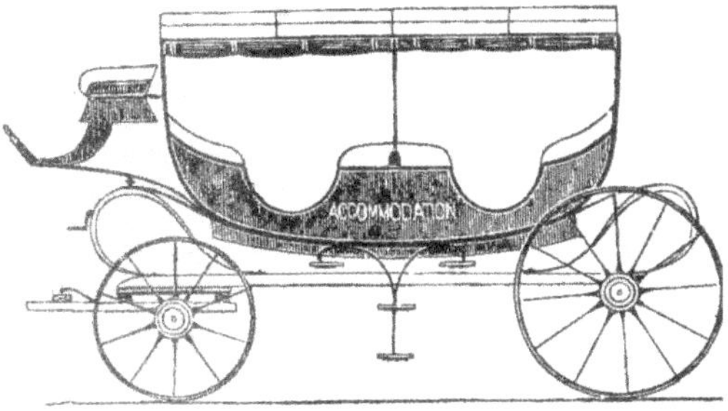

Fig. 4.2 Adam Brower's "Accommodation" in 1827

Fig. 4.3 Adam Brower's "Sociable" in 1829

Knowledge of Brower's first two omnibus coaches comes from a self-published history of animal-drawn vehicles by Ezra Stratton written in 1878. She obtained information and drawings of the two vehicles from John Stephenson, who would become the nation's leading builder of omnibus coaches and streetcars. She was a contemporary of Stephenson, and obtained the information through personal contact.

By 1931, Brower's service was so popular, that he greatly expanded, adding a number of coaches, now built by John Stephenson, who had established a coach-making business after considerable success in making rail cars. The Stephenson coaches are credited with being the first to be designed and manufactured as omnibuses, and set the standard in the fledgling industry.

Brower's service expanded to cover Broadway from the Battery to Bond Street. The coaches operated along a fixed route and charged a flat fare of one schilling (12.5 cents). Initially, the fare was collected by a boy (child labor laws were not yet in effect) standing on a narrow platform near the rear steps of the omnibus. Later, this practice was eliminated, and drivers collected the fare using a variety of mechanisms, usually a small box that would be handed back and forth through a small opening between the coach and the driver's seat.

Fig. 4.4 John Stephenson's Omnibus in 1831

Through 1931, it was technically illegal to pick up and discharge passengers on a city street. Early omnibus drivers were therefore subject to frequent arrest. Many fought this by chaining themselves to their coaches. In 1832, the Stage Carriage Act was passed which required drivers and conductors to take out licenses. This set the legal precedent for the regulation of local public transportation.

By 1835, there were over a hundred omnibuses operated by a variety of organized companies and independents, running up and down Broadway and other streets of New York. Competition was fierce, particularly among the independents. Riders frequently complained about being cheated by drivers – who returned less than the appropriate change in many cases. Dangerous driving was also a prevalent hazard as individual drivers jockeyed for position and speed on the crowded New York streets.

Despite the arrival of street railways, horse-drawn omnibus service on Fifth Avenue survived until the introduction of motor buses in 1905. Thus, there was an omnibus service still running on the day the city's first subway opened in 1904.

Fig. 4.5 An Omnibus operating on Broadway, circa 1860

4.3 The Street Railways

Horse-drawn (and even human-drawn) vehicles on rails had existed for over a century, mostly in mining applications for the extraction of various natural materials. In the 1830s, they would spawn a new type of urban public transportation service.

The idea of the street railway evolved from the early demonstration of a steam engine hauling three railroad cars over the Mohawk & Hudson Railroad between Albany and Schenectady. John Mason, a wealthy New York banker and merchant, believed that it was only a matter of time before a functional passenger railroad would extend all the way to the city. He and a group of associates organized the "New York and Harlem Rail Company," to provide a NYC terminus for such a railroad. New York State gave a charter to the company which included the right to build a single- or double-track street railroad or way between the Harlem River and 23rd Street along any route approved by the Common Council of the City of New York.

The Common Council and the Mayor quickly agreed on 4th Avenue, and an authorizing ordinance was enacted and signed on December 22, 1831. Within days of the ordinance, a rush of investors had provided the $350,000 of investment capital to initiate construction.

The route between the Harlem River and 23rd Street was sparsely inhabited, so the idea quickly arose to continue the line through the built-up areas between 23rd Street and lower Manhattan. The Common Council quickly authorized the extension, allowing the road to be extended along 4th Avenue and the Bowery to Walker Street.

New York and Harlem commissioned John Stephenson to provide two rail coaches to inaugurate the service. The design was similar to a stage coach, but was larger. Each coach was separated into three compartments, each with its own entry and exit doors. Interiors were opulently upholstered, and the exteriors were embellished with attractive paintings.

Fig. 4.6 The "John Mason," built by John Stephenson in 1832

Despite a minor accident when the new service was demonstrated in the Fall of 1832, the line began operation later in the year to generally good reviews and booming business.

The horse-drawn railed vehicle had many technological advantages over the omnibus:

- Passengers received a relatively smooth ride over the rails, as opposed to the decidedly bumpy ride provided by omnibuses operating on wooden wheels over mostly cobblestone streets.
- Because of the lower friction of railed vehicles, they traveled faster than omnibuses, about 8 mph compared to 5 mph.
- Again, because of the lower friction of railed vehicles, a horse-streetcar could carry three times the load of an omnibus for the same investment of time and labor, that is, 30–40 passengers compared to 12–15 passengers.

John Stephenson was convinced that the street railway was the wave of the future. When a fire destroyed his shops in March of 1832, he built a much larger shop on Elizabeth Street, near Bleeker Street. Unfortunately, the original line was not joined by competitors for almost a decade. The financial collapse of 1837 left Stephenson bankrupt, and he paid off his investors at a rate of $0.50 on the dollar, and in some cases paid off his debts by building horse-cars at no cost. He managed to survive the period, and his business rebounded strongly in the 1840s. The standard form of the streetcar changed over time. Eventually, the three compartments with transverse seating were replaced by a car with a single compartment and longitudinal seating. Platforms at the front and rear accommodated the driver and conductor, and a canopy was extended over both.

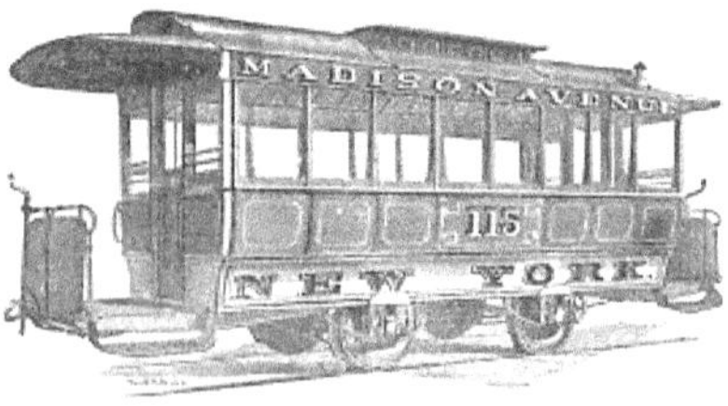

Fig. 4.7 A typical John Stephenson streetcar of the 1840s

Horse-streetcars were an enormous success. On July 3, 1854, the first Brooklyn streetcar line was opened along Myrtle Avenue, traveling from the Fulton Ferry to a former stagecoach stable at Marcy Avenue. Operated by the Brooklyn City Railroad, it was soon followed by the Fulton Street line from Fulton Ferry to City Hall, the Greenwood (later Court Street) line, and the Greenpoint line.

(a) Third Avenue railway streetcar at Park Row Broadway; driver feeding and watering his team, circa 1906

(b) Last horse-streetcar line on Bleeker Street and in 1917

(c) Brooklyn City Railroad, Greenpoint Line, 1910

(d) Brooklyn City Railroad, Fulton Street Line, 1888; Fulton St. Elevated rumbles overhead

(e) Central Park and East River Railroad open summer car in 1890

(f) Streetcars, omnibuses, and elevated trains approach the South Ferry terminal, circa 1895

Fig. 4.8

By 1860, Manhattan's fourteen horse-streetcar lines carried over 38 million passengers per year, in addition to those served by Manhattan's 29 omnibus lines using 671 vehicles.

While omnibus service had little impact on the development of Manhattan, more popular street railways had a much more profound impact. Between 1832 and 1860, the northern boundary of intense development had moved northward from Houston Street to 42nd Street. This northward growth was more than twice the total northward movement of the city's boundaries in the previous 200 years!

By the 1890s, it was clear that the age of horse-powered public transportation was coming to an end. The difficulties involved in keeping horses healthy and the environmental problems of animal waste products eventually became barriers to effective or profitable operation. One line on Bleeker Street continued to operate until July 26, 1917, but it was reported to collect only 30 cents in fares per day during its final year.

4.4 The Age-Old Problem of Traffic Congestion

By 1850, it was clear that the mix of pedestrians, private carriages, omnibuses, and horse-streetcars were making most of Manhattan's streets impassable during the peak hours of the day. The limits of horse-drawn vehicles on city streets had clearly been reached, and the northward expansion of the city had basically stopped dead in its tracks, even though some horse-drawn services would exist for another 67 years.

In August of 1852, Jacob Sharpe, a horse-streetcar and ferry operator, stood on Broadway at Chambers Street and counted the number of omnibuses that passed by. During a 13 h period, approximately 470 omnibuses per hour (with an average headway of 13 seconds) were observed. In 1867, a reporter the repeated the experiment, counting over 1300 in all – more than doubling the 1852 observation!

By 1860, it was clear that New York needed a different and more efficient means of public transportation. It needed to be faster, have significantly higher capacity, and had to relieve street congestion. Discussions of a *rapid transit* system – an urban railroad – were already under discussion. They would soon lead to the elevateds, and by the early 20th century, the subway.

Horse-streetcars eventually gave way to trolleys; trolleys later gave way to buses. As such, they remain a critical part of our public transportation system, and help make the subway system more effective and efficient.

4.5 Regulation

While omnibuses and street railways were built and operated by private enterprise, they (of necessity) operated on city streets that had been built using public tax revenues. Since the city owned the streets, and since omnibuses and street railways obviously represented an opportunity for profit, and therefore, for plunder, the city offered operating franchises to private enterprises to be permitted to use its streets. Given their relative success, various companies strenuously competed to obtain such franchises, granted by the Common Council. The losers frequently sued. Eventually, one suit resulted in a decision that the city did not have the right to limit private enterprise in this way, but the state and courts soon provided new ways in which to do so.

The franchise system continues to this day, making it very difficult to change bus routes for any reason. In the early days of city transit, the franchise system created a system of graft and pay-offs, perfected by the infamous William "Boss" Tweed, who (with his city and state cronies) raked in millions in pay-offs from omnibus and streetcar operators. This would become quite relevant in the development of NYC's transit system, as Tweed and his Tammany Hall cronies long opposed attempts to initiate rapid transit in any form, as it posed a significant threat to a major source of income.

References

1. Stratton, E.M.: The World on Wheels. Published by the author, New York (1878)
2. Hood, C.: 722 Miles: The Building of the Subways and How They Transformed New York. The Johns Hopkins University Press, Baltimore (1993)
3. Bobrick, B.: Labyrinths of Iron: A History of the World's Subways. Newsweek Books, New York (1981)
4. Greller, J., Watson, E.: Brooklyn Trolleys. N.J. International Inc., Newark (1995)
5. Walker, J.: Fifty Years of Rapid Transit (1918), `http://www.nycsubway.org`
6. Miller, J.: Fares Please. Dover Publications Inc., New York (1960)
7. Gould, P.: The Horse Bus,1662–1932, `http://www.petergould.co.uk/localtransport-history/generalhistories/general/horsebus.html`
8. The John Stephenson Car Co., `http://www.midcontinent.org/rollingstock/builders/stephenson1.html`

Figure Sources

4.1: *Originally published in* Stratton, Ezra M., *The World on Wheels*, published by the author, New York NY, 1878, pg 370.

4.2: *Originally published in* Stratton, Ezra M., *The World on Wheels*, published by the author, New York NY, 1878, pg 432.

4.3: *Originally published in* Stratton, Ezra M., *The World on Wheels*, published by the author, New York NY, 1878, pg 432.

4.4: *Originally published in* Stratton, Ezra M., *The World on Wheels*, published by the author, New York NY, 1878, pg 438.

4.5: *Courtesy of* The New York Transit Museum.

4.6: *Originally published in* Stratton, Ezra M., *The World on Wheels*, published by the author, New York NY, 1878, pg 440.

4.7: *Originally published in* The John Stephenson Car Company brochure.

4.8: *Courtesy of* The New York Transit Museum, Lonto–Watson Collection (a – f).

Chapter 5
The Age of Trolleys

Trolleys were the bridging technology between the horse-drawn streetcar and the motor bus. The former operated into the 1890s in abundance, with one straggler lasting as long as 1917. The latter became a major part of the public transportation system through the 1930s and 1940s.

Trolleys never had the marketplace all to themselves. Throughout their history, they were forced to compete with horse-drawn omnibuses and street railways, the elevateds, the subways, and eventually the motor bus.

Despite this, there was a time when trolley lines covered Manhattan, the Bronx, Brooklyn, and Queens like a blanket, and one could travel from almost anywhere to anywhere else in the city by trolley. Staten Island had trolleys too, but services reflected the lower densities that existed in that borough.

Trolleys had an intercity cousin as well: the interurbans. These were generally larger trolleys covering longer distances connecting sometimes isolated small towns to each other. At one time, a rider could travel from New York to Chicago or Boston using a series of interconnected interurban trolleys.

5.1 Introduction

The trolley system was never really operated as a coordinated entity, nor designed to be a rational system. Rather, a plethora of trolley operators sprung up, many from street railway operators, and operated a rather loose system of competing, often duplicative, routes. The trolley industry in New York never enjoyed corporate stability, nor sustained financial success. For every trolley operator that made a substantial profit, there were two that went bankrupt. The corporate history of trolley operations in New York is a dizzying sequence of small companies bought up by larger ones, massive consolidations followed by subsequent break-ups, and a significant number of simple abandonments of routes and franchises that ceased to make money.

There was also little standardization across the industry, even within companies. While trolleys were propelled by electric traction motors of various types, their source of power varied widely from battery packs that had to be recharged (a process taking 8–12 h) each night, to overhead wires providing power through "trolley poles," to a center power rail just under the surface of the street, to third rails similar to rapid transit on private rights-of-way. The parallel technology of cable cars also shared the stage, and several of these ran in New York as well.

R.P. Roess and G. Sansone: The Wheels That Drove New York, STTT 1, pp. 63–87.
springerlink.com

5.2 What's in a Name?

In 1885, an entrepreneur by the name of Leo Daft built a short electrically-powered street railway line in Orange, New Jersey. He employed a new system for delivering electricity to the cars: two overhead wires and a small four-wheeled carriage that rolled along the wire, connected to the car by a flexible cable. The little carriage was called a "troller," which later became "trolley," the name subsequently applied to virtually all types of electrically-powered streetcars.

The trolley also played a major role in one of New York's most famous sports franchises. Brooklyn, even before incorporation into the city in 1898, had a very dense trolley network. Brooklyn residents were often referred to as "trolley dodgers" because crossing any major street involved dodging the frequent trolleys as they passed. Brooklyn's baseball team, originally named the *Suburbas*, changed their name to the *Trolley Dodgers*. Later, the term "trolley" was dropped, and the Brooklyn Dodgers were born.

5.3 The Beginnings of Electric Streetcars

Various inventors had been toying with the use of electricity to power vehicles since the late 1870s. The first successful electrically-powered streetcar line was in Montgomery, Alabama, where it began operations in April of 1886. James Gaboury was an engineer and promoter who built a number of street railways throughout the south, using mule-power, as was the southern custom. In the summer of 1885, he attended the Toronto Exposition, where he observed a small electric vehicle running back and forth on a small track with no visible means of power other than a thin overhead wire. The model had been built by Charles Van Depoele, who had been trying to sell electric operation to a variety of streetcar operators across the country. Gaboury was so impressed that he offered to let Van Depoele convert the Capital City Street Railway in Montgomery, in which he held a controlling interest, to electric power.

The system started using converted horse-cars, using a single electric motor mounted on the front platform of the car. Power was transmitted to the wheels using a chain, and was collected from the overhead wire by a small two-wheeled carriage (called a "monkey") connected to the car by a flexible cable. Beginning on the Court Street Line, the experiment was so successful that electrification was extended until the system had 18 cars running over 15 miles of track. Montgomery was the first city to have a city-wide transit system powered by electricity.

Thomas Edison actually built a small electrically-powered railroad extension in Menlo Park, New Jersey in 1880. The locomotive was said to operate so fast as to "scare the wits" out of riders. He soon lost interest, however, and turned to the problem of electric lighting. One of his employees, Frank Sprague, would go out on his own and revolutionize the area of electric traction. Sprague was convinced that the most lucrative market for electricity would not be in lighting, but in the powering of transit vehicles. In 1884, he left Edison to set up the Sprague Electric Railway and Motor Company.

Fig. 5.1 Van Depoele's trolley system in Montgomery, Alabama, 1906

His first major opportunity came when he was commissioned to electrify the Union Passenger Railway of Richmond, Virginia. He was only 29 years old at the time. Upon reviewing the route of the new line, he was aghast at a steep grade on Franklin Street, and planned to build a steam-driven cable to help cars get up the hill. An associate, however, encouraged him to test a regular electric car on the grade before doing this. Sprague himself drove the test car. As he approached the grade, he was sure that the car wouldn't make it. Unexpectedly, the grade was successfully negotiated, but one of the motors short-circuited in the process. In dark of night, after all of the curious onlookers had departed, the car was hauled back to the shop by mules. Nevertheless, Sprague had demonstrated that an electrically-powered car could in fact negotiate the grade, and work commenced to begin operating the system.

A long series of difficulties ensued, not the least of which was the problem of designing a system to connect the car to the overhead wires on a continuous basis. Eventually, a rigid "trolley pole" was mounted on a swivel base that would allow it considerable flexibility to follow the car on curves. A rope arrangement was designed to allow the pole to be easily reattached to the power wire when it became dislodged.

The line eventually opened on February 2, 1988, but experienced a number of operating difficulties. On the first day, many trolleys suddenly stalled as their gears locked. Gears had to be manually reset, and in some cases, entire trolley cars had to be moved off the tracks to allow following vehicles to pass through. Improper oiling was eventually found to be the culprit. For many months, motor parts kept short-circuiting, and had to be returned to New York for rebuilding. Motor brushes sparked frequently, often welding themselves into uselessness. Eventually, however, Sprague improved the gears and motors, and the service was soon extremely successful. Sprague actually lost $75,000 on his initial contract, but the technical knowledge gained, and the publicity for the system was worth it.

The system was visited by a number of operators of streetcars elsewhere in the country, including Boston and Chicago, which soon followed in electrifying their streetcars. The rapid development of trolley systems throughout the U.S. had begun.

In 1890, a census of street railways indicated that there were 1260 miles of electrified trolley track in the U.S., compared to 5700 miles of horse-powered railways. Twelve years later, in 1902, the numbers were 22,000 miles of electrified street railways and only 250 miles of horse-railway.

5.4 The Power of the Franchise

From the earliest days of public transportation, the city government recognized that operation of for-profit businesses using public thoroughfares provided an opportunity for taxation. Needless to say, it also offered endless possibilities for graft and corruption as well.

The mechanism for controlling the growing public transportation business was the franchise. For each specific public transportation line, the city would issue a "franchise" to the operating company. Initially, these were easily obtained, and did not involve a revenue stream for the government, though the same could not be said for government agents who dispensed them.

Over time, various requirements were introduced into each franchise agreement. Horse-railway operators, for example, who operated primarily on unpaved streets, were required to pave the street between the tracks to protect it from damage caused by horses' hooves. When double and triple horse teams came into use, that requirement was expanded to include two feet on each side of the tracks as well. Later, streetcar operators were required to clear snow from the streets. When electric traction was introduced, franchise operators attempted to remove the paving requirement, but city officials continued to enforce it and include it in new franchises.

Other forms of taxation included the requirement to carry police and fire personnel for free, and eventually, a percentage tax on revenues. At one point, under the Cantor Act, the city made all franchises available based upon the bidder offering the highest percentage of gross revenues. In one of the wilder bidding contests, two paper companies got into a bidding war over a new route to be built in the Bronx. As the bidding progressed, the bids exceeded 100% of gross revenues. The winner settled at a rate of 97% of gross revenues, but, to no surprise, never built the route.

Most franchises were term-specific, usually for a period of 10 years. Some, however, were granted in perpetuity. In these cases, the city often tried to recoup right to reissue by requiring changes in the perpetuity clauses of existing franchises whenever the holder requested a new franchise for another route.

The franchise system is still in place today. Many of the bus routes in Manhattan have their roots in streetcar franchises from the turn of the century. The system also explains why it is so hard to change bus routes. Because each route has a separate franchise, any change must be accompanied by a change in the franchise, which must be negotiated with the city.

5.5 The Trolleys of Manhattan and the Bronx

Electrification of streetcar lines was off and running with the success of Frank Sprague's Richmond system. In 1888, however, a major weather event changed the course of electrification in Manhattan. Through 1888, all electric and phone service in New York was provided by overhead wires strung along a veritable forest of "telephone" poles, some with as many as 40 crossties. The blizzard of 1888 caused massive failures as ice-covered lines collapsed all over Manhattan and other areas of the city. Street traffic, phone service, and electric service were disrupted for months as the damage was slowly repaired. As a result, the city established an ordinance requiring that all Manhattan utilities be located underground. This edict applied to trolley wires as well.

A new system was developed for use in Manhattan. A slot was located between the running rails. Within it were two rails carrying power. Each car would have a "plow" that moved with the car in the slot to draw power. Unfortunately, the cost of laying track with a center power slot was 3 to 10 times more expensive than the cost of normal rails with an overhead trolley wire.

5.5.1 Corporate History

The corporate history of Manhattan's streetcar and trolley lines is indeed tortured. From 1832, when the New York and Harlem Railroad began operating the first horse-drawn streetcar along Broadway, the number of such lines in Manhattan and the Bronx quickly multiplied. By 1865, there were twelve companies operating a variety of streetcar lines in Manhattan, a number which reached 20 by 1886.

Fig. 5.2 Broadway Line, near 10th Street, of the Metropolitan Street Railway in 1911; center rail for power is clearly seen

A group of Philadelphia investors incorporated the Metropolitan Traction Company on February 19, 1886. Essentially a holding company, it began to acquire a number of Manhattan streetcar operations, including the:

Broadway and Seventh Avenue Railroad (June 1886)
Houston, West Street and Pavonia Ferry Railroad (June 1886)
Chambers Street and Grand Street Ferry Railroad (June 1886)
South Ferry Railroad (January 1889)
Twenty-Third Street Railway (March 1890)
Broadway Railway (October 1890)
Metropolitan Cross-Town Railway (March 1891)

On August 4, 1892, a second-generation Metropolitan Traction Company of New York was formed with twice the capitalization of the original, which had been incorporated in New Jersey. The new holding company continued to acquire streetcar properties, adding the:

Central Park, North and East River Railroad (August 1892)
Forty-Second Street and Grand Street Ferry Railroad (March 1893)
Thirty-Fourth Street and Eleventh Avenue Railroad (April 1893)
Columbus and Ninth Avenue Railroad (May 1893)
Lexington Avenue and Pavonia Ferry Railroad (May 1893)
Fulton Street Railroad (October 1895)
Twenty-Eighth and Twenty Ninth Streets Crosstown Railroad (September 1896)
Central Crosstown Railroad (May 1897)

If this were not complicated enough, the Metropolitan Traction Company of New York began to leasing its subsidiaries to each other. The Metropolitan Street Railway was formed on November 12, 1895 from a collection of the leased subsidiaries. It added two more railroads not owned by the Metropolitan Traction Company:

Eighth Avenue Railroad (November 23, 1895)
New York and Harlem Railroad (June 11, 1896)

On September 16, 1897, the Metropolitan Traction Company was formally dissolved, transferring all stock and subsidiary holdings to the Metropolitan Street Railway Company. It then acquired a lease on the:

Second Avenue Railroad (January 28, 1898)

At this point, there was only one other railway company operating in Manhattan and the Bronx, that is, the Third Avenue Railroad. It held crosstown lines on 42nd Street and 125th Street, two north-south lines on Third Avenue and Broadway, the entire street railway network in the Bronx, and a number of lines in Westchester County. The rapidly rising cost of electrification drove the company into bankruptcy in 1900, and the Metropolitan Street Railway Company acquired a majority of its stock in March of 1900. The third version of the Metropolitan now controlled all of the streetcar lines in Manhattan and the Bronx.

Corporate stability didn't last long. On November 25, 1901, the Interurban Street Railway Company was incorporated to take over the bankrupt North Mount Vernon Street Railway. By 1902, the Metropolitan was failing, burdened by overcapitalization, and a veritable Ponzi scheme of interlocking leases that brought with it the obligation to pay dividends to a huge number of stockholders. The Interurban Street Railway leased the Metropolitan on February 14, 1902. In 1904, the Interurban changed its name to the New York City Railway Company.

In January of 1906, yet another giant conglomerate was formed. Beginning in 1902, the directors of the Metropolitan realized that the new subway being built by August Belmont and the IRT (Ch. 8) would be a major threat to its streetcar lines. The Metropolitan countered by offering to build a second subway under Lexington Avenue, with the specific intent of having its streetcar lines serving as feeders to the new subway. Belmont responded by buying out the Metropolitan, forming the Interborough-Metropolitan Company, which now owned the IRT Subway, the elevated lines of the Manhattan Railway Company (which Belmont had acquired in 1895), and all of the streetcar and trolley lines in Manhattan and the Bronx.

The consolidation was a ploy by Belmont to eliminate a potential competitor in the subway business, and it succeeded, at least until the BRT entered the picture and precipitated the Dual Contracts (Ch. 9). The street railway portion of the conglomerate did not, however, benefit from the arrangement, and by 1907, the New York City Railway Company, the operator through lease of all Metropolitan street railways, went into receivership.

Following the failure of the New York City Railway Company, its leases and operating agreements were cancelled by the courts, and its properties turned over to the receivers of its subsidiaries. The break-up left a number of independent operating companies, all in receivership, including the Third Avenue Railroad, the Metropolitan Street Railway, the Central Park, North and East River Railroad, the Twenty-Eighth and Twenty-Ninth Streets Crosstown Railroad, and the Second Avenue Railroad. Remaining lines were consolidated into the New York Railways Company, incorporated on December 30, 1911. This consolidation itself went bankrupt on March 20, 1919.

From 1907 on, unprofitable streetcar lines were abandoned regularly. The process continued until the middle 1930s when buses began to replace remaining trolley lines. The New York Railways Corporation, the successor to the New York Railways Company, regained control of its remaining trolley lines from the receiver on May 1, 1925, and it operated with mixed success until conversion to buses.

5.5.2 Trolley Lines after the Metropolitan Failure

After the crash of the Metropolitan, surface transit in Manhattan fell into one of three categories: (1) lines operated by the consolidated New York Railways, (2) lines operated by the Third Avenue Railway Company, and (3) lines operated by a number of surviving independents. New York Railway Company held the largest system, and began operations in 1912. The new company was still mired in a maze of leases and joint holdings from the Metropolitan. It operated both wholly-owned properties, and leased properties under its umbrella. Its wholly-owned properties included the:

Metropolitan Street Railway (the operating company, not the then-defunct holding company)
Broadway Railway
Chambers Street and Grand Street Ferry Railroad
Columbus and Ninth Avenue Railroad
Houston, West Street and Pavonia Ferry Railroad
Metropolitan Cross-Town Railroad
Lexington Avenue and Pavonia Ferry Railroad
Central Crosstown Railroad

Its leased companies included the:

Bleeker Street and Fulton Ferry Company
Broadway and Seventh Avenue Railroad
Christopher and 10th Street Railroad
42nd Street and Grand Street Ferry Railroad
23rd Street Railway
Sixth Avenue Railroad
Fort George and 11th Avenue Railroad
34th Street Crosstown Railroad

In some cases, several different entities would use the tracks of one or another of the companies. When used by a line operated by a company other than the owner of the tracks, the five cent fare collected had to be shared.

In attempting to rationalize the network of lines it inherited, the New York Railways Company made several critical mistakes. It initially cast off several of the leased companies to independent operation, only to realize that they were among the more profitable properties. Thus, lines of the New York and Harlem Railroad, the Ninth Avenue Railroad, and the Eighth Avenue Railroad had to be re-leased and brought back into the fold at significant expense.

In attempting to modernize its system, New York Railway Company designed a new trolley car for ease of access and egress, intended to speed the process and increase the average speed of operation. The trolleys featured a single set of low-slung doors at the center of the car. Unfortunately, both the doors and the aisles were too small, and access and egress was actually slowed. Moreover, to collect the fare systematically, the cars had to be manned by two persons. By the time the company figured out these problems, they had ordered 293 of the cars.

The company also looked at its route structure, and adopted a variety of approaches to marginal services. Such services were electrified, replaced by battery-operated trolleys, continued as horse-drawn lines, or abandoned. This didn't work out too well either, as cars converted to battery operation came from the 293 center-door fleet, which required a two-man crew. The four lines that were converted to battery operation had to be abandoned in 1919, but not before an order for 70 additional battery-powered cars had been placed, essentially with no place to use them. Ironically, one of their leased lines, the Bleeker Street and Fulton Ferry Company, was the last horse-drawn streetcar line to operate, lasting until July 26, 1917.

Due to its financial problems, intensified by their unfortunate decision-making, the New York Railways Company went into receivership in 1919, and re-emerged as the New York Railways Corporation in 1925, when it became commonly known as the Green Lines due to the color of its trolleys. The Green Lines, however, were owned by the Omnibus Corporation, a subsidiary of General Motors, whose intent was to replace all trolleys by motor buses.

5.5.3 A Sampling of Manhattan/Bronx Trolley Cars

Over the years, the New York City transit system has used a wide variety of rolling stock in its rapid transit system. Despite a few critical errors, however, the rapid transit fleet was carefully developed, and subjected to serious design considerations and testing. By comparison, trolley rolling stock varied widely, with little cooperation between operators. With so many companies vying for ridership, a plethora of trolley vehicles was developed and operated. The lack of standardization made operation and maintenance difficult and expensive. Further, given the number of failures in the business, used trolleys were forever being transferred from one property to another. Many were reconditioned, but to a very variable quality standard.

A common trolley was the "single-truck" design. The trolley car rested on one motorized truck unit with four wheels. Single-truck trolleys were used primarily on routes with relatively low to moderate demand, and typically seated 25–30 passengers in a vehicle that was about 30 ft long. Figure 5.3 shows a group of single-truck trolleys of the Third Avenue Railway.

Fig. 5.3 Single-truck trolleys in the foreground on the Third Avenue Railway's Avenue B Line at Delancey and Essex Streets in 1923. Congestion created by the mix of trolleys and cars is evident.

For more densely-populated areas, "double-truck" trolleys were most frequently used. Double-truck trolleys, which were supported by two trucks each having four wheels, were much longer and carried many more passengers. Double-truck trolleys could be up to 50 ft long, and typically carried as many as 70 passengers. Figure 5.4 shows a double-track trolley of the NYC Interborough, circa 1910; Figure 5.5 shows the interior of a typical two-truck trolley.

Fig. 5.4 A two-truck trolley of the NYC Interborough, circa 1910

Fig. 5.5 Interior of a typical two-truck trolley manufactured by J.G. Brill

Open-sided trolleys were popular, but could only be used in the late spring and summer seasons. Operating these trolleys became a financial burden, as they were out of service for the majority of the year. Figure 5.6 shows such a summer car of the NYC Interborough circa 1910. Open cars had a side safety bar to keep passengers from falling off the trolley, although the bar was kept in the "up" position on the boarding side of the vehicle, as seen in Figure 5.6.

Fig. 5.6 An open summer trolley of the NYC Interborough, circa 1915

One solution was a "combination" trolley in which half the seats were enclosed, and half were open. These proved to be uneconomic as well, as passengers would only use the enclosed half during winter and the open half during summer. Figure 5.7 shows a J.G. Brill combination trolley manufactured for the Metropolitan Street Railway in 1897.

Fig. 5.7 A J.G. Brill combination trolley in 1897; this view, taken at the J.G. Brill factory, was manufactured for the Metropolitan Street Railway

Eventually, trolleys that could be "converted" were built, allowing open operation on warm days, and closed operation during colder weather. Side panels were removed and stored during the summer season.

For some Manhattan streetcar lines that were lightly travelled, the costly conversion to electric power (using the required center power rail) was not justifiable. In a number of these cases, literal horsepower was replaced by battery-powered electric trolleys. This eliminated the need to construct a power rail. Battery-operated trolleys were, however, somewhat inefficient, as the battery packs would have to be recharged each evening, a process taking 8–12 h. In battery-operated trolleys, the batteries were generally located under the seats.

Fig. 5.9 Unique photo shows the last horse-car operated on the Third Avenue Railway's Avenue B line in 1913; behind it is the first battery-powered trolley on the line

Marginal lines often got some unique treatments. The Third Avenue Railway leased the 28th and 29th Streets Crosstown Railway, but its lines quickly decreased in importance. The Third Avenue actually ran a limited number of trolleys using an internal combustion engine, beginning as early as 1909. These were eventually, however, converted to battery operation.

A single problem plagued many early trolley designs. They required two-man operation, with a driver and a conductor to collect the fare. This was because the designs permitted entry and exits through both front and rear doors. Eventually, most designs were converted for one-man operation. For one-man operation, all entries had to be through the front door, necessitating a double front-door design, with one for entering passengers and one for exiting passengers. Some companies simply required that all entries were through the front door and all exits were through the rear door. Figure 5.10 shows a J.B. Brill trolley designed for one-man operation.

Fig. 5.10 A J.G Brill trolley designed for one-man operation.

Figure 5.11 shows a New York Railway low-step entry car traveling on Broadway in 1920. They were designed to speed entry and egress, which was slowed by the high steps necessary to board and alight from the trolley. The design featured a low-step middle door that, it was hoped, would speed the entry and exit process. The cars were colossal failures due to several significant design flaws: the doors were too narrow (and so were the aisles), so that access and egress was actually slower, not faster; the extended ends became an accident hazard, as vehicles often clipped them when trying to maneuver around the trolley cars. Finally, the design necessitated inefficient two-man operation.

Fig. 5.11 New York Railways low-step trolley in Times Square; crowds are watching the ticker describing the 1920 World Series

Figure 5.12 shows New York City's one and only double-decker trolley. Owned and operated by New York Railways on its Broadway line, it was popular with riders and ran from 1912 through 1922. Its single-door entry/exit system shared the flaws of its single-deck brethren. The car was specially designed so that it could operate on trolley lines located under elevated rapid transit lines, and was frequently referred to as the "Broadway Battleship."

Fig. 5.12 New York Railway's "Broadway Battleship"

5.5.4 *The End Is Here*

Manhattan was the first borough to end its trolley services. The great congestion of Manhattan, coupled with the exponential growth in motor traffic, placed trolleys in a hazardous environment for all involved. By the early 1930s, Mayor Fiorello LaGuardia had made it clear that he wanted to replace all of Manhattan's trolley lines with buses. The fact that the New York Railway Corporation (the Green Line) and its subsidiaries was by then owned by General Motors clearly accelerated the process, as the Corporation was interested in motorizing its lines as fast as possible. The last trolley line in Manhattan (not counting services operating over the East River bridges) was the 86th Street crosstown route. It was motorized in 1936. Trolley lines in the outer boroughs, and over several bridges, would survive until the mid-1950s.

5.6 Trolleys in Brooklyn and Queens

The history of trolleys in Brooklyn and Queens mirrors that of Manhattan and the Bronx. Their business history is a maze of small and large operators, consolidations of smaller companies into bigger ones, and a number of bankruptcies that forced breakups and abandonments.

The leaders of trolley companies in Brooklyn and Queens were perhaps a bit more aggressive, and were clearly better organized, but their end would be similar, namely, a conversion of remaining viable trolley lines into bus lines. It is interesting to note that the first electrification of a trolley line in New York occurred on a line that operated in Queens. Also of note is that a number of lines, particularly in Queens, ran on exclusive rights-of-way.

5.6.1 *Corporate History*

As was the case in Manhattan, the corporate antecedents of trolley operators were horse-streetcar companies. The first streetcar line was built and operated by the Brooklyn City Railroad, which incorporated in 1853. The company originally planned to begin operations with four lines. By July of 1854, it only had enough equipment to operate one. On July 3, 1854, it initiated the first streetcar line in Brooklyn along Myrtle Avenue using 15 streetcars, all hauled by teams of four horses. By the end of October of 1854, the Brooklyn City Railroad was operating all four of its intended lines. The company continued to expand its service, and by the end of the decade had lines serving the outlying communities of East New York, Bushwick, Ridgewood, Williamsburg, Greenpoint, and Flushing. Thus, its services extended into Queens.

Gradually, a number of competitors began operating competing lines, and lines into previously unserved areas. The major competitors were the:

Broadway Railroad
Atlantic Avenue Railroad
Coney Island and Brooklyn Railroad

A number of smaller companies also began to operate one or two lines, but these were gradually merged into the larger operations.

In 1884, the first Brooklyn elevateds began construction. Two major companies were involved: the Brooklyn Union Elevated System, and the Kings County Elevated Railroad Company. This is noted here, as the elevated companies and the trolley companies would eventually fall under a single broad corporate umbrella.

The first electrification of a streetcar line occurred on Jamaica Avenue in Queens County in 1887, before the great success of Frank Sprague's system in Richmond, Virginia. It used a power system developed earlier by Van Depoele. A second electrified line opened in Coney Island in 1890. On November 11, 1891, the Brooklyn Common Council authorized all streetcar companies to convert to electric operation. Unlike Manhattan, there were no restrictions on the use of overhead wires, and most of Brooklyn's trolley lines used an overhead wire system.

On March 13, 1893, a large holding company known as the Long Island Traction Company was incorporated. It immediately purchased a small one-line trolley company called the Brooklyn Heights Railroad (BHRR).The holding company then arranged a 999-year lease of the Brooklyn Railroad Company and its subsidiaries through the BHRR, giving it control over 27 trolley lines. The conglomerate then built the then-largest car barn in the world at 3rd Avenue and 58th Street, with a capacity of 1000 cars.

By the end of 1895, the Long Island Traction Company was forced to reorganize. It emerged as the Brooklyn Rapid Transit Company (BRT). By 1900, the BRT had gained control of all elevated rapid transit lines, and all trolley lines in Brooklyn, with the exception of five small independents. It would eventually acquire two of the independents by 1912.

The BRT empire crashed into bankruptcy on December 31, 1918, as its entry into the subway business as part of the Dual Contracts (Ch. 9) proved to be unprofitable. The company remained in receivership until 1923, when it emerged as the Brooklyn-Manhattan Transit Company (BMT). In 1929, the BMT rationalized its trolley lines by creating a subsidiary to operate them, that is, the Brooklyn and Queens Transit Corporation (B&QT). While its term of operation was relatively short, the B&QT was one of the more forward-looking operators in the business. The corporation played a major role in the development of the President's Conference Car (PCC), which was an attempt to standardize the design of trolleys throughout the country. The "President" was the President of the Electric Railway Association.

By 1940, unification of the city's transit system had occurred, and the NYC Board of Transportation took over operation of the B&QT and began to motorize trolley lines. By 1951, only three trolley lines remained in Brooklyn, and in 1956, the last Brooklyn trolley line ceased operation. Ironically, it was the same year that Brooklyn lost its Dodgers to L.A., the team that got its name from Brooklyn's trolleys.

While the consolidated B&QT ran a number of lines in Queens, there were a number of other smaller operations that existed in Queens and in Nassau County as well. These included the:

New York and Queens County Railway: Operated a number of lines in Long Island City, Woodside, Astoria, Flushing, College Point, North Beach, and Jamaica. It also operated a line over the 59th Street Bridge into Manhattan. It owned a private right-of-way for its Flushing-Jamaica line through Kissena Park and Flushing Meadows. NYC eventually took over this right-of-way to build the Grand Central Parkway, and by 1937, all of the companies trolley lines (with the exception of the 59th Street Bridge service) to buses.

Manhattan and Queens Traction Company: This company was the successor to the South Shore Traction Company that operated primarily in Suffolk County. When reorganized as the Manhattan and Queens, it operated primarily along Queens Blvd., serving Woodside, Elmhurst, Forest Hills, and Jamaica. By 1914, it had a convenient transfer to the Long Island Railroad's Jamaica Station. The company was reasonably successful, but when the Independent Subway was built under its primary routes, it could not compete. Recognizing the problem, its lines were abandoned in 1937, even though their ridership at the time was as high as it had ever been.

New York and North Shore Traction Company: This was a most unique company. Its lines were more like interurbans than trolleys, and covered substantial distances. It provided service between Flushing, Queens, to a number of North Shore communities in Nassau County, including Mineola, Roslyn, and Port Washington. It also had a line running to the current location of the Whitestone Bridge, in Queens. The company never made money, but managed to struggle along until 1920, when its lines were abandoned.

New York and Long Island Traction Company: This was one of the larger traction companies operating out of Jamaica. Its routes traveled east from Jamaica to various Nassau County destinations, including the Belmont Park, Elmont, Bellerose, Creedmore, Floral Park, New Hyde Park, Mineola, Garden City, Hempstead, Freeport, Lynbrook, Valley Stream, Baldwin, Oceanside, Rosedale, and Rockville Center. The company went bankrupt in 1924, despite the highest revenues in its history. In one of the strangest abandonments on record, at midnight on April 4, 1926, motormen simply deserted their cars in mid-run and left.

Long Island Electric Railway: This was yet another company providing service from Jamaica to some of the inner communities in Nassau County. It also served a number of Queens communities, including a line that originated in downtown Brooklyn (the current Borough Hall area). It provided service to communities such as Ozone Park, Woodhaven, Hollis, Far Rockaway, and Belmont Park. Both the LIRR and the IRT considered it to be an important feeder, and were heavily invested in the company. Its biggest money-maker was the summer-only runs to Far Rockaway. It suffered with the poor condition of the roads used in Nassau County, and from its aged rolling stock. In 1924, the company spent $63,000 to purchase 25 third-hand trolleys built in 1898, but two major banks called in their bonds shortly thereafter, effectively killing the company.

There were also a number of smaller companies operating isolated services in Nassau County, primarily connecting small communities to LIRR stations and/or beach recreation areas. In Queens, as bankruptcies and abandonments accelerated, two successor companies, Steinway Lines, and the Jamaica Central, attempted to rescue the more important lines. Although they each had some success, both efforts failed. Steinway Lines abandoned its last services in early 1939, the Jamaica Central in December of 1933.

5.6.2 The Trolleys of Brooklyn and Queens

In the early days of the Brooklyn Rapid Transit Company, most of the Brooklyn trolleys were of the single-truck variety. BRT actually maintained two sets of cars, open for summer and warm weather operation, and closed for winter and cooler weather operation. This made for a somewhat inefficient operation, but it kept the customers happy. The BRT ran a number of services to Coney Island amusements such as Dreamland, Luna Park, and Steeplechase Park.

Most of the open cars, however, were retired by 1928.

Later, most companies moved to larger, double-truck cars, but the practice of having open and closed fleets continued. "Convertible" cars were eventually developed; windows could be removed and stored during the summer, and installed during the colder parts of the year. One of the earliest convertible cars, called the "Chamberlain" after the BRT equipment superintendent who designed it in 1900, featured a unique interior with single cane-back seats that could swivel towards the outside of the car or inward towards the aisle. The arrangement was not considered practical, and the use of this design ended in 1915. The general use of convertibles, however, continued to increase.

In a move parallel to their Manhattan brethren, the BQ&T also moved towards a low-step middle-door arrangement. It worked better in Brooklyn, where the rider crunch was a little less intense, and because larger center doors were used. The car still required two men to operate it, however, and most were later converted to one-man operation by adding a front door for entries only, where the fare was paid, using the center doors for exiting only. In Brooklyn, some of the busier lines began running multiunit trolleys, with up to three individual units. Special trailer cars were built to make up these short trolley-trains.

In many ways, the Brooklyn traction companies were more aggressive in developing new trolley cars than their Manhattan counterparts. After WWI and the reorganization of the BRT, the Brooklyn City Railroad ordered a new series of trolley's referred to as the "Peter Witts" because they were thought to be inspired by a Peter Witt design for his Philadelphia Rapid Transit Company. They were also known by their series number, as the 8000s. The 8000s were large two-truck trolleys set up for one-man operation, with a front door for entry, and center doors for exit. Originally, two interiors were built, one with upholstered seats, the other with wooden slat seats. Riders were polled as to which they preferred. Remarkably, they chose the wooden seats, and the remainder of the fleet was outfitted with them. The cars were all steel, except for the end platforms, which were wood and steel, and they were equipped with pneumatically operated doors. These trolleys were extremely durable, and operated for a period of over 30 years.

Fig. 5.13 A Peter Witt trolley of the BQ&T, circa 1935

Because the BRT was relatively forward-looking and innovative, it is no surprise that it was an early supporter of the Electric Railway President's Conference Committee work to design a new and standard trolley car. The cars that resulted from the committee's work were thereafter referred to as PCC cars. The BQ&T allowed the PCC staff to use part of its Ninth Avenue Depot as a testing facility, and two early PCC cars were tested in Brooklyn in 1931. The BQ&T purchased 100 PCC cars from the St. Louis Car Company in 1935, but political pressure to motorize trolley lines across the city prevented the company from ordering additional cars. The cars were extremely popular, and ran for twenty years, until trolley service in Brooklyn was virtually ended.

Fig. 5.14 A PCC Trolley in Brooklyn, circa 1940

While the BQ&T and its corporate parent, the BRT, were frequently at the forefront of trolley technology, the more scattered Queens and Long Island traction companies were most often forced to use second- and third-hand reconditioned trolleys along its diverse and more sparsely-populated routes. Because Queens was not as "urban" in character as Brooklyn, many of its lines took on the character of interurbans, and a number of them ran on separated rights of way. Figure 5.15 shows a section of the New York and Queens Railroad's main line between Flushing and Jamaica. The line ran on a private right of way except near its termini in Flushing and Jamaica. Parts of its R.O.W. were used in the construction of the Grand Central Parkway. The Manhattan and Queens Railroad's Queens Boulevard route ran in a separated median of the highway, as shown in Figure 5.16.

Fig. 5.15 New York and Queens Railway trolley traverses a private right of way passing through Kissena Park, St. Mary's cemetery is seen on the right

Fig. 5.16 Manhattan and Queens Railway trolley in the median of Queens Boulevard

The pictures illustrate one of the principal difficulties facing Queens and Nassau County trolley companies. The majority of their mileage served areas that were positively rural in character, making it difficult to sustain an economically viable ridership.

In the end, Brooklyn and Queens trolleys suffered the same fate as the Manhattan and Bronx carriers. With the drastic increase in motor traffic, the operation of trolleys on city streets created numerous conflicts, delays, and potential accident situations. Further, motor buses were free to move within the traffic stream, not tethered to a single path, which, if blocked, disrupted service. Motor buses introduced increased air pollution, and it would take more buses to provide service than the larger trolleys, but buses could cover a broader area, and (even with franchise regulations) could respond more quickly to changes in usage patterns. The last trolley in Queens ran in 1939, while a few trolley lines in Brooklyn lasted until 1956. Neither, however, held the distinction of being the last trolley service in New York City.

5.7 Over the Rivers

New York was able to grow beyond Manhattan only by conquering the substantial barriers created by the rivers that surround it. To reach the Bronx, the Harlem River had to be defeated; to reach Queens and Brooklyn, the East River had to be crossed. Throughout its history, City entrepreneurs and planners used a variety of transportation services to provide the necessary connections. Originally, ferry services were the only services, but with the construction of the East River Bridges, crossing would become possible in a number of ways. The Brooklyn, Manhattan, and Williamsburg Bridges, connecting Manhattan to Brooklyn, could all be crossed by four different modes at one time or another: by walking, by private motorized or horse-drawn vehicle, by rapid transit, and by trolley. Further uptown, the Queensborough (59th Street) Bridge served the same mix. These four bridges were very busy places in the early 1900s.

When the Brooklyn Bridge opened in 1883, it was the first solid connection between Brooklyn and Manhattan. While the elevateds in both Manhattan and Brooklyn were operating by then, the weight of the steam engines that propelled them was too heavy to permit them on the bridge. The first cross-bridge public transit service was provided by a cable railroad, which simply crossed back and forth over the bridge, connecting with the plethora of horsecar services on both sides.

With the advent of electrification at the turn of the century, the BRT (and later the BQ&T) revised many of its Brooklyn trolley lines to travel over the bridge, providing direct service to Manhattan. When the BRT began operating some of its elevated trains over the bridge, the trolley services were not discontinued. Because of the heavy trans-river demand, both the elevateds and the trolleys carried substantial numbers of passengers. While the electrified elevated trains connected directly to Brooklyn elevated lines, all trains terminated in a Manhattan-side terminal, and passengers had to transfer to Manhattan elevated trains.

Fig. 5.17 Brooklyn Bridge Cable Railway leaving the Manhattan terminal in 1885. Cables are clearly seen in the center of the tracks

Fig. 5.18 After electrification, elevated trains and trolleys can be seen approaching the Sands Street terminal on the Brooklyn side the bridge, circa 1898

Trolley service over the Manhattan Bridge was far less successful, largely due to a costly franchise dispute between competing companies. On December 30, 1909, the Manhattan Bridge Three-Cent Line was incorporated specifically to run a trolley service over the then just-opened bridge. It took three years for the company to achieve a franchise to operate from the LIRR's Brooklyn terminal to the Hudson River ferry terminal on Manhattan's lower west side. The large traction companies – New York Railways, the Third Avenue Railway, and the BRT – wanted a piece of the action, and formed the Brooklyn and North River Railroad, obtaining substantially similar franchises. The two services competed both in court and on the street for passengers. Unlike the Brooklyn Bridge services, once rapid transit trains started crossing the bridge, there was not sufficient demand to keep the trolley services viable. The Brooklyn and North River Railroad ceased operation in 1919, while the Three Cent Line staggered along until November of 1929.

The Williamsburg Bridge was outfitted similarly to the Manhattan Bridge, with a set of rapid transit tracks, and two sets of separate trolley tracks on either side of the structure. On the south side of the bridge, New York Railways and the BRT jointly operated the Bridge Local. The service was eventually a BRT service when New York Railways decided to join the Third Avenue Railroad in operating a service on the north side of the bridge. Neither service did well, and service on the north tracks was quickly abandoned. When the Bridge Local was suspended as well, the city began to operate it as a municipal service, the first city-run trolley line. Eventually, the BRT resumed control and operation of the line.

The last East River crossing to be completed was the Queensborough Bridge. It included four trolley tracks, two on separated lanes on each side of the bridge, and two that shared vehicular lanes. The trolley served a unique connection: the municipal hospital complex on what is now Roosevelt Island (formerly Welfare Island, even more formerly Blackwell Island). The trolley stopped mid-bridge, where passengers would take an elevator between the bridge level and surface. For years, it was the only connection (other than boat) between the mainland and the island. For this reason, the Queensborough Bridge trolley was the last operating trolley service in New York. In its final years, it was operated by the Queensborough Bridge Railway, as all other trolley companies had ceased to exist, or converted to bus companies. The trolley ran until April of 1957, when a small vehicular bridge was opened between the island and Queens. In 1976, the Roosevelt Avenue Tramway from Manhattan was opened. Today, there is a major housing complex on Roosevelt Island, which is served by the small bridge connection to Queens, and by an IND station in the new (relatively) 63rd Street Tunnel, in addition to the tramway.

Fig. 5.19 The last trolley line in New York over the Queensborough Bridge, 1955

5.8 The Trolley-Bus in Brooklyn

When it became clear that New York City was going to force the conversion of all trolley lines to motor buses, the BRT experimented with a hybrid vehicle called the trolley-bus. It was basically a bus that was electrically powered from overhead wires. Tracks could be removed, and the vehicle, while tethered to the overhead wire, was more maneuverable in traffic, and cheaper to operate than a traditional motor coach.

The trolley-bus was not unique to New York, and was not initiated in New York. The first commercial trolley-bus line in the U.S. occurred in 1910, when the Laurel Canyon Utilities Company installed a 1.5-mile trolley-bus route in 1910. In 1913, another line was operating in Merrill, Wisconsin.

Several trolley-bus companies had formed even earlier, as the technology for combining the best of rubber-tired vehicles with cheap electric traction was extremely promising. The American Trackless Trolley Company was formed out of the previous Eastern Trackless Trolley Company, which had been trying to demonstrate a successful line since the late 1890s. In 1903, the company finally succeeded in demonstrating an 800-ft line in Scranton, Pa. The success of the demonstration did not result in a flood of U.S. services, but was sufficient to spark the interest of English engineers, who initiated service in both Bradford and Leeds (England) in June of 1911.

By the 1920s, interest in trolley-buses was sufficient to spark General Electric and other major corporations to develop technology. Eventually, systems would be operated in a number of cities, including Philadelphia, Toronto, Salt Lake City, Newark, Honolulu, Cincinnati, Chicago, and others.

In Brooklyn, the BMT converted several trolley lines to trolley-bus operation beginning in July of 1930. The trolley-bus in Brooklyn did not survive the trolleys themselves for very long. The last trolley in Brooklyn was discontinued in 1956; the last trolley-bus line was closed in July of 1960. The trolley-bus was never the found to be a reliable replacement for trolleys. The freedom afforded the trolley-bus in traffic was also its chief operating problem: drivers would frequently stray too far from the overhead wire, disengaging the trolley pole. Reattachment was a major problem, as the vehicle had to be pushed or pulled into place, and the pole manually replaced. Further, removal of the trolleys was not only due to the problems of tracked operation; the maze of overhead wires was considered unsightly, and their removal was a priority for city politicians. While they only served a few lines, the trolley-buses did run for almost exactly 30 years, and are part of Brooklyn's mass transit heritage.

Modern trolley-buses survive, and many routes throughout Europe and Asia continue to run using new up-to-date equipment. Seattle and San Francisco operate trolley-buses in the U.S., as they are more efficient than buses or tracked vehicles in hilly operation. Vancouver, B.C. also has extensive trolley-bus service, including articulated trolley buses.

Fig. 5.20 A trolley-bus in downtown Brooklyn; unused trolley tracks are still visible, circa 1955

5.9 Onward

By the end of 1960, all trolley and trolley-bus services in New York had ended. At this point, the mass transportation system, with a few exceptions, had reached the configuration we still operate, namely, a combination of rapid transit, heavy rail services, and buses. The trolleys never had the playing field to themselves, and in the end simply could not overcome the combination of competition and aesthetic concerns. Over the past several decades, trolleys have been reborn in numerous cities across the U.S. under the title "light rail." The primary difference between newer services and the historic trolleys is that light rail almost always is provided with a private right of way, crossing streets only at intersections. They remain an attractive option for cities that cannot create the demand densities required for heavy rail operation, but where buses alone cannot satisfy demand effectively.

The history of trolley services, however, was also affected by the byzantine corporate environment in which they operated. Lack of planning and coordination certainly doomed many trolley services to financial ruin. New York has not joined many other parts of the world in reintroducing light rail services into the mix of modern public transportation options.

References

1. Miller, J.: Fares Please. Dover Publications, New York (1960)
2. Meyers, S.: Lost Trolleys of Queens and Long Island. Arcadia Publishing, Charleston (2006)
3. Meyers, S.: Manhattan's Lost Streetcars. Arcadia Publishing, Charleston (2005)
4. Geller, J., Watson, E.: Brooklyn Trolleys. NJ International Inc. (1995)

5. Diehl, L.: Subways. Clarkson Potter Publishers, New York (2004)
6. Sandler, M.: Straphanging in the USA: Trolleys and Subways in American Life. Oxford University Press, New York (2003)
7. Rowsome Jr., F.: Trolley Car Treasury: A Century of American Streetcars – Horsecars, Cable Cars, Interurbans, and Trolleys. McGraw-Hill, New York (1956)

Figure Sources

5.1: *Courtesy of* Library of Congress, Prints and Photographs Division, Detroit Publishing Collection.
5.2: *Courtesy of* New York Transit Museum, Lonto–Watson Collection.
5.3: *Courtesy of* New York Transit Museum, Lonto–Watson Collection.
5.4: *Courtesy of* New York Transit Museum, Lonto–Watson Collection.
5.5: *Courtesy of* New York Transit Museum, Lonto–Watson Collection.
5.6: *Courtesy of* New York Transit Museum, Lonto–Watson Collection.
5.7: *Courtesy of* New York Transit Museum, Lonto–Watson Collection.
5.8: *Courtesy of* New York Transit Museum, Lonto–Watson Collection.
5.9: *Courtesy of* New York Transit Museum, Lonto–Watson Collection.
5.10: *Courtesy of* New York Transit Museum, Lonto–Watson Collection.
5.11: *Originally published in* New York Times, October 12, 1920.
5.12: *Courtesy of* New York Transit Museum.
5.13: *Courtesy of* Brooklyn Historical Railway Association.
5.14: *Courtesy of* Brooklyn Historical Railway Association.
5.15: *Courtesy of* New York Transit Museum, Lonto–Watson Collection.
5.16 : *Courtesy of* New York Transit Museum, Lonto–Watson Collection.
5.17: *Courtesy of* New York Transit Museum.
5.18: *Courtesy of* New York Transit Museum.
5.19: *Courtesy of* New York Transit Museum, Lonto–Watson Collection.
5.20: *Courtesy of* Motor Bus Society.

Chapter 6
To "El" and Back: The Era of the Elevated Railroad

The story of New York's elevated rail rapid transit system is a critical juncture in the history of the city. Surrounded by water on a long, narrow island, the city needed to expand, and it needed a transportation system to enable it to do so. The elevated railway would provide the only answer for the thirty-four year period between 1870 and 1904, when the first subway was opened. The story of the "Els" is one of great innovation and invention, dedicated (even if often misguided) engineers and scientists, and both political and financial intrigue involving some of the most well-known names in the city's history. Always controversial, the elevated railroads paved the way for New York to rapidly expand, particularly to the north, while creating significant environmental problems virtually from their very beginnings.

6.1 A City in Need

By the middle of the 19th century, New York was choking on virtually intolerable congestion. Poorly-paved streets were filled with pedestrians, private carriages, horse-drawn omnibuses, horse-drawn streetcars, and merchants selling their wares. Congestion was particularly bad between the Hudson River docks and the city center on downtown Broadway, as all goods to and from ships were hauled on an endless stream of horse-drawn trucks.

Fig. 6.1 T. Benecke's Sleighing in New York depicts the chaos of winter traffic on Broadway in 1855

R.P. Roess and G. Sansone: The Wheels That Drove New York, STTT 1, pp. 89–138.
springerlink.com

During most of the day, it was impossible to safely cross Broadway, and accidents and injuries were commonplace. Robert Reed (1), in "The New York Elevated," notes:

> *"Traffic was so terrible on lower Broadway in the commercial district that it was dangerous, if not impossible for pedestrians to cross it during the day. Mr. Genin, an enterprising hat merchant occupying a busy spot across from St. Paul's Church offered to erect a bridge across the street to allow shoppers to cross Broadway to his store."* (Ref. 1, pg 16)

The proposed bridge was the subject of an article in *Gleason's Pictorial Drawing Room Companion* (2), and was enthusiastically endorsed by the paper. The proposed bridge was not built, although a less ornate bridge was built somewhat later.

Fig. 6.2 Genin's proposed Broadway bridge

Conditions on streetcars and omnibuses were poor at best, and the problem received much attention from newspapers. Overcrowding, discomfort, and extremely slow progress made most public transportation trips an unpleasant experience. An article in *Scientific American* (3) speaking of the conditions on omnibuses and streetcars in 1863 noted:

> *"Today, mildly speaking, some of our lines of omnibuses are a disgusting nuisance. They are infested with rowdies, drunken men and other objectionable characters who ought never be allowed to enter…the public now generally prefer to walk rather than be submitted to the degradation and discomfort as attends a ride in an omnibus for any distance."* (Ref. 3, pg 169)

The use of horse-drawn vehicles was a major health concern in its own right. At their peak, omnibus and streetcar operators used over 11,000 horses. On average, a horse would discharge 10 lbs of fecal material each day, much of it directly on the city's streets. They also literally drenched the streets with urine. City streets produced horrid odors, and carried the potential for transmittal of serious disease, including cholera and tetanus, which were, at the time, almost always fatal. When the New York Metropolitan Board of Health was created in 1866, one of its first projects was the removal of 160,000 tons of horse manure from vacant lots throughout the city.

Fig. 6.3 "Horrors of the Horsecar" as depicted in Frank Leslie's Illustrated News in 1865

The economics of horse-powered transit were also difficult. The cost of a cart-horse was about $150.00. Such a horse could be kept in service for only about four years before being sold for less strenuous service. Because most streetcars required two horses, and because horses had to be changed frequently, most companies owned 4 to 10 horses per cart owned. Horses were the single most expensive investment for a streetcar company, and, as an asset, they depreciated quickly, and were subject to total loss from disease. In 1872, when an epidemic virus among horses broke out on the east coast, 18,000 horses in New York were either killed or disabled, literally crippling the public transportation system.

Horse-car companies were also frequently targeted for criticism of their treatment of horses. While they typically ignored such criticism, they were criticized for their treatment of passengers as well.

If the streets of New York were virtually clogged with traffic, the harbor was almost as bad. Before the Brooklyn Bridge opened in 1883, the only travel between Manhattan and Brooklyn, Staten Island, or New Jersey was via ferry. Those affluent enough to flee Manhattan started a number of communities in these areas, and used ferries to commute back and forth to work. The harbor had traffic problems of its own.

Unlike many European cities, New Yorkers had long been resistant to allowing steam railroads to enter the heart of the city. The environmental and health concerns related to steam engines running on or over city streets were feared, and ordinances limited steam railroad access to 42nd Street and north. These ordinances resulted in Grand Central being located at 42nd Street, and most railroads ended at either the Hudson River or the East River.

Nevertheless, by mid-century, most New Yorker's had begun to think about a system of elevated railroads, perhaps using small steam engines for power. While some proposals were made as early as the 1820s, they were ignored by politicians and financiers for over 40 years. As the Civil War approached, and eventually ended, the number of proposals mushroomed, and both the public and the politicians began to seriously contemplate the immediate need for a rapid transit system in New York City.

6.2 A Flood of Proposals, Great and Small

The first elevated railroad was proposed by Col. John Stevens in 1812, and he was granted the first railroad charter in North America in 1815. He proposed a wooden elevated structure between Lake Erie and Albany, NY, to be powered by small steam locomotives. He argued that building an elevated structure would remove the cost of grading and reinforcement of the surface, and speed the construction process greatly. In 1812, the only steam locomotives existed in England, and thus his idea was considered quite radical.

By 1826, however, Stevens had built and demonstrated a small circular steam railway on his homestead in Hoboken. His steam locomotive operated at a speed of 12 mph. In 1830, he proposed an elevated steam railway from the Battery along Greenwich Street to around 34th Street, where it would cross the Hudson River on a bridge and connect to the Erie Canal system in Hoboken. The railway would carry passengers during the day and freight, including water and coal, at night. The proposal, however, was never implemented.

While the first elevated railway in New York would not appear until 1868, ideas for urban rapid transit were developing for decades. The first patent for an elevated railway in New York was filed by Henry Sargent in 1825. His proposal was for a "suspension" or "single-rail" railway on Broadway. It was an early version of a monorail. It was described in an article in the *Mechanics' Magazine, and Journal of the Mechanic's Institute* (4) in 1834. The article touts the benefits of the design in terms of its costs, stability, and general unobtrusiveness of its design. As a monorail, a relatively slender structure could be used while providing much-needed transit service. The mode of power is not specified, but it is conceivable that a cable-system was envisioned.

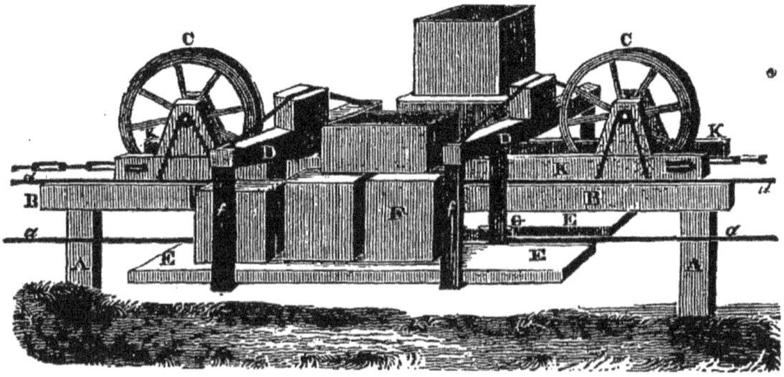

Fig. 6.4 Truck Design for the Sargent suspended railway

A similar railway was patented in England by Henry Palmer in 1821, but it is not clear whether Palmer or Sargent knew about each others' work.

Between 1825 and 1899, the U.S. patent office received 125 patent requests for rapid transit designs. In 1867 alone, over 40 drawings of elevated systems were submitted to the New York State Legislature. Some of the early plans called for subways as a more environmentally acceptable alternative to the elevated, and the discussions of subways vs. elevated systems waged on throughout this period.

One of the journals that published many of the proposed schemes was *Scientific American*, which was run by Alfred Ely Beach, who would himself play a pivotal role in the development of rapid transit. There was much discussion of the general safety and practicality of elevated structures. Most plans involved one of three options for motive power:

Small Steam Engines: Technically, steam was the most practical system, as it had already been demonstrated on intercity railroads. It, however, involved known environmental problems related to the release of hot steam into the immediate surroundings.

Cable Systems: Railcars would be propelled by a continuous moving cable (or rope). The cable or rope would be propelled by stationary steam engines at periodic locations. Some form of grip was used to detach the car from the cable to stop, and to reattach to the cable to start.

Atmospheric or Pneumatic Power: Railcars would be "pushed" by forced air pressure and "pulled" by creation of a vacuum.

In 1837, Uri Emmons (5) patented a monorail system in which the car is placed above and around the single rail. Emmons claimed that his system provided better stability for both the car and the structure. Again, the power mode is not specified.

Fig. 6.5 Uri Emmon's single-rail, supported monorail system

In 1946, a proposal for an elevated railroad along Broadway was reported in *Scientific American* (6). The article discussed the form of the structure to be used. From Figure 6.6, it is obvious that a steam-powered conventional railway was envisioned. The proposal called for a single track, with trains running once per half-hour. Presumably, some side-tracks would have been needed to accommodate the passing of trains in the opposite direction. It is interesting to note that the structural design bears an uncanny resemblance to a modern cable-stay bridge.

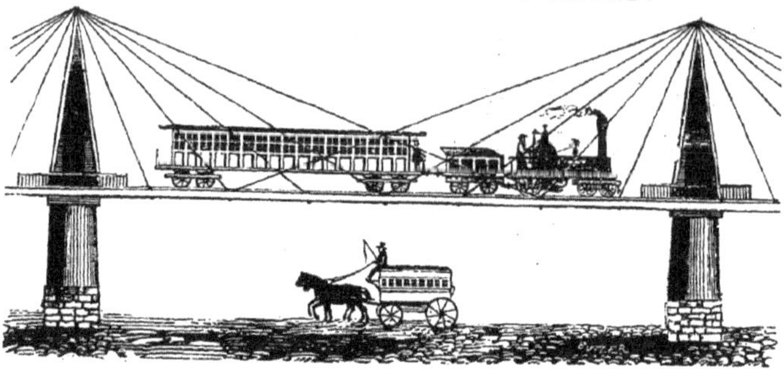

Fig. 6.6 Proposed Broadway elevated railway structure

In 1847, Varick DeWitt and Thomas Randall presented competing plans for a rope-cable elevated along Broadway. The former was reviewed in an article in *The Literary World* (7) in 1847. Dewitt's plan was for a suspended railway in which the bottom of the passenger cars would be 9 ft above the surface. With one structure on each side of the street, the appearance of the railway is described as "*...ornamental architectural style, light and airy in its appearance.*" (Ref. 7, pg 461)

By 1850, ongoing debates were not concerned with the need for or inevitability of some form of transit system, but the details, namely, should the railway be located below the street, on the street, or above the street; if above, should it be above the sidewalk, or over the street itself; how should intersection crossings be handled; etc.

In 1853, William Deitz (8) postulated a system using two 4 ft wide cars, each running on a single rail (one in each direction) stabilized by elevated tracks above. Power would be supplied by a rope-cable system. The cars themselves ran at ground level, and thus did not address the existing congestion on Broadway.

Fig. 6.7 Deitz' proposed system for Broadway

In the same year, James Swett (9) proposed an elevated railway for Broadway that featured a steam engine running above the rails, with a passenger car suspended below them.

Fig. 6.8 Swett's elevated railway proposal for Broadway

T.M. Brennan proposed an elevated, pneumatic railway for Broadway, also in 1853, but the idea did not gain much support. Pneumatic or atmospheric railways, however, became a significant part of the rapid transit discussion. The basic principals of pneumatic or atmospheric power are discussed in great detail in Ch. 7.

In 1854, J.B. Wickersham suggested an elegant elevated terrace along Broadway, with a horse-railway on the outside and a pedestrian promenade on the inside. The plan attracted much attention, and was described in both *Scientific American* (10) and *Gleason's Pictorial Drawing-Room Companion* (11). The elevated terrace would be located over the sidewalk, and would also provide a "protective canopy" without blocking light from the middle of the street. While the proposal attracted attention, it was not taken very seriously, as it ignored the problems of horse-drawn transportation and the need for greater speed and convenience. The lack of a solution for the liquid and solid wastes deposited by horses on an elevated structure was considered to be a critical deficiency in the proposal. Wickersham did allude to the potential for a steam railway at some point in the future, but had no specific plan for using steam power. Further, in conjunction with an adjacent pedestrian way, a steam railway would pose a significant hazard.

In 1865, P. Andrew (12) proposed a complex rope-cable elevated that had cars operating both above and below the rails on a single structure located over the outside edge of the sidewalk. There were four tracks in total, two above and two below. There would be a track in both directions on each level.

Fig. 6.9 J.B. Wickersham's elevated horse-railway/promenade

Fig. 6.10 Andrew's elevated railroad

In 1869, Richard Morgan (13) patented a proposal for a steam railroad running on a graceful arched structure. Several designs focused on this type of arched structure, but none was ever built in this manner. The proposed structure was actually borrowed from a somewhat earlier proposal from Charles T. Harvey.

In 1867, a grand plan called the Arcade Railway (14) was introduced for Broadway. Designed by civil engineer B.B. Nowlan, the plan was supported by many of New York's leading entrepreneurs. *Scientific American* literally gushed in praise of the plan. Its description of the plan underscored its ambitious objectives:

"The plan, as seen, is not merely to tunnel under the street, but to remove the street itself, block by block, from wall to wall, and construct another street at the depth of 15 feet, supporting the present street level on arches, and making stores in what are now the basements and sub-basements of buildings. Below all are the sewers, with tunnels for the passage of carts which remove the offal etc that naturally finds its way to the lowest level. The lowest portion is to be of the best masonry, strengthening the foundation walls of the buildings and giving ample support to the superstructure. Light, daylight, is afforded the sub-street or arcade by areas enclosed with iron railings between the upper roadway and sidewalk, ample room for which will be found when the upper street will be relieved by the construction of the lower. At convenient intervals, flights of steps will lead from these areas to the street below, and the pedestrians as well as those who ride in the cars, will be sheltered from the storms of winter and the sun of summer." (Ref. 14, pg 91)

Fig. 6.11 The Arcade Railway

One of the problems with the Arcade Railway was the lack of specificity on the mode of power to be used. While Figure 6.11 clearly shows a steam engine, other portrayals of the railway do not show any source of power, and written plans did not specify one. The Arcade Railway survived for over a decade as an active plan, including one bankruptcy, and had the support of many prominent citizens, including William Barclay Parsons, who would eventually help build the city's first operational subway of significant length. In the end, despite generating a lot of interest and discussion, it was never implemented.

In 1870, J.M. Hannah (15) proposed an elevated railway structure in which two tracks would be supported by a single row of steel columns in the middle of the street. As an alternate, for narrow streets, two single rows of steel columns would be located at each curb line, each supporting a single track. His proposal focused on the structural aspects of the railroad, and was less specific concerning the power system and car design.

In 1871, New Jersey wine merchant Alfred Speer produced a spectacularly innovative proposal for a continuously-moving walkway on Broadway (16). The moving sidewalk would be a very long loop, traveling up and down Broadway, with benches and seating areas for users. A number of sketches depicting the proposal appeared over time, with occasional differences in the details. Over time, the concept of moving walkways would arise in the transit debate, occasionally elevated, occasionally underground. The Speer design, however, was not implemented.

Fig. 6.12 Speer's moving sidewalk proposal for Broadway

There was no lack of ideas or proposals in the period between 1825 and the 1870s, only the lack of political and financial will to make rapid transit in New York a reality. Some schemes were quite innovative and some completely impractical. Reality, however, began to set in during the 1860s. The need for a transit system of some kind became clearer and clearer as the problems of congestion became worse and worse. A gathering consensus that something simply had to be done developed throughout this period. While so many of the proposed systems focused on New York's main street, that is, Broadway, it would be one of the last streets to acquire rapid transit services, as part of the Dual Contracts (see Ch. 9).

Rapid transit proponents had a number of major political adversaries. Boss Tweed and his cronies were becoming wealthy on the graft extorted from streetcar and omnibus operators. They vigorously fought any proposal for alternative forms of public transportation for decades. Property owners along proposed routes were generally opposed to elevated systems, fearing that the environmental conditions they would create would significantly reduce the value of their real estate. The merchants along Broadway were particularly vocal in their opposition. Property owners would have been perfectly happy to own land one block away from an elevated system, but no one wanted to own property right next to one. Businesses that catered to and serviced the omnibus/streetcar system were also opposed. Stable operators, veterinarians, feed sellers, etc., all stood to see their livelihoods seriously disadvantaged by new systems that did not employ literal horse-power.

Eventually, need would win out over political resistance. In the end, two great pioneers, Charles T. Harvey and Dr. Rufus Gilbert would initiate elevated rail rapid transit in New York. Neither would see the financial benefits of their innovations, as unscrupulous financiers would destroy them both. Interestingly, neither man wanted to use steam as the propulsion system, although both wound up employing it as the only practical propulsion system for large-scale elevated systems.

6.3 Prelude to Chaos: Hugh B. Willson's Subway Plan

Before either Harvey or Gilbert would begin to build their elevateds, railroad engineer Hugh B. Willson submitted a plan to the NYS legislature for a subway to be built under Broadway from the Battery to Central Park. Willson was a railroad magnate who had been in London during the construction of their underground, and was convinced a similar system would be ideal for New York City. In 1864, he incorporated the Metropolitan Railway Company, and accumulated over $5,000,000 of financing, much of it from John Jacob Astor, who strongly supported his plan. The legislature postponed consideration of the plan for a year.

Willson enlisted the aid of A.P. Robinson, a well-known engineer, to draft improved plans for the legislature. Robinson proposed a unique construction technique, called "inverted cut-and-cover," in which the tunnel is built from the top down, providing a way to restore the street surface while construction of the tunnel continued. The legislature passed the plan, which was promptly vetoed by Governor Fenton, a Tammany Hall regular. Willson tried once more, but was derailed in the legislature by Alfred Craven, chief engineer of the Croton Aqueduct Board. After previously stating that the construction of the proposed subway would not encounter a significant problem in moving underground water lines, he testified that the costs due to disruption and replacement of water lines would be "uncontrollable." Willson's subway proposal was again rejected by the legislature.

After defeat of the subway proposal, Senator Samuel Ruggles encouraged Governor Fenton to form a special committee to study proposals for rapid transit in New York City, both elevated and subway, and to make a recommendation on which system and location would provide the best solution to New York's critical need for public transportation.

The "Select Senate Committee on Rapid Transit" included three state senators, the mayor of New York City, and Alfred Craven. The committee solicited plans, among which was a subway plan that was remarkably similar to Willson's. It was submitted by a former supporter of Willson's plan, Orgin Vanderburgh. Willson accused Vanderburgh of stealing his plan, and Vanderburgh eventually made a settlement offer of $112,500, which Willson rejected.

The committee approved Vanderburgh's plan, but it was subsequently revealed that each committee member had accepted $10,000 to support the proposal. The ensuing scandal effectively killed the plan. Thus, the first truly viable proposal for a New York City subway was defeated first by lobbying interests for omnibus and streetcar operators and others, and finally by official corruption. Because of the scandal surrounding its considerations, the committee did not formally endorse any plan. It did, however, recommend that another submission get serious consideration, that is, a plan for an elevated railroad powered by cables and stationary steam engines, submitted by Charles T. Harvey. While it was considered to be a minor proposal for only a small test facility, it would finally result in New York's first functioning elevated railroad.

6.4 Charles T. Harvey and the Ninth Avenue El

In 1866, Charles T. Harvey incorporated the West Side and Yonkers Patent Railway, along with a number of financial backers. Its name encapsules his vision: an elevated railway along the westside of Manhattan serving the entire length of the island, providing service through rail connections all the way to Yonkers.

The legislature, after receiving a favorable review of his proposal from the Select Committee, approved his plan. Harvey had secured the approval by proposing a short test section, and agreeing to finance its removal if it was deemed unsuccessful.

Harvey came to the transit world with a substantial reputation as an engineer and innovator. He was best known for being the chief engineer on the "Soo" project, a canal joining Lake Superior with Lake Huron. Working under a strict time limit imposed by the construction contract, he ran into a severe problem during the winter of 1854. The official survey showed a sandbar along the canal's route which turned out to be a granite reef. Harvey had no time to transport equipment to the site to break up the granite. He improvised a drop-hammer made out of a propeller shaft of an old steamship, driven by a hoisting engine. The invention saved the day, allowing workers to break up the granite and haul it away. The President of the canal company was Erastus Corning, who became a life-long supporter of Harvey. Corning's influence helped get Harvey's elevated plan approved.

Fig. 6.13 Portrait of Charles T. Harvey

6.4.1 The System Begins

Harvey's plan was for a cable-operated, one-track elevated system. The cable was driven by underground steam engines located every 1500 ft along the route. The cable-car had a grip that was released to stop the car, and reengaged to start again.

The initial test section was built along Greenwich Street, between the Battery and Morris Street. On December 7, 1867, Harvey piloted a one-person car along the one-fourth mile test section, primarily for his financial backers. It was the first operation of an elevated railcar in New York City. Remarkably, this historic event was captured by a photographer. His backers were sufficiently enthused to authorize him to complete the initial system to Cortland Street, another one-fourth mile.

Fig. 6.14 Harvey tests his elevated cable system on Greenwich St. on December 7, 1867

By June of 1868, the line had been completed to Cortland Street, well ahead of the legislative deadline for its completion. The Mayor and Governor each took separate trips on the completed El, and declared it a success. On July 1, 1868, the Rapid Transit Commission made its final report, fully supporting its success. The Governor agreed, and vested the company with full power to extend the line all the way to Spuyten Duyvil in upper Manhattan.

Immediately after receiving permission to proceed, Harvey and his directors took a test run on their new cable car. They reached a speed of 12 mph, which exceeded their initial expectations by several mph. Of interest is the fact that the car was propelled by a cable manufactured by John Roebling in his Trenton, N.J. works. The cable design was basically the same as would be used by Roebling on the Brooklyn Bridge.

Fig. 6.15 Harvey's first cable car

Harvey had invested much of his personal fortune to get the elevated project implemented. He had spent $200,000 directly in construction, and posted a $500,000 bond against possible removal of the line had it been deemed unsuccessful. Instead of reclaiming his $200,000 from investors, he accepted one-fifth of the company's stock in return for his continued services, use of his patents, and repayment of his $200,000 investment.

In 1868 and 1869, the company was able to provide sporadic passenger service, while advancing the line to 30th Street, where a connection to the New York Central railroad could be made. The line used three cable cars, and was restricted to a single track, with periodic turn-outs to allow cars operating in opposite directions to pass each other.

Fig. 6.16 Greenwich Street el in 1869 at 12th Street, turning onto Ninth Avenue

However, to continue construction to 30th Street, Harvey was forced to mortgage the railroad to obtain the right to issue $750,000 in corporate construction bonds. The bonds were issued by two prominent investment houses: Clark, Dodge, and Company; and Lockwood & Company. The bonds were to provide an initial investment of $200,000 to finance additional construction.

6.4.2 Jay Gould and Black Friday

Just as Harvey's elevated seemed poised to be a big success, he was snared in the "Black Friday" financial collapse of September 24, 1869. Jay Gould, one of the best-known of the so-called "robber barons" of Wall Street, implemented a scheme to corner the gold market in the U.S. which led to the collapse of the markets.

In an era when there was virtually no regulation of markets, Gould and several cronies started buying large amounts of gold, gradually forcing the price up. After buying throughout the summer of 1869, Gould and his cronies flooded the market with gold, causing a precipitous drop in the price. Gould started buying again when the price bottomed out. He made an estimated $11,000,000 on the deal, even though he did not achieve his goal of controlling the entire gold market in the U.S.

While Gould got richer, general economic panic ensued. Wall Street brokerages went bankrupt, railroad stocks shrank, and business was generally paralyzed. Unfortunately for Charles Harvey, one of the banks that failed was Lockwood & Company. When it failed, Harvey lost the much-needed capital to continue building his elevated.

Because of the need for capital, Harvey and his directors obtained a $200,000 loan from a powerful group of investors. They, as a condition of the loan, required that a majority holding of stock in the company be transferred to them until the loan was repaid. The investors, however, had more in mind than financing new construction. They intended to manipulate the stock by having the West End and Yonkers Railway appear to be failing for some time; as the stock price fell, they would buy up as much as possible. The railroad would then remarkably begin to look like a success again, and the stock would be resold at an enormous profit. Harvey refused to participate in this process, even though he stood to gain significantly from it. Since Harvey no longer owned a controlling majority of the stock shares, Charles T. Harvey was dismissed from the company, and the investors went on with their plan.

6.4.3 The Elevated Fails

The new owners didn't fare very well in the final analysis. Without Harvey, the elevated began to experience many operating and safety problems. Cable failures were common, and passengers would be stranded for hours, and often had to be evacuated using ladders. Public sentiment and support for the elevated began to erode. By the Fall of 1870, operation of the elevated ceased, except for the running of occasional test cars to keep the system ready for renewed operations. After rusting for several months, the elevated, now bankrupt, was sold at a sheriff's auction for a grand total of $960 to Francis Tows. Tows became the proud owner of the structure, which now stretched from the Battery to 31st Street, three cable cars, and all cable mechanisms, including the steam engines that moved the cables. Within a year, the railway was auctioned off again to a new set of investors.

6.4.4 Reorganization and Conversion to Steam Power

The new owners reorganized the West Side and Yonkers Patent Railway, renaming it the New York Elevated Railway, and began test runs using small steam engines to power the system.

They faced a number of daunting obstacles. A plethora of lawsuits from businesses and land-owners along the route slowed progress significantly. Emerging from these, the new railway faced perhaps the most daunting opponent of all: a proposal for a massive eastside elevated built on huge stone arches, known as the "Viaduct Plan." The plan proposed a large stone viaduct along the sides of and over the tops of new buildings on the eastside.

The plan, which would have leveled virtually all property east of 3rd Avenue, was backed by some of the city's most prominent and distinguished citizens, including John Jacob Astor, Horace Greeley, Oswald Ottendorfer, Peter Cooper, August Belmont, Charles Lamont, Levi Morton, Charles Tiffany, John Agnew, Simeon Chittenden, and James Gordon Bennett Jr.

Fig. 6.17 The Viaduct Plan

The plan was the brainchild of A.T. Stewart, one of the most prominent businessmen on Broadway. His "Marble Palace" was one of the first department stores in the nation. He, with many allies in the business community, wanted to prevent any form of rapid transit from being implemented on Broadway. The Viaduct Plan would insure that rapid transit would occur elsewhere.

One of his political allies was Boss Tweed. Tweed doubtless envisioned the Viaduct Plan as a vehicle for massive graft. Tweed succeeded in getting a bill through the State Senate calling for the removal of the elevated, the only viable rival for rapid transit in NYC at that point in time. In response, Charles Harvey made his last appearance on the transit scene, calling on his old friend and supporter, Erastus Corning, to use his influence to save the elevated. Corning, whose influence in the legislature was second only to Tweed's, succeeded, and the bill was eventually defeated by an overwhelming margin.

It was about this time that Boss Tweed was brought under indictment and eventually convicted. His influence waned quickly, and the Viaduct Plan quietly disappeared from the scene, despite its impressive list of backers.

With the defeat of the Viaduct Plan, the New York Elevated prospered. It ran regular steam-powered service between the Battery and 30th Street using small "dummy" engines, that is, small steam engines disguised to look like a regular railcar. When usage increased enough to buy additional railcars, the "shadbelly" design was adopted. The centers of shadbelly coaches were lowered to provide both the feeling and reality of additional stability. Because the floors were lowered, but ceilings were not, the shadbelly cars had a very spacious feel, and immediately became popular with riders.

Fig. 6.18 Ninth Avenue El train at 59th Street; "Dummy" engine is pulling three cars, two of which are of the "shadbelly" type, circa 1876

By the summer of 1875, the elevated had been extended to Central Park. In May and June of 1875, the elevated carried over 170,000 passengers; and during the same two months of 1876, over 340,000 passengers.

In 1876, Cyrus Field bought out the New York Elevated Railway. Within several years, he doubled the fare from five to ten cents, and rebuilt the line with two tracks (one on each side of the street) through 61st Street. The Ninth Avenue El was off and running.

6.5 Dr. Rufus Henry Gilbert and the Gilbert Elevated

Dr. Rufus Henry Gilbert had a unique vision for elevated railways in New York City. He came to the world of rapid transit after a distinguished career as a doctor and surgeon, often serving the urban poor. After his wife's early death and his own exhaustion, he went to Europe for rest and recuperation. While there, he toured many European cities, particularly their crowded tenement areas. He became convinced that tenement living, with all of its obvious environmental difficulties, was a chief cause of illness and early death among urban poor populations. He returned to the U.S. to serve in the Civil War as a surgeon, and become the superintendent of all U.S. Army hospitals by the war's end.

When he returned from the war, he switched careers, joining the New Jersey Central railroad, where he gained a substantial reputation by remodeling the railroad. In 1867, he resigned from his position in order to become a full-time advocate for rapid transit.

6.5.1 Gilbert's Vision and Plan

Gilbert had become convinced that rapid transit was the cure for the problems of the urban poor. Good transit, he reasoned, would allow the urban poor to move out to more desirable, more sparsely-populated areas, where their health would substantially improve. This vision, however, was flawed. Even when the elevateds eventually flourished, the urban poor simply moved to new tenements located further from the center of the city.

Gilbert developed a plan for an elevated pneumatic railway, powered by atmospheric pressure. Though aesthetically pleasing, the system was believed to be impractical and too expensive. While he obtained a charter to build in 1872, he had difficulty attracting financial backers. The financial panic of 1873 further complicated the raising of capital.

Fig. 6.19 Portrait of Dr. Rufus Gilbert

Fig. 6.20 Gilbert's elevated pneumatic railway concept

6.5.2 The Husted Act of 1875 and the Rapid Transit Commission

In 1875, the New York State legislature passed the Husted Act. This landmark act allowed the Mayor of New York City to appoint a Rapid Transit Commission to study alternatives and recommend the best systems and routes for the city. For the first time, it shifted power over rapid transit from the state to the city.

The Commission studied various proposals for a year. Early on, it decided that elevated railways should be adopted as the rapid transit system of choice. Routing issues were more difficult, as various city ordinances had created a number of obstacles, the most serious of which was that no elevated structure could be built over, or even cross, Broadway south of 35th Street.

While the Commission studied the problem, Gilbert took on new financial backers. Because of his desperate need to raise capital to build his el, Gilbert took on William Foster Jr. as a partner. Foster convinced Gilbert to at least temporarily alter his plans to steam power to attract additional money. Together, Foster and the unsuspecting Gilbert reorganized the company, formed a board of directors, and issued stock. Foster convinced Gilbert to turn over some of his stock for sale to others. Unfortunately, Foster had no intention of selling the stock, and kept all of it, purchasing it himself. He also convinced Gilbert to accept the backing of the New York Loan and Improvement Company, led by Jose Navarro, a crony of Boss Tweed. Navarro exchanged stock in the soon-to-be defunct viaduct enterprise for Gilbert's stock, with the understanding that the stock could be recovered after the elevated was completed.This was, unfortunately, part of the scam: the transfer left Foster and Navarro in complete control of the Gilbert Elevated.

When the Commission completed its study, it approved plans for the following:

- The New York Elevated was permitted to build a loop from the Battery to South Ferry, providing a means to connect westside and eastside elevated service.
- The Gilbert Elevated was granted a franchise to build elevated lines on 6th Avenue and 2nd Avenue.
- The New York Elevated was given a franchise to build an elevated line on 3rd Avenue, with a connecting link to Grand Central Terminal along 42nd Street.

This landmark plan became the blueprint for New York City's elevated rapid transit system. A unique feature of the plan was that the 6th Avenue El would turn west at 53rd Street (it could not run over or through Central Park), joining the 9th Avenue El in a single system further north. At 110th Street, the combined elevated would turn east, continuing northbound on 8th Avenue.

The Commission also created the Manhattan Railway Company. The Commission worried that if either or both of the private elevated companies were to go bankrupt, the city would instantly lose its public transportation system. The Manhattan Railway Company was created to provide a back-up. It would take over either or both railways in the event of their financial demise.

6.5.3 Gilbert Loses His Elevated

The 6th Avenue El opened to great fanfare on June 7, 1878. The next day, Gilbert was voted off the company's directorate, and a few days later dismissed from the company altogether. The company was renamed the Metropolitan Elevated Company, removing any connection to Gilbert. Gilbert spent seven years and most of his money in court trying to regain the rights to the railroad he had created. He failed, and died at the age of 53, a virtual pauper.

Fig. 6.21 The 6th Ave. elevated opens on June 7, 1878

The 6th Avenue El was physically different from the 9th Avenue system. Its structures were more substantial, it was built as a two-track system from the start, and its cars were far more luxurious, and heavier.

On February 9th , 1878, *Harper's Weekly* (17) published a pictorial article extolling the virtues of the new elevated rapid transit system being built in New York. The article opened with:

> *"The sketches, diagrams, and plans presented in our opening pages this week will serve to show our readers in the city and country the progress which has been made by the two Elevated Railway companies to confer upon the citizen's of New York the benefits of rapid transit, of which they have so long been deprived by the jealousy and opposition of the horse-railroad companies."* (Ref. 17, pg 114)

Fig. 6.22 The Gilbert Elevated at 6th Ave. and 34th Street

The 6th Avenue El also attracted more attention than the 9th Avenue El. The 9th Avenue El was on the far westside of the city, relatively remote from the main business areas around Broadway and 5th Avenue. The 6th Avenue El went right through the virtual heart of Manhattan.

Harper's Weekly followed up on May 18th with an illustrated description of the rolling stock for the Gilbert Elevated. Both the sketches and descriptions somewhat overstate the degree of comfort provided by the cars.

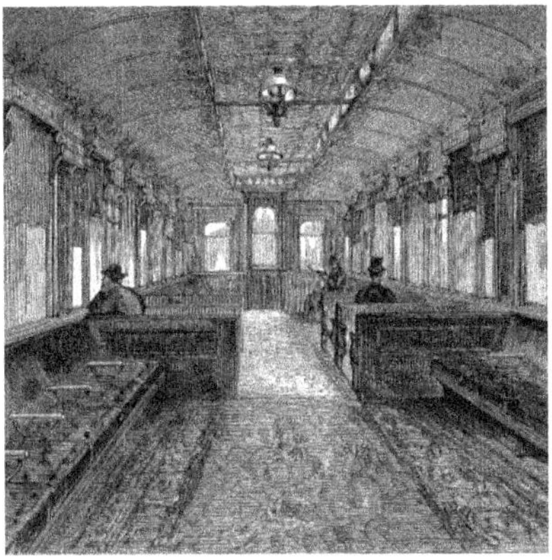

Fig. 6.23 Exterior and interior of Gilbert Elevated Railcar, as depicted in Harper's Weekly in 1878

The Gilbert Elevated cars and structures were heavier and more substantial than those of the New York Elevated. The cars were undoubtedly comfortable, but were not as luxurious as depicted in the *Harper's* illustrations.

Despite the fact that both Harvey and Gilbert were "relieved" of their interests in their inventions, as the 1880s approached, New York had its long-awaited rapid transit system, in the form of steam-powered elevated railways.

6.6 The Manhattan Railway and Completion of the Elevated System

On September 1, 1879, the New York and Metropolitan Elevateds were both leased to the Manhattan Railway, the creation of the Rapid Transit Commission. At the time of the leases, the principal directors on all three company boards were more or less the same. The leases, at the time, amounted to little more than a series of lucrative stock transfers. The President of the Manhattan Railway was Jay Gould, who would later concoct yet another financial scam that would leave him in control of the entire system.

The merger did allow for greater efficiency by reducing costs of duplication on the separate railways. It also simplified the problem created by the sharing of service by the 9th and 6th Avenue Els north of 53rd Street.

6.6.1 The 6th Avenue and 9th Avenue Els Move Northward

Meanwhile, elavateds continued to expand northward. The westside elevated was extended first to 110th Street, and again to 125th Street. At 110th Street, the combined lines turned east, continuing their uptown journey on 8th Avenue. The turn at 110th Street was the highest point in the system, and provided a dramatic view for passengers. It became popularly known as "suicide curve," not due to any accidents at the site, but due to the number of people who supposedly committed suicide by jumping off the structure.

Fig. 6.24 Downtown elevated train approaches "Suicide Curve" onto 9th Ave. from 110th Street, circa 1880

By December of 1879, the westside els were operating to 155th Street, the northern-most regions of Manhattan. A few years later, connections at 155th Street to the New York and Northern Railroad, which serviced the Bronx and Westchester, were made available.

(a) 9th Ave. El, looking north from Perry St. Single track is braced on the opposite curb. circa 1873

(b) 9th Ave. El approaching Battery and the South Ferry Terminal

(c) 9th Ave. and 53rd Street junction of the 6th and 9th Ave. Els, looking north

(d) 6th Ave. and 53rd Street; the 6th Ave. El turns west to 9th Ave.

Fig. 6.25. Views of the westside elevateds (6th and 9th Ave. Els)

6.6.2 The 3rd Avenue El

The New York Elevated opened its Third Avenue Elevated in August 1878 from South Ferry to 42nd Street, where it turned west to link with the Grand Central Terminal. Once the mainline had been opened further north, shuttle trains were run to provide the link to Grand Central. By the end of 1878, the el ran as far north as 129th Street.

In March of 1879, the el completed another branch connecting with the west terminal of the Brooklyn Bridge, which was then under construction. This was a far-sighted development. When the bridge opened in 1883, a cable-rail service was provided across it, permitting direct connections to the elevated lines of Brooklyn. The bridge was not deemed strong enough to support steam engines, necessitating a cable-car system when it first opened. When the els were electrified, continuous service from Manhattan to Brooklyn would be provided over the bridge.

Another spur at 34th Street connected to the East River ferry services operated by the Long Island Railroad.

(a) 3rd Ave. El along the Bowery in downtown Manhatttan; note the single-track design at both curb lines

(b) 3rd Ave. El at Coenties Slip in lower Manhattan

(c) 3rd Ave. El on the Bowery; new three-track segment under construction to replace the original two tracks in 1915

(d) 3rd Ave. El at Chatham Square

Fig. 6.26 Views of the 3rd Ave. Elevated

6.6.3 *The 2nd Avenue El*

The last elevated to be built was the 2nd Avenue El. Construction began in 1879, and by the end of 1880, the line was operating to 129th Street. The 2nd and 3rd Avenue Els shared common routing between South Ferry and Chatham Square. Because of this, 2nd Avenue trains ran to South Ferry, while 3rd Avenue trains ran to City Hall, using the City Hall branch. In 1882, an interlocking switch was installed, which allowed trains from either el to travel either path.

The Chatham Square station was a critical transfer point. Passengers could access trains to South Ferry or to City Hall and the Brooklyn Bridge, and could travel uptown via 2nd Avenue or 3rd Avenue. Initially, anyone wishing to transfer from one el to the other had to go to the street level and walk to the other station. After a few years, an overhead pedestrian bridge between the North Chatham Square station (3rd Avenue El) and the South Chatham Square station (2nd Avenue El) accommodated these transfers. Another critical junction was the South Ferry terminal, where many of the eastside and westside elevated trains converged.

(a) Chatham Square, looking north. 2nd Ave route departs to the right, 3rd Ave. to the left

(b) Chatham Square, looking south. South Ferry train to the left; City Hall train to the right

Fig. 6.27 Chatham Square Junction of the 2nd and 3rd Ave. Elevateds

As in the case of the joint 6th/9th Avenue Elevateds, connections directly serving the Bronx would also eventually be made. When the 59th Street Bridge was built, the 2nd Avenue Elevated initiated a spur across it which connected with other elevated lines in Queens.

6.6.4 *Jay Gould and the Hijack of the Elevateds*

In 1881, Jay Gould implemented a scheme to take over control of the elevateds. By this time, his empire included a major newspaper, and close connections to (and control of) key judges and legislators. Using his newspaper empire, stories began to appear that the Manhattan Railway was insolvent, and its directors corrupt. The stock price fell from 57 to 15 ½ within months. Meanwhile, Gould, along with Russell Sage and Cyrus Fields, began to buy shares. To insure that the price did not spike up too soon, Gould arranged to have a New York Supreme Court judge prepare a petition of bankruptcy for the company. Once Gould accumulated enough stock to guarantee full control of the railway, the bankruptcy petition was withdrawn, and the negative news reports ceased, replaced by more positive pieces. The stock price swiftly rebounded to 48, leaving Gould even more wealthy and powerful, and in complete control of New York City's entire rapid transit system.

By 1880, the 9th, 6th, 3rd, and 2nd Avenue Elevateds in Manhattan were completed. Under the coordinated control of the Manhattan Railways Co., they comprised a system that would accelerate the development of New York in northern Manhattan and the Bronx.

6.7 Elevated Railways in the Bronx

On October 19, 1880, the Suburban Rapid Transit Company was incorporated, with the purpose of building elevated railroads in the Bronx. The Rapid Transit Commission (of the Bronx) laid out four routes, all using private rights-of-way that had to be acquired at significant expense before construction could begin. Under its charter, it was required to provide links to existing trunk railroads terminating in the Bronx, and with existing eastside elevateds in Manhattan, specifically, the 2nd Avenue El and the 3rd Avenue El. Its first construction project was the building of a bridge across the Harlem River, so that the link to eastside elevateds could be provided. The bridge took three years to build, and was completed in 1886.

While building the bridge, the company moved ahead with the construction of its four elevated lines, while the Manhattan Railway constructed a joint station at the end of the 2nd Avenue El. The Bronx els began operations on May 17, 1886, with a completed connection to the 2nd Avenue El.

With its service complete to 145th Street, the company was interested in extending its lines without having to purchase private rights-of-way. A second company, the New York, Fordham, and Bronx Railway Company, had also secured a charter to build elevated lines in the Bronx. Its charter, however, allowed it to build over public roads. While the New York, Fordham, and Bronx Railway never built any elevateds, its franchise rights were valuable. On March 17, 1886, the Suburban Rapid Transit Company leased the New York, Fordham, and Bronx, and formally merged with it the following month. With the merger, the Suburban expanded its lines, abandoning most of the original franchise routes (except where rights-of-way had already been purchased), and proceeded to build over public streets as far north as 169th Street.

In 1891, the Manhattan Railway Company leased all of the railway lines, equipment, property, and franchises of the Suburban, unifying control of the elevateds in Manhattan and the Bronx. The Manhattan Railway Company continued to expand the elevateds further north to 177th Street, Fordham Road, and the Bronx Botanical Gardens.

6.8 An El Grows in Brooklyn

As Manhattan became congested, the independent City of Brooklyn began to expand as well. In the latter quarter of the 19th century, the "city" portion of Brooklyn was restricted to what is now called "Downtown Brooklyn," in the Borough Hall district. The area became an attractive alternative to middle- and upper-class New Yorkers seeking better, less congested living areas. Many commuted to Manhattan daily, using a plethora of ferry services to and from work. When the Brooklyn Bridge opened in 1883, many walked to and from Manhattan, while others rode a cable-car across the bridge.

At the other end of Brooklyn, the Coney Island area began to develop as a recreational destination Several grand hotels and beaches were opened, and a number of prominent Brooklynites built private mansions on or near the beach. The area developed steadily throughout the 19th century:

6.8.1 *The Coney Island Resorts and Attractions*

In 1829, the first seaside resort, the *Coney Island House*, was opened. It was built by the Coney Island Road and Bridge Company. The company also built a shell roadway across the creek that separated Coney Island from the mainland. At the time, there were no public conveyances to Coney Island, and the hotel catered to an upscale clientele who reached the hotel in private carriages to access the sea and salty air environment.

In 1844, the *Pavilion* was opened on the western end of Coney Island, and became the first amusement attraction on Coney Island. It was primarily a large wooden floor covered by a tent, and it was used primarily as a dance floor and for musical shows. By 1847, a steamboat service had been initiated between the Pavilion location and Manhattan.

In the 1860s, the politician, Mike "Thunderbolt" Norton, built a hotel in the same area. The area became known as "Norton's Point." Unfortunately, Norton had a substantial reputation as a crooked politician, and the area around his hotel rapidly became a magnate for gambling and prostitution, resulting in an overall seedy environment which quickly became a deterrent to those seeking the more genteel benefits of sun, beach, and sand.

In 1873, Austin Corbin, a successful investor and President of the Long Island Railroad, moved his family to Coney Island. Doctors thought the sea environment would help his son, who was generally sickly. Corbin soon recognized that an opportunity existed to develop Coney Island as a major resort area. He formed the Manhattan Beach Improvement Company, and in 1877 built the upscale and opulent *Manhattan Beach Hotel* on eastern end of Coney Island. This was soon followed by the *Oriental Hotel* in 1880. In between, William Engeman opened the *Brighton Beach Hotel* in 1878, west of the Manhattan Beach Hotel. Because his hotel was closer to the "bad" part of Coney Island, the Brighton Beach Hotel became a popular destination for the upper-middle class, while the Manhattan Beach and Oriental Hotels attracted a wealthier clientele. Ten years after the Brighton

Beach Hotel opened, however, beach erosion began to threaten its foundation. It became necessary to move the entire hotel 600 ft inland. The hotel was literally jacked up and placed on rail flatcars, moved, and placed on a new footing 600 ft from its original position.

(a) Manhattan Beach Hotel in 1904 (b) Oriental Hotel in 1903

Fig. 6.28 Austin Corbin's Eastern Coney Island Hotels

(a) Brighton Beach Hotel in 1904 (b) Moving of the Brighton Beach Hotel 6 ft Inland in 1888

Fig. 6.29 William Engeman's Brighton Beach Hotel

In 1882, one of the most unique attractions was built on the western end of Coney Island. The *Elephant Hotel*, also known as the Elephantine Colossus, was built in the form of an elephant. It was thought to be used as a brothel, giving rise to a local phrase, "going to see the elephant." The Elephant Hotel, however, was never financially successful. It was eventually abandoned, and was destroyed in a fire in 1896. A year after the opening of the Elephant, the *Sea Beach Palace Hotel* was opened immediately to the east.

With the opening of hotels and beaches, other amusements soon made their way to Coney Island.

In the 1870s and 1880s, horse racing was introduced, at first for the upper classes. The first race tracks were located on the eastern portion of Coney Island for their convenience. Racing, however, became a substantial draw for middle-class and other patrons as well, and caused the development of a number of smaller, less costly hotels near the tracks.

Fig. 6.30 Western Coney Island in 1890; the Elephant Hotel is seen towards the upper-center of the photograph; the Sea Beach Palace Hotel is shown immediately to the right

In 1882, George and Peter Tilyou opened the *Surf Theater*, which became a familiar venue for many of the vaudeville stars of the era. It was built on an alleyway running between Surf Avenue and the ocean. The alleyway was paved with wooden planks, and was also home to a number of beer saloons, clam bars, and bathhouses.

In 1884, the world's first amusement roller coaster, the *Switchback Railway*, was opened by LaMarcus Thompson. The design was said to be based upon a coal mining train that had been converted to an amusement in 1827. The car was manually pushed off a high point, reaching a maximum speed of 6 mph. When it reached the other end (another high point), the car was switched to a return track, where the process was repeated.

Fig. 6.31 The Switchback Railway in 1885

In June of 1895, Paul Boynton opened Coney Island's first amusement park, that is, Sea Lion Park, on 16 acres located immediately behind the Elephant Hotel. Boynton was an international celebrity for numerous swimming exploits featuring a rubber suit of his design. The park was outdoors, but enclosed by a fence. The park's two signature attractions were the *Flip-Flap* and the *Shoot-the-Chutes*. The former was one of the world's first roller coasters that turned a complete loop, taking the occupants upside-down. While popular for a while, few were brave enough to actually take the ride. The Shoot-the-Chutes was a waterslide that landed the occupants of its boats in the surf of the beach. The park was never a real financial success, and a summer of bad weather in 1902 led to its closure and sale.

(a) The Flip-Flap (b) The Shoot-the-Chutes

Fig. 6.32 Signature rides of Sea Lion Park

In 1897, Coney Island's most famous and enduring amusement park opened. *Steeplechase* was built by George Tilyou. It was burned down in a fire in 1907, but later rebuilt. It remained in operation until 1964. It originally included a large pool and swimming pavilion, but featured a variety of indoor amusements that always attracted a crowd. One of its most famous rides was the *Parachute Jump*, which was only added after the 1939 World's Fair, from which it was obtained. Declared a national historic site, the structure of the parachute drop is the only remaining artifact of Steeplechase. Its signature ride, however, was the *Steeplechase* (for which the park was named), a set of mechanical horses on rails that careened around a track at speeds approaching 60 mph. With no safety belts, riders hung onto poles for dear life. Steeplechase closed in 1964, due to waning interest and a rapidly-deteriorating neighborhood. Several attempts to redevelop the site were tried with no success. Eventually, it became the site of a minor league baseball stadium housing the *Cyclones*, a New York Mets affiliate.

Fig. 6.33 The Steeplechase ride

The last of the great Coney Island amusement parks was opened by Frederic Thompson and Elmer Dundy in 1903. Thompson and Dundy had designed a popular ride called "A Trip to the Moon" for the Pan-American Exposition, held in Buffalo, New York in 1901. George Tilyou brought the ride to *Steeplechase* for the 1902 season. Thompson and Dundy, however, purchased the site of the former Sea Lion Park and opened their own facility, which they named *Luna Park*. Luna Park operated until 1944, and was reincarnated in a modern format in 2010.

(a) Entrance to Luna Park in 1903 (b) Loop-the-Loop ride in 1903

Fig. 6.34 Luna Park in 1903

Luna Park retained the popular "Shoot-the-Chutes" ride from Seal Lion Park, but replaced the "Flip-Flap" with its own looping roller coaster, the "Loop-the-Loop." Like its predecessor, however, more people wanted to watch than ride. Luna Park also featured a full circus, and was well-known for its bright lights at night.

6.8.2 The Coney Island Railroads

The Brooklyn rapid transit system grew out of two needs. The first was the need to transport people to and from the hotels and amusements of Coney Island from the more populated areas, which were clustered around the ferries to Manhattan, and after 1883, the Brooklyn Bridge.

The story of the various railroads built to transport people to and from the growing resorts in Coney Island could fill a book. As usual, cutthroat competition among competing developers took its toll. In virtually all cases, the railroads were built by or at the behest of resort owners, each trying to make their facilities the easiest to get to, and therefore, the most profitable.

The first railroad to serve Coney Island was the *Brooklyn, Bath, and Coney Island Railroad*. In 1864, it began operations as a horse-railroad. It was extended to Coney Island in 1867, by which time it had converted to steam operation. In 1885, the line was reorganized into the *Brooklyn, Bath, and West End Railroad*. The current West End Line still uses part of the original right-of-way.

Fig. 6.35 The Brooklyn, Bath, and Coney Island Terminal in 1876

Fig. 6.36 The Brooklyn, Bath and West End Railroad, Bensonhurst Station, in 1892

The *Prospect Park and Coney Island Railroad* opened on June 19, 1875 as a steam railroad operating between Prospect Park, 9th Avenue and 20th Street, to Gravesend Avenue and Neck Road. Originally a single-track line, it was double-tracked in 1878. It was built by Andrew Culver, and was referred to as the "Culver Line," a name that still used today for an elevated that travels the same route. The Coney Island terminal, called Culver Terminal, was the predecessor to the present Stillwell Avenue complex.

The *New York and Sea Beach Railroad* opened on July 18, 1877 as a traditional steam railroad. The line eventually served the Sea Beach Palace Hotel, and was operated as a seasonal service only. Originally a single-track railroad, it was double-tracked in 1885. The current Sea Beach line runs along much of the original right-of-way of the railroad.

The *Brooklyn, Flatbush, and Coney Island Railroad* began operations on July 2, 1878 from Flatbush and Ocean Avenues and the Brighton Beach Hotel. In August, service was extended to the Long Island Railroad terminal at Atlantic and Flatbush Avenues. After a bankruptcy, the railroad was reorganized as the *Brooklyn and Brighton Beach Railroad* in 1887. The current Brighton Beach line uses open cuts originally constructed for the railroad.

Fig. 6.37 The Prospect Park Terminal on the Brooklyn, Flatbush, and Coney Island Railroad in 1878

Fig. 6.38 An isolated-looking Kings Highway Station on the Brooklyn and Brighton Beach Railroad in 1888

Perhaps the most unique operation of all was the two-week demonstration of Eben Boynton's Unicycle Railway during the summer of 1890. Using an abandoned section of the Sea Beach and Brighton Railroad, the experimental service used a single-railed locomotive and cars stabilized by an overhead rail. Both the locomotive and the cars were double-decked, but were, of course, quite narrow. The service operated continuously throughout the period, going from Gravesend to Coney Island, a distance of approximately two miles.

Fig. 6.39 Drawings of the Boynton Unicycle Railroad

Fig. 6.40 The Actual Boynton Engine used in the Coney Island experimental service, 1889

While technically not a Coney Island railway, another steam railroad would become part of the Brooklyn rapid transit system. The *Brooklyn and Rockaway Beach Railroad*, which opened in 1865, connected the Long Island Railroad in East New York with a pier at Canarsie Landing. It remained a single-track line until 1894, when a second track was added. The line was run by the Long Island Railroad until it was taken over by the Brooklyn Rapid Transit Company in 1906.

6.8.3 Brooklyn Elevateds

While the Coney Island railroads were providing options for travel from the inhabited areas of Brooklyn to Coney Island, a second set of rapid transit lines was developing with a different objective, namely, carrying people living in Brooklyn to the ferry lines, and after 1883, the Brooklyn Bridge. The ferries and the bridge were the lifelines to Manhattan. Early connections via horse-streetcars were popular, but were eventually replaced by trolleys, and later by elevated rapid transit lines.

6.8.3.1 The New York and Brooklyn Bridge Railway

On May 24th, 1883, the Brooklyn Bridge opened. Because there was some thought that the bridge could not bear the weight of steam engines, rail traffic across the bridge was provided by a cable train service operated by the *New York and Brooklyn Bridge Railway Company*. The cable system predated any of the Brooklyn elevateds, and facilitated crossings for passengers arriving primarily on horse-streetcars.

The cable-trains ran from a Brooklyn terminal on Sands Street to a Manhattan terminal near City Hall. After electrification of the elevateds, Brooklyn trains provided direct service over the Brooklyn Bridge. The Manhattan terminal, however, was never physically connected to the Manhattan Elevated system, and passengers had to transfer to the City Hall terminal of the 2nd/3rd Avenue Els.

6.8.3.2 The Lexington Avenue El (Old Main Line)

The first elevated railway to operate in Brooklyn opened in 1885. Known as the *Old Main Line* (also known as the *Lexington Avenue El*), it was operated by the Brooklyn Elevated Railroad from Washington and York Streets to Gates Avenue and Broadway. By the end of 1885, it had, in a series of steps, been extended west to Fulton Ferry, and east to Van Siclen Avenue in East New York. The line was eventually extended as far as Jamaica, in Queens. The small section of the el that operated over Broadway is the oldest section of the current rapid transit system in current operation.

6.8.3.3 The Myrtle Avenue El

The year 1888 saw a rapid expansion of Brooklyn elevateds. The *Myrtle Avenue El* was built by the Union Elevated Railroad Company, but was leased to the

Brooklyn Elevated Railroad for operation before opening. It began operations in 1888 from Adams Street to Grand Avenue, where it joined the Old Main Line to Dekalb Avenue. In 1889, the Myrtle Avenue El was extended (down Myrtle Avenue) to Grant Ave. Later in the year, the Myrtle Ave El was extended to Wyckoff Avenue, crossing over the Old Main Line at Broadway, but no track connection between the two lines was provided. In 1906, the line was extended to Metropolitan Avenue in Queens. A portion of the route ran at-grade through a right-of-way purchased from the Lutheran Cemetery. The last two stations on this line remain at grade to this day, while its remainder was elevated in 1915.

(a) The Manhattan Terminal of the NY & BB Cable Railway as depicted in *Harper's Weekly*

(b) A train leaves the Manhattan Terminal in 1885. The cable can be seen in the center of the tracks

(c) The Manhattan Terminal after electrification. Passengers are walking to the City Hall Terminal to access the 2nd and 3rd Avenue Els

Fig. 6.41 Views of the New York and Brooklyn Bridge Cable Railway

(a) Lexington Avenue train joins Myrtle Ave El at Grand Avenue

(b) Train leaves Broadway El to join the Lexington Avenue El

(c) Engine ramp at Grand Ave. and Park Ave.; One of the few existing pictures showing any portion of the Park Avenue El

(d) Lexington Avenue train weaves its way through Brooklyn

(e) Cypress Hills Terminal of the Lexington Ave. El in 1915; Broadway El trains shared the trackage at that time

Fig. 6.42 Views of the Lexington Avenue El in Brooklyn

In 1943, a new station was built at Jay Street on the Myrtle Avenue El. This was in preparation for the 1944 abandonment of through service from the Myrtle Avenue El over the Brooklyn Bridge to Manhattan (indeed, all elevated service over the Brooklyn Bridge was terminated in 1944) From 1944 through October 4, 1969, the station served as the Brooklyn terminus of the Myrtle Avenue El.

(a) Myrtle Ave. El near Borough Hall in 1940

(b) Myrtle Ave. train near Fresh Pond Road at-grade on the Lutheran Cemetery section in 1939

(c) Jay Street station under construction in June 1943

(d) Jay Street terminus of the Myrtle Avenue El in 1944

Fig. 6.43 Views of the Myrtle Avenue El

In Figure 6.43(c–d), the GEM Razor factory is seen. It is the future home of Polytechnic Institute of Brooklyn, now Polytechnic Institute of NYU, the authors' Alma Marta.

6.8.3.4 The Broadway El

Later in 1888, the Union Elevated opened and operated the *Broadway El*, which extended north from the portion of the Old Main Line over Broadway to Driggs Avenue, and a bit later to Broadway Ferry. The Brooklyn Elevated and the Union Elevated merged in 1890, consolidating their operations, and retaining the

Brooklyn Elevated name. Eventually, the Broadway El would be extended over the Williamsburg Bridge. The Broadway El shared tracks with the Old Main Line south of the intersection of Myrtle Ave. and Broadway.

(a) Broadway El train, circa 1935

(b) Broadway El approaching the Williamsburg Bridge; tracks to the right go over the; tracks to the left go over the Broadway Ferry, circa 1910

Fig. 6.44 Views of the Broadway Elevated

6.8.3.5 The Fifth Avenue Elevated

Also built by the Union Elevated and leased to the Brooklyn Elevated, the *5th Avenue Elevated* opened on November 5th, 1888. It started at Park and Hudson Avenues and connected the Old Main Line to Atlantic and Flatbush Avenues. This el had a scattered history, with an accident occurring between a 5th Ave. El and a Myrtle Ave. El train two days after opening. The accident occurred at the intersection of Hanover Street and Myrtle Avenue, which was an at-grade elevated crossing of the two lines. Service on the line was interrupted for several months, although several extensions were added through 1893, bringing the line down Brooklyn's 5th Avenue in several steps. At 5th Avenue and 38th Street, the line turned west, finishing its run on 3rd Avenue to 65th Street in Bay Ridge.

Fig. 6.45 The Fifth Avenue El at Flatbush and Atlantic Avenues; entrance to the Long Island Railroad Atlantic Avenue Terminal can be seen across the far street

6.8.3.6 The Fulton Avenue El

The last of the Brooklyn elevateds began operations in late 1888. Built and operated by the Kings County Elevated Railway Company, the *Fulton Street Elevated* ran from Fulton Ferry to Sackman Street. In 1893, it was extended to Grant Avenue, then to the City Line (of Brooklyn) and finally to Lefferts Blvd. in Queens. The segment to Lefferts Blvd. is still used as part of the A-train service.

(a) Fulton Street El at Borough Hall, circa 1898

(b) Typical Fulton Street El train and station,circa 1912

Fig. 6.46 Views of the Fulton Street Elevated

6.8.3.7 The Brooklyn Elevated System

While the Brooklyn elevateds involved three companies (Brooklyn Elevated, Union Elevated, Kings County Elevated), they shared trackage and routes, particularly over Broadway east of Lexington Avenue. There were also transfer points when a passenger could transfer from one to another without paying an additional fare (Broadway and Myrtle, Broadway Junction). Thus, the Brooklyn elevateds formed a functioning system that was reasonably coordinated from their very beginnings.

Of historical note, the first abandonment of an elevated railroad in New York came on December 30, 1890. The Old Main Line (Lexington Avenue) link between Myrtle and Hudson Avenues was closed, as its service duplicated that of the extended Myrtle Avenue El. The abandoned portion of the el, which ran over Park Avenue, was also been known as the "Park Avenue El." An interesting footnote is that this was the only section of New York City's elevated system to be removed without ever being converted to electric operation.

Figure 6.47 shows a map of the Brooklyn elevated lines as they existed in the early 1900s.

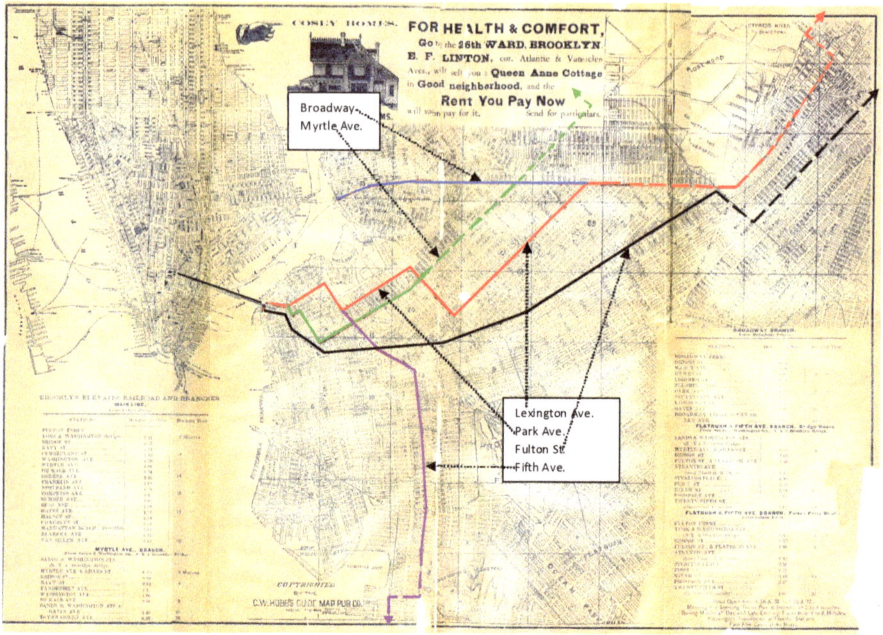

Key for Map:

Red = Old Main Line (Lexington Avenue El)
Green = Myrtle Avenue El
Blue = Broadway El
Purple = Fifth Avenue El
Black = Fulton Street El
Dotted Black = Brooklyn Bridge Cable RR
Solid Lines = Original Sections
Dashed Lines = Extensions after opening

Fig. 6.47 Map of the Brooklyn Elevated System, circa 1910

6.8.4 The BRT Emerges

In January of 1896, the Brooklyn Rapid Transit Corporation was created as a holding company, specifically set up to acquire the properties of other transit systems in Brooklyn. Prior to incorporation, it had begun the process by acquiring most of the street-car operations in Brooklyn. Because many of the railroads were experiencing financial difficulties, the BRT had acquired all of the Coney Island Railroads and all of the elevated lines by the end of 1900, except for the Brooklyn and Rockaway Beach Railroad (the current Canarsie Line), which was not acquired until 1906.

Figure 6.48 is a map of the BRT system in 1912, just prior to the Dual Contracts (See Ch. 10). It shows a coordinated network consisting of surviving Brooklyn elevateds and the Coney Island excursion railroads that had been taken over by the BRT.

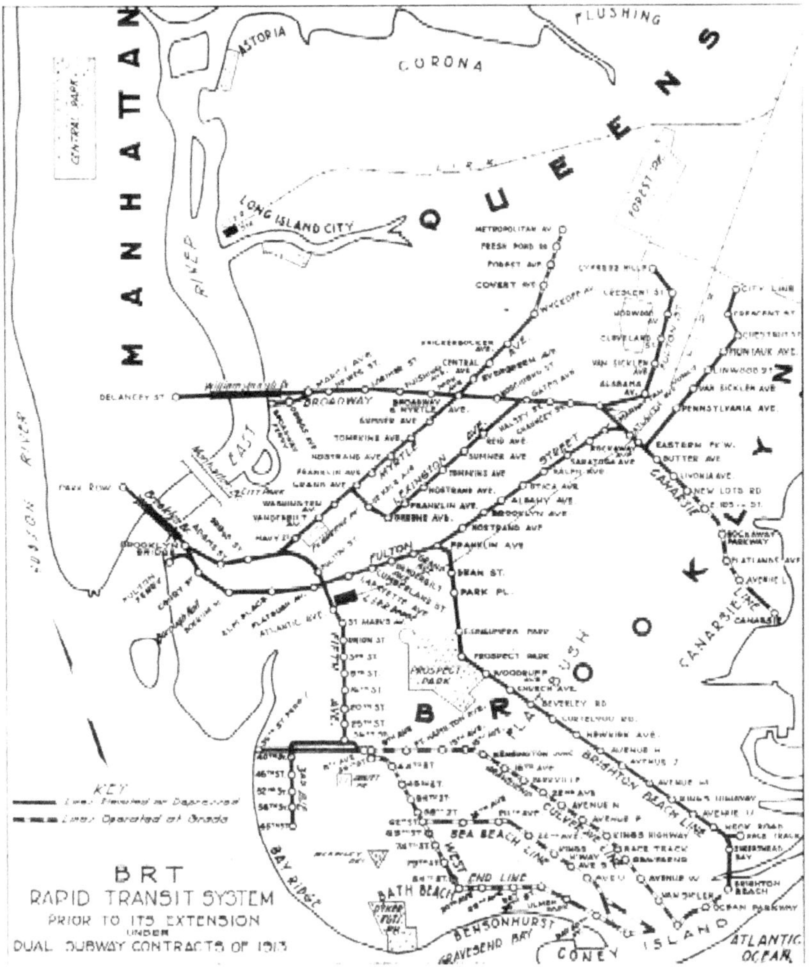

Fig. 6.48 Map of the Brooklyn Rapid Transit System in 1912

The map shows that the Fifth Avenue El was connected to the West End, Sea Beach, and Culver Lines, using rights of way originally owned by the Coney Island excursion railroads. The Brighton Beach Line, was directly connected to the Fulton Street Line. While in Manhattan, coordination of the elevateds was achieved primarily by financial skullduggery on the part of Jay Gould and others, Brooklyn was being serviced by a more comprehensive system that was unified under the BRT for the express purpose of designing and operating a functional transportation system.

6.9 Electrification

From their very inception, elevated railways in New York were considered to be at best a necessary evil. The need for transportation to allow the city to grow was evident. Just as evident, however, were the environmental issues created by the elevateds. Some of the worst complaints concerned the noise, soot, and heat generated by steam operation.

By the 1880s, New York was becoming an electric city. By 1885, telegraph, telephone, and lighting services were available to most in Manhattan, and to some in Brooklyn. It, therefore, was not startling when public support began to develop for conversion of the els to electric traction operation.

Like most significant technological changes, those most affected, namely, the operators of the elevateds, were generally opposed to the idea, which (in the mid-1880s) had not been developed or demonstrated.

The first demonstration of an electric traction engine in New York was on the Ninth Avenue El in 1885 (1). Under cover of darkness, a Daft-electric traction engine was tested, the Benjamin Franklin, pulling two cars. The engine created huge electric arcs every time contact was made, and threw a steady stream of sparks over the side of the el and onto the street. While it reached a speed of 25 mph, the owners were not impressed, as their steam engines could hit almost 50 mph while pulling up to four cars.

Later that year, Frank Sprague, Thomas Edison's former associate, demonstrated an electric traction car to the bigwigs of the Manhattan Railway, including its President, Jay Gould. (1) Unfortunately, when Sprague started the train, it lurched forward and a fuse blew out in a blinding explosion. Several people had to prevent Jay Gould from jumping off of the el in fear.

Stephen Field demonstrated his electric locomotive in 1887 on the 34th Street spur of the 3rd Avenue El, pulling one car uphill at 8 mph. Daft reappeared in 1888 with a second test run on the Ninth Avenue El pulling four cars (19). The moguls of the els remained unconvinced that replacing steam engines with electrically-powered locomotives would be of benefit to their companies, despite growing complaints concerning the environmental problems of the steam engines.

Fig. 6.49 Demonstration of the Daft Electric Locomotive on 9th Avenue El, Nov. 27, 1888

Sprague, of course, went on to demonstrate electric traction power elsewhere with great success, electrifying the entire Richmond, Virginia streetcar system. He followed this by converting the Boston streetcar system, the largest in the country, to successful electric operation.

The final piece for nationwide adoption of electric traction power occurred in 1895. Sprague developed a means to operate multiple-car units from a single controller. Now, each car could be separately electrified, and each car would be supplying its own traction power, all controlled from either end of the train by a single operator. The system permitted operation of trains in any length, and provided great flexibility in operations. Chicago was the first major system to adopt this mode of operation.

Interestingly, electrification of the New York elevateds began in Brooklyn. The Brooklyn Rapid Transit Company began converting its operation to multiple-unit, electric traction power in 1898, the same time that London abandoned steam operation of its Underground for electric traction.

Eventually, in Manhattan, economics won out. While conversion was an expensive project, the Manhattan Railway could sell off 225 locomotives. The savings in fuel would also be enormous, that is, 924 tons of coal and 535 million gallons of water per year. Electric service on the 2nd Avenue El was initiated with great fanfare on January 9, 1902. Full service was established on 2nd Avenue in March, on 3rd Avenue in August, on 6th Avenue in November, and finally, on 9th Avenue in the spring of 1903 (1).

Electrification enabled operators to run longer trains, which eased overcrowding significantly. While the environmental hazards of steam operation were eliminated, those associated with the elevated structures themselves, for example, blocking of light, constant dropping of steel dust, and noise, were not. However, without electrification, it is doubtful that the elevateds would have survived long into the 20th century.

They did, however, provide a vital force for growth at a time of exploding population in the city. In 1898, the great consolidation of the so-called "outer boroughs" into New York City occurred. At their height, the elevateds carried New Yorkers throughout all of the boroughs, except Staten Island, which to this day remains unconnected by rapid transit to the rest of the city.

6.10 The Ideas Keep Coming

As the steam-powered elevateds provided increased outreach for all New Yorkers, the crescendo of technological proposals for alternative systems continued before it became clear that subways were both feasible and desirable for rapid transit. As before, some of the ideas were quite innovative.

In 1886 in Boston, Massachusetts, Joseph Meig proposed an elevated, steam-powered, cylindrical monorail system (20). The monorail system allowed for lighter, less imposing structures. Switching became more complex however. Meig actually built a demonstration track and vehicle, but never got it running.

L.A. Thompson proposed a "gravity system" which did away with locomotives entirely in 1888 (21). His proposal was just the latest version of what had been referred to as "undulating railways" since the 1830s. The gravity system worked much like a roller coaster. Periodically, trains would be pulled up a grade by cables or cogs (driven by stationary steam engines). They would then coast down gentle grades to the next station.

In 1889, Gibard advanced his "sliding railway," which he had been working on since 1852 (22). Instead of wheels, his system "slid" along flat tracks with flat wheels. Claiming significant benefits in reduced friction, the system required that a thin film of water, oil, or other lubricant be maintained on the rails.

In 1890, T. C. Clarke proposed a system that utilized electric power (23). He proposed a system in which two elevated tracks were supported on single columns. Directly below the el, two trolley-tracks would be placed. A combined electric supply powered both systems, thus claiming to double the capacity of a traditional el, which (Clarke claimed) "wasted" the space directly below.

(a) Meig's cylindrical monorail concept

(b) Thompson's gravity railway

(c) Details of Gibard's sliding truck

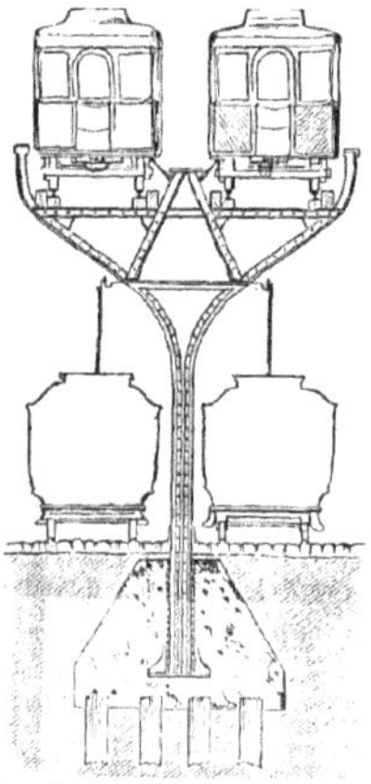

(d) Clarke's combined elevated-trolley scheme

(e) Anderson's propeller-powered el

(f) The Chase–Kirchner aerodromic el

Fig. 6.50 Additional ideas on elevated rapid transit

A. Anderson proposed an elevated with cars driven by aircraft-type propellers on the front and back of railcar (24). The proposal did not address possible noise created by the propellers, which would have been driven by electric motors.

In 1894, a design surfaced that claimed to allow speeds of up to 125–150 mph using stacked "wings" to provide lift, thereby reducing friction (25). To accommodate such speeds, however, the proposers, Chase and Kirchner, believed that the route would have to be perfectly straight, with no curves.

As the 20th century approached, despite all of the intriguing proposals, some brilliant, though some totally impractical (often at the same time), it had become clear that a traditional railway with standard-gauge tracks was the logical system upon which to build a rapid transit system. By the turn of the century, arguments over the mode of power would be nearing a conclusion, with electric traction power being the overwhelming winner.

6.11 The Age of Subways Dawns

The elevateds survived well into the 20th century, and indeed many continue to operate today. The age of subways, however, was about to begin as the century turned. Unresolved environmental issues, and the building of the subway system caused most of the Manhattan elevateds to be removed in the 1930s and 1940s. The 3rd Avenue El was the last to operate in Manhattan; its last train left the station on May 12, 1955. Portions of the 3rd Avenue El continued to operate in the Bronx through the mid-1970s.

Many, but not all, of the elevateds were replaced by subways. In Manhattan, the 9th Avenue El was replaced by the 8th Avenue Subway, and the 6th Avenue El was replaced by the 6th Avenue Subway. However, the 2nd and 3rd Avenue Els were also removed, and were not replaced. The 2nd Avenue Subway remained a dream for almost 80 years; the first stage of a new 2nd Avenue Subway is now under construction and scheduled to open in December of 2016. There is no guarantee that remaining stages will be completed. The Eastside of Manhattan continues to struggle without adequate rapid transit service.

In Brooklyn, the Fulton Street El was replaced by the Fulton Street Subway, and the 5th Avenue El was replaced by the 4th Avenue Subway. The Myrtle Avenue El was demolished between Broadway and Jay Street, replaced only by buses. The surface trip down Myrtle Avenue remains long and difficult.

Subways, of course, did not displace all of the elevateds. The Bronx, Brooklyn, and Queens still have many miles of elevated rail transit, some of it built after 1900.

6.12 Impacts

Table 6.1 illustrates the importance of the elevateds to New York. It shows annual ridership on the system from 1872 through 1939, the last year for which separate data is available.

Table 6.1 Ridership on the elevateds

Year	Annual Ridership	Year	Annual Ridership
1872	242,000	1899	183,000,000
1875	910,000	1905	266,000,000
1880	60,000,000	1910	294,000,000
1885	103,000,000	1921	374,000,000
1890	186,000,000	1931	327,000,000
1895	188,000,000	1939	169,000,000

The expansion of the city to uptown Manhattan and into the Bronx was fueled by the Manhattan elevated rail system. Rufus Gilbert's dream of providing a better living environment for the poor never happened, and thus the poor simply moved to new tenements located further away from downtown. The coordinated system of rapid transit serving Brooklyn, together with direct service over the Brooklyn Bridge, fueled significant development there as well, and eventually into Queens.

Fortunes were made on real estate, as the Wall Street tycoons, using what is now considered to be "inside information," bought up properties along proposed elevated lines, reselling at huge profits. The rich indeed got richer. Jay Gould, in particular, used the elevated railroad system as a key cornerstone of his wealth, and earned much of his "robber baron" reputation by manipulating elevated railroad company stocks.

After the turn of the century, however, subways would be the focus of rapid transit development, eventually leading to one of the world's largest and most effective public transportation networks.

References

1. Reed, R.C.: The New York Elevated. A.S. Barnes & Co., Cranberry (1978)
2. Genin's Bridge. Gleason's Pictorial Drawing-Room Companion, vol. 3(26), Boston, MA, December 25 (1852) (American Periodicals On-Line)
3. Abuses in Public Conveyances. Scientific American, March 14, vol. VIII(11). Munn & Co., New York (1863) (American Periodicals On-Line)
4. Suspension Railway. Mechanic's Magazine and Journal of the Mechanical Institute, vol. 3(1). D. K Munn & George C. Schaeffer, New York, NY (1834) (American Periodicals On-Line)
5. Mechanic's Register. Journal of the Franklin Institute of the State of Penn-sylvania for the Promotion of the Mechanical Arts, vol. 19(6), Philadelphia, PA (American Periodicals On-Line)
6. Broadway Elevated Railroad. Scientific American, January 1, vol. 1(16). Munn & Co., New York (1846) (American Periodicals On-lLine)
7. Dewitt's Elevated Railway and Suspension Car. The Literary World, New York, NY, December 11, vol. 2(15) (1847) (American Periodicals On-Line)
8. (Title Unknown). Scientific American, November 26, vol. IX(11). Munn & Co., New York (1853) (American Periodicals On-Line)

9. Swett's Elevated Railroad for Broadway. Scientific American, October 15, vol. IX(5). Munn & Co., New York (1853) (American Periodicals On-Line)
10. Elevated Promenade and Railroad for Broadway. Scientific American, January 21, vol. IX(19). Munn & Co., New York (1854) (American Periodicals On-Line)
11. Elevated Railroad. Gleason's Pictorial Drawing-Room Companion, Boston, MA, April 1, vol. 6(13) (1854) (American Periodicals On-Line)
12. Elevated Railway for Broadway. Scientific American, September 9, vol. XIII(11). Munn & Co., New York (1865) (American Periodicals On-Line)
13. Reeves, W.: The First Elevated Railroads in Manhattan and the Bronx in the City of New York. New York Historical Society, New York (1936)
14. The Arcade Railway. Scientific American, February 9, vol. XVI(6). Munn & Co., New York (1867) (American Periodicals On-Line)
15. Hannah's Patent Metropolitan Railway. Scientific American, May 13, vol. XXIV(20). Munn & Co., New York (1871) (American Periodicals On-Line)
16. Endless Traveling or Railway Sidewalks. Scientific American, April 20, vol. XXVI(17). Munn & Co., New York (1872) (American Periodicals On-Line)
17. Rapid Transit. Harper's Weekly, February 9, New York, NY (1878) (American Periodicals On-Line)
18. Christiero, G.J.: An Early History of New York City's First Elevated Railway, http://www.nycsubway.org
19. Running Elevated Railroad Trains by Electricity, December 8, vol. LIX(23). Munn & Co., New York (1888) (American Periodicals On-Line)
20. The Meigs Elevated Railway. Scientific American, July 10, vol. LV(2). Munn & Co., New York (1886) (American Periodicals On-Line)
21. Thompson's Gravity System for Rapid Transit in Towns and Cities. Scientific American, September 8, vol. LIX(10). Munn & Co., New York (1888) (American Periodicals On-Line)
22. Gibard's Sliding Railway. Scientific American, November 23, vol. LXI(21) (1889) (American Periodicals On-Line)
23. New Elevated Railway. Scientific American, May 10, vol. LXII(19). Munn & Co., New York (1890) (American Periodicals On-Line)
24. An Improved Elevated Railway. Scientific American, April 2, vol. LXVI(14). Munn & Co., New York (1892) (American Periodicals On-Line)
25. The Chase-Kirschner Aerodrome Railway. Scientific American, May 5, vol. LXX(18). Munn & Co., New York (1894) (American Periodicals On-Line)

General References

1. Hood, C.: 722 Miles: The Building of the Subways and How They Transformed New York. The Johns Hopkins University Press, Baltimore (1993)
2. Bobrick, B.: Labyrinths of Iron: A History of the World's Subways. Newsweek Books, New York (1981)
3. Reeves, W.: The First Elevated Railroads in Manhattan and the Bronx of the City of New York. J.J. Little and Ives Company, New York (1936)
4. Cunningham, J.: Second Avenue El in Manhattan. NJ International Inc., Hicks-ville (1995)

5. Geller, J., Watson, E.: The Brooklyn Elevated. NJ International Inc., Hicks-ville (1995)
6. Feinman, M.: Early Rapid Transit in Brooklyn, 1878 to 1913, `http://www.nycsubway.org`
7. Walker, J.: Fifty Years of Rapid Transit (1918), `http://www.nycsubway.org`

Figure Sources

6.1 *Courtesy of Wikimedia Commons*, T. Beneche, *Sleighing in New York*, 1855, from Library of Congress Prints and Photographs Division.
6.2 *Origninally published in* "Genin's Bridge," *Gleason's Pictorial Drawing Room Companion*, Vol.3, No. 26, December 25, 1852, Boston, MA, pg 416.
6.3 *Originally published in* Frank Leslie's Illustrated News, New York, NY 1865.
6.4 *Originally published in* "Suspension Railway," *Mechanic's Magazine, and Journal of the Mechanic's Institute*, Vol 3 No. 1, January 1834, D.K. Munn and George C. Schaeffer, New York, NY, pg 30.
6.5 *Originally published in* "Mechanics Register," *Journal of the FrankliInstitute of the State of Pennsylvania for the Promotion of the Mechanical Arts*, Vol. 19, No 6, June 1, 1837, Philadelphia PA, pg 480.
6.6 *Originally published in* "Broadway Elevated Railway," *Scientific American*, Vol. 1, No. 16, January 1, 1846, Munn & Co., New York, NY.
6.7 *Originally published in Scientific American*, Vol. IX, No. 11, Munn and Company, New York, NY, Nov. 26, 1853, pg 30.
6.8 *Originally published in* "Swett's Elevated Railroad for Broadway," *Scientific American*, Vol. IX, No. 5, October 15, 1853, Munn & Co., New York, NY, pg 33.
6.9 *Originally published in* "Elevated Railroad," *Gleason's Pictorial Drawing-Room Companion*, Vol. 6, No. 13, April 1, 1854, New York, NY, pg 201.
6.10 *Originally published in* "Elevated Railway for Broadway, *Scientific American*, Vol. XII, No. 11, September 9, 1865, Munn and Co, New York, NY, pg 159.
6.11 *Courtesy of* Library of Congress, Division of Prints and Photographs.
6.12 *Courtesy of* Library of Congress, Division of Prints and Photographs.
6.13 *Courtesy of* New York Historical Society, William F. Reeves Collection.
6.14 *Courtesy of* Library of Congress, Prints and Photographs Division, LC USZ62 4471.
6.15 *Courtesy of* New York Historical Society, William F. Reeves Collection.
6.16 *Courtesy of* New York Historical Society, William F. Reeves Collection.
6.17 *Originally published in* Clark, Thomas, "Rapid Transit in Cities," *Scribner's Magazine*, May/June 1892, pg 567.
6.18 *Courtesy of* New York City Transit Museum.
6.19 *Courtesy of* New York Historical Society.
6.20 *Originally published in* Frank Leslie's Illustrated News, March 18, 1871, pg 1.

6.21 *Originally published in* Frank Leslie's Illustrated News, May 25, 1878, pg 201.
6.22 *Originally published in* "Rapid Transit," *Harper's Weekly*, February 9, 1878, New York, NY, pg 112.
6.23 *Originally published in* Harper's Weekly, May 18, 1878, pg 388.
6.24 *Courtesy of* Library of Congress, Prints and Photographs Division, LC USZ62 956799.
6.25 *Courtesy of* New York City Transit Museum (a, c, d); *Courtesy of* Library of Congress, Prints and Photographs Division, Detroit Publishing Collection, LC D4 155942L.
6.26 (a) *Courtesy of* Library of Congress, Prints and Photographs Division, LC-USZ62 66774, (b) *Courtesy of* Library of Congress, Prints and Photographs Division, August Loeffler Collection, LC USZ62 96201, (c,d) *Courtesy of* New York Transit Museum.
6.27 (a) *Courtesy of* Wikimedia Commons, (b) *Courtesy of the* New York Transit Museum.
6.28 *Courtesy of* Library of Congress, Division of Prints and Photographs, August Loeffler Collection, LC USZ62 66522 and LC USZ62 66523.
6.29 (a) *Courtesy of* Library of Congress, Division of Prints and Photographs, August Loeffler Collection, LC USZ62 66521, (b) *Courtesy of* New York Transit Museum.
6.30 *Courtesy of* New York Transit Museum.
6.31 *Courtesy of* Wikimedia Commons.
6.32 *Courtesy of* Library of Congress, Division of Prints and Photographs, (a) Detroit Publishing Collection, LC DH 9026, (b) August Loeffler Collection, LC USZ62 90966.
6.33 *Courtesy of* Library of Congress, Division of Prints and Photographs, LC USZ62 102896.
6.34 *Courtesy of* Library of Congress, Division of Prints and Photographs, Detroit Publishing Collection, LC D4 16699 and LC D4 9180.
6.35 *Courtesy of* New York Transit Museum, Lonto–Watson Collection.
6.36 *Courtesy of* New York Transit Museum, Lonto–Watson Collection.
6.37 *Courtesy of* New York Transit Museum, Lonto–Watson Collection.
6.38 *Courtesy of* New York Transit Museum, Lonto–Watson Collection.
6.39 *Originally published in* "Unicycle Railway," *Scientific American*, Munn & Company, New York, NY, March 28, 1891, pg 66.
6.40 *Courtesy of* New York Transit Museum, Lonto–Watson Collection.
6.41 (a) *Originally published in* Harper's Weekly, 1885; (b,c) *Courtesy of* New York Transit Museum.
6.42 (a–e) *Courtesy of* New York Transit Museum.
6.43 (a–d) *Courtesy of* New York Transit Museum, (c) Lonto–Watson Collection.
6.44 (a–b) *Courtesy of* New York Transit Museum.
6.45 *Courtesy of* Library of Congress, Division of Prints and Photographs, Detroit Publishing Collection, LC D4 72114.

6.46 (a–b) *Courtesy of* New York Transit Museum (a) Lonto–Watson Collection.
6.47 Base map prepared for E. F. Linton, a realtor, in 1888. Elevated lines drawn by the authors.
6.48 *Courtesy of* New York Transit Museum.
6.49 *Originally published in* "Running elevated Railroad Trains by Electricity," *Scientific American*, Vol. LIX, No. 23, December 8, 1888, Munn & Co, New York, NY, pg 361.
6.50 *Originally published in* (a) "The Meigs Elevated Railway," *Scientific American*, Vol. LV, No 2, Munn & Co., New York, NY, July 10, 1886, pg 21, (b) "Thompson's Gravity System for Rapid Transit in Towns and Cities," *Scientific American*, Vol. LIX, No 10, Munn & Co., New York, NY, September 8, 1888, pg 149. (c) "Gibard's Sliding Railway," *Scientific American*, Vol. XLI, No 21, September 8, 1888, pg 327, (d) "New elevated Railway," *Scientific American*, Vol. LXII, No. 19, Munn & Co., New York, NY, May 10, 1890, (e) "An Improved Elevated Railway," *Scientific American*, Vol. LXVI, No. 14, Munn & Co., New York, NY, April 2, 1892, pg 211, (f) "The Chase-Kirchner Aerodrome Railway," *Scientific American*, Vol. LXX, No. 18, Munn & Co, May 5, 1894, pg 276.

Chapter 7
Alfred E. Beach and New York's First Subway

On February 26, 1870, while Harvey, Gilbert and others were working on a practical elevated railroad, Alfred Ely Beach introduced a one-block demonstration subway to the public, charging $0.25 fare per ride. All of the proceeds were donated to the Union Home for Orphans of Soldiers and Sailors. The subway ran for one block under Broadway, between Warren and Cedar Streets, and was powered by air pressure. Thirty-four years would pass before another subway opened its doors in New York.

The story of Beach and his unique subway is the stuff of legend, because Beach built his subway by stealth of night over a period of 58 days, supposedly with no one, other than the workers, knowing what was occurring. There is, however, more myth than truth in that version of the tale, although the truth is still pretty amazing stuff.

In building his subway, Beach demonstrated the viability of subways, proved that they could be built without compromising the structural integrity of adjacent buildings, that travelers would indeed travel underground without fear, and that even the political opposition of the powers that be could be confounded.

If Alfred Beach's only accomplishment was his subway, he would be mostly forgotten as a footnote in transit history. He was, however, far more than that, and his leadership and thought was critical in bringing the eventual reality of rapid transit to New York, even if his demonstration proved to be impractical for larger-scale operations.

7.1 Who was Alfred E. Beach?

Alfred Ely Beach's pedigree was impressive. He could trace his lineage back to Pilgrim elder William Brewster, and to Elihu Yale, who founded Yale University. His father was Moses Yale Beach, who founded and published the *New York Sun*. Working his way through the newspaper in various capacities, Alfred and his brother Moses were given full control in 1848. At the age of 22, Alfred Beach was already a very prominent New Yorker.

R.P. Roess and G. Sansone: The Wheels That Drove New York, STTT 1, pp. 139–154.
springerlink.com

Fig. 7.1 Alfred Ely Beach

While Beach was interested in the publishing business, he was more interested in invention, and in promoting invention. In 1846, Beach, together with his friend Orus Munn, had purchased a newly-launched, but struggling magazine, renaming it the *Scientific American*. He became its publisher, dividing his efforts between the magazine and the newspaper. As editor, he focused *Scientific American* on inventions and inventors, and began publishing detailed articles and discussions on emerging technologies, including plans for rapid transit systems in New York and elsewhere. In 1853, he started another publication, *People's Journal*, which was a more accessible version focusing on agriculture, mechanics, and chemistry. It was the forerunner of *Popular Mechanics*. Both publications and the newspaper prospered under his leadership.

In 1852, wishing to devote himself fully to promoting invention, he left the *Sun* in the hands of his brother. In 1846, he and Munn had established a patent agency, that is, Munn & Company, dedicated to helping inventors obtain patents for their work. It was an immediate success, and became the most important business of its kind in the U.S. Thomas Edison first demonstrated his phonograph in Beach's office. Alexander Graham Bell, Samuel Morse, John Ericsson, and others, great and small, consulted him. At his own expense, Beach would travel to Washington several times a month to promote his clients' interests. It could be argued that Beach did more for American inventors than anyone in history. In 1846, approximately 600 patents were issued in the U.S. Fifty years later, when Beach died, more than 20,000 per year were being issued.

Beach was a successful inventor in his own right. He held many patents, including a typewriter for the blind. It is generally considered to be the forerunner of the standard office typewriter, and it was awarded first prize at the Crystal Palace Exhibition in New York in 1856.

7.2 An Early Proposal for Transit in New York

From the day he became editor of *Scientific American*, Beach used the journal to publish descriptions of many of the proposals for rapid transit that were being generated for New York and for other cities. In 1849, as editor of the *Sun*, he proposed a tunnel running the length of Broadway. He proposed two tracks for a horse-drawn railway, and two promenades for pedestrians.

7.3 Early Development of Pneumatic Railways

Beach believed that pneumatic (or atmospheric) power was the answer to an environmentally viable subway. Beach was not the first person to propose and build a pneumatic railway. Beginning in the 1830s, several had already been developed and demonstrated in Europe.

Such railways were also referred to as "atmospheric" railways because they made use of air pressure for propulsion. In concept, atmospheric propulsion was relatively simple: using large fans and a piston tightly enclosed in a tube, air pressure could be brought to bear behind the piston; moreover, using a fan in reverse, a vacuum could be created in front of the piston. The piston could then be connected to a railcar, or any other device needing propulsion.

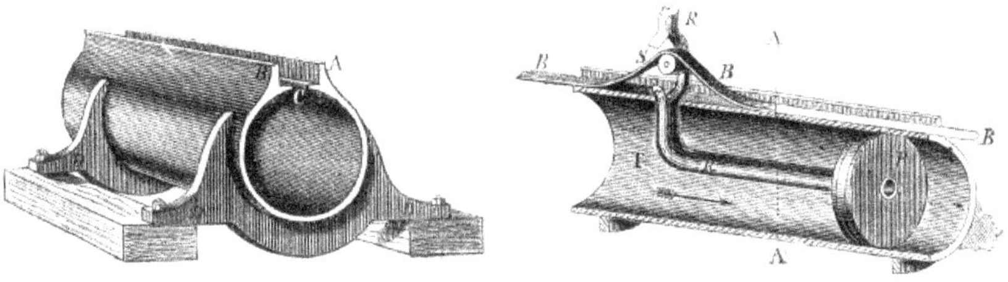

Fig. 7.2 The pneumatic piston concept

One of the earliest discussions of pneumatic railways appeared in *Mechanic's Magazine, and Journal of the Mechanics' Institute* in 1835 (1). Concerning the early steam railroads of the day, it notes that experience:

> "*...has at the same time disclosed the inefficiency of the construction by its utter inability to withstand the violent action and tremendous concussions of the ponderous engine, which is also destroyed by its own force, and the imperfections of the locomotive system by the incompetency of the engine to work with advantage except upon a level.*" (Ref. 1, pg 97)

It extols the virtues of pneumatic power created by stationary steam engines which do not have to carry the load of moving their own dead weight, and claims substantial operating and construction cost savings over "conventional" railroads (themselves only in their infancy).

Fig. 7.3 Sketch of a pneumatic railway proposal

Figure 7.3 shows drive wheels located in a 30–40 inch diameter cylinder, located under the center of the tracks, attached to a piston driven by air pressure. The drive wheels are connected to an "engine" pulling a passenger carriage or carriages. The invention was put forth by G. Ralston, and drew a great deal of interest from other inventors and scientists of the time.

One atmospheric railway operated in Ireland between 1844 and 1854, and another in France between 1847 and 1860. In both cases, a small pneumatic tube was built alongside the track. Both were relatively short railways on inclines of less than two miles. Each only used pneumatic propulsion upgrade, using gravity alone to move cars down the incline (2).

Longer lines were built in England, and both placed the pneumatic tube between the running rails. The longest ran between London and Croydon, about 7.5 miles, and achieved a remarkable speed of 70 mph during a test run in September 1845. Both of the English lines, however, were short-lived, each operating for less than two years (2)

In 1860, Thomas Rammell patented a pneumatic system in which the entire car was placed in a larger tube. The car itself was the piston. The system was primarily developed to carry mail and packages between rail terminals and post offices. A

Fig. 7.4 Rammell's concept demonstrated at Battersea, 1861

prototype using 30 inch diameter tubes was built on open ground at Battersea, London in 1861. The London Pneumatic Dispatch Company put two similar tubes into service in 1863 and 1865, but ended service by 1873. While technically sound, the costs of operation of the tubes exceeded what the company could charge for parcel delivery.

7.4 The American Institute Fair of 1867

While *Scientific American* was publishing a significant volume of material concerning rapid transit systems, Beach was developing his own concept. He was convinced that steam power would cause significant environmental problems, particularly if run in tunnels, where the enclosed surroundings would trap the steam and make adequate ventilation almost impossible. London's experience with underground steam railways was decidedly mixed, and a number of horror stories involving passengers sickened by engine effluents had been published, both in England and in the U.S. He began to advocate a pneumatic system of propulsion for a subway in New York.

In 1867, Beach provided two pneumatic exhibits for the American Institute Fair, which was held in Armory Hall on 14th Street. One was a small pneumatic system for delivery of mail and packages. The second, however, was a 6-foot diameter vacuum tube, 107 ft in length, which carried passengers. The tube was suspended from the ceiling of the armory. By the closing of the fair in November 1867, 75,000 people had ridden through the demonstration railway. This impressive accomplishment earned both notoriety and praise for the system and its potential for rapid transit use.

Beach's intentions were clear. He wrote a pamphlet for the American Institute Fair outlining a system of pneumatic subways that would carry passengers from City Hall to Madison Square in five minutes, to Central Park in eight minutes, and all the way to Washington Heights in twenty minutes.

Fig. 7.5 Beach's American Institute Fair Exhibit 1867

7.5 Beach's Pneumatic Subway

In 1864, Hugh B. Willson's plan for a New York subway under Broadway had been confounded by a tough political environment. The opposition of business interests to anything built under or over Broadway, the general discomfort with the idea of steam railroads in tunnels, and a bribery scandal sealed the proposals demise. Beach did not believe that a bill for passenger rail service could be practically achieved. Instead, Beach submitted a bill to allow construction of small pneumatic tubes for carrying mail and parcels across the North (Harlem) and East Rivers. The bill passed under the radar of the politicians, and Boss Tweed may have even supported it, thinking that it would keep Beach out of the growing rapid transit debate. The Senate passed the bill on May 4, and it was signed into law by Governor Fenton on June 2, 1868. Under the auspices of this act, the Beach Pneumatic Subway Transit Company was incorporated on August 28, 1868. Under the charter granted by the state, the company had to first construct a pneumatic parcel system from the main Post Office to a point no further north than 14th Street, and operate it successfully for three months before constructing any additional tubes. No details of routing were contained in the charter, but Broadway was a reasonable expectation.

In May 1869, the company's charter was revised in two ways. The amendment authorized the construction of a parcel tube along Broadway in the event that the postmaster rejected a connection with the Post Office. Rarely reported, the postmaster had already rejected the connection months earlier, citing the lack of authority to allow the tube to be constructed across a public sidewalk. A second provision was critical: the original charter limited the size of the tubes to be constructed to a 4.5-ft diameter. As part of the requested amendment, Beach claimed that the most efficient system to build two small packet tubes would be to insert them into a single larger tube. No mention, of course, was made that the larger tube would indeed be large enough to house a passenger car.

Most historians believe that Beach always intended to build a passenger railway under Broadway, and that the packet system was simply a ruse to get a charter to build. Beach did, however, begin by building a small packet tube before embarking on his passenger demonstration subway. The packet tube may have been built first to give Beach some experience in tunneling. Whatever the case, Beach did proceed to build a one-block passenger system without having any charter to do so.

The company took a five-year lease in the basement of Devlin's Clothing Store, and used it as a base for construction of the tubes. While legend has the passenger tube built in complete secrecy, this was not entirely correct. It was well-known that packet tubes were being built, and news reports throughout the construction period appeared. The "secret" was that Beach was also building a passenger system under the noses of the politicians, lacking a charter to do so.

Nevertheless, construction did proceed with a great air of secrecy around it. On December 29, the *New-York Tribune* (4) published an article on progress that specifically notes the lack of public information concerning the project:

> *"It is about twelve months since the basement of the store on the corner of Warren St. and Broadway was rented by the managers of the projected Broadway Pneumatic Railway. Operations were immediately commenced looking to the boring of that thoroughfare, but from that day to the present the utmost secrecy regarding the work has been maintained. The laborers are all sworn to secrecy as a condition precedent to their employment. Press reporters are denied admittance, and the public are, therefore, kept in comparative ignorance of the real nature and progress of the work."* (Ref. 4, pg 2)

At the time, however, the *Tribune* was most interested in a slight sagging of the Broadway sidewalk, and was trying to find out why this had occurred and whether it remained safe. The project itself was well-known, but its exact nature and intent were kept well under wraps. The *Tribune*, however, was clearly going to persist in investigating the project.

On January 8, the *Tribune* published a reply to several of its articles from the Beach Pneumatic Company. The response focused on the perceived lack of public information, and on the apparent sagging of a portion of Broadway (5):

> *"In reference to the ridiculous stories that have been circulated about our men being sworn to secrecy, and the doors being closed to all persons, there is no truth in them. Our work has been carried on under the constant supervision of the officers of the Croton Aqueduct Department, where all of our plans are matters of public record. It has been alleged that the surface of the pavement on Broadway has settled slightly in consequence of the building of our tunnel. If it appears that we have so disturbed the pavement, it will become our duty to make the needful repairs at our own expense, that being one of the stipulations expressed in our charter. As to the danger from caving, the Chief Engineer of the Croton Aqueduct Department has reported officially to the Mayor that the street is perfectly safe, and that our works are constructed in a substantial manner of iron plates and brick masonry."* (Ref. 5, pg 5)

7.5.1 Beach's Tunneling Shield

In order to build his tunnel, Beach patented several improvements to Brunel's basic shield. He was the first to use a cylindrical form, predating James Henry Greathead's design by several months. After trying a small version of the shield building a packet tube, he used the full-size version to construct his passenger tube. The shield could be operated by six men. Horizontal shelves with sharpened edges were located across the front of the shield. Eighteen hydraulic rams (or pistons) were located at the rear and along the sides of the shield. Each piston had its own control, so that the direction of the shield, that is, left or right, up or down, could be precisely established. It dug out 16 inches of earth in each forward movement (3).

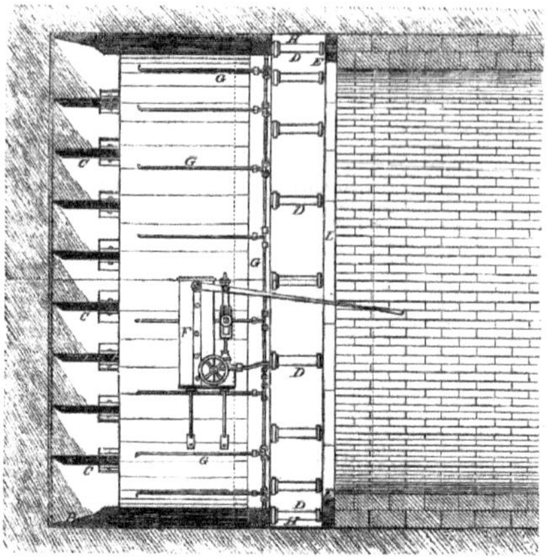

Fig. 7.6 Beach's Tunneling Shield

Fig. 7.7 Workers in the Beach Tunneling Shield

7.5.2 The Subway Is Built

The passenger tunnel was built with considerable stealth. Constructing out of Devlin's basement, much of the work was done at night. While some historians believe all of the work was done at night, there are no direct accounts that indicate this. Given that the soil along the route was soft, except for a stone foundation from an old Dutch fort that was encountered, the work was completed in 58 days.

At completion, the tunnel was 312 ft long. It had a cylindrical shape, was lined with brick, and contained a single set of tracks. One cylindrical car was lavishly furnished. The tunnel ran along a straight route under Broadway, except for a curve as the subway approached the Warren Street station. Beach wanted to avoid the claustrophobic nature of a tunnel, and did so by painting its inner walls white, and brightly lighting it using a new invention, Zircon oxygen lamps.

The waiting room for passengers was lavishly decorated and furnished to provide the utmost comfort. Even the station platform itself, several steps down from the waiting room, was invitingly designed.

At the Warren Street end of the line, in a hidden alcove, the systems power source was located. It was a 100 horsepower fan weighing 50 tons, and commonly referred to as the "Western Tornado." The fan provided the atmospheric pressure to push the car to Warren Street. When an air valve reversed the fan, the car was "sucked" back to its origin.

Without permission to link a packet service to the Post Office, Beach fulfilled the letter of his charter by also building 1000 ft of 8-inch diameter tubes that could carry letters and papers around the premises at remarkable speeds. It was connected to a hollow lamp post drop on the street above.

On January 11, 1870, just three days after the *Tribune* ran the response of the Beach Pneumatic's superintendent, it published a complete description of the project, after one of its reporters, supposedly disguised as a workman, gained entry into the work area (6). The reporter states that:

> *"The workmanship appears first-class, with no signs by which anyone could imagine the probability of its caving in or doing any damage to Broadway."* (Ref. 6, pg 8)

This statement belies many of the previous articles run by the *Tribune*, which had alleged that the work was compromising the structural integrity of Broadway. Even more amazingly, although he had viewed the full-size tunnel, he did not challenge the fact that it would be divided into two smaller tubes, as had been stated in the Pneumatic's January 8th reply. Referring to that reply, the reporter noted:

> *"This is probably the fact, but at the present time there is no sign of a wall."* (Ref. 6, pg 8)

The *Tribune*, on the other hand, also published a negative editorial attacking the whole idea (of a pneumatic railway) as impractical.

Beach responded by opening his subway to invited public officials on February 26, 1870, and to the public several days later on March 1st. The pneumatic subway became an immediate success.

The official opening to invited dignitaries was covered by virtually all of New York's daily and weekly newspapers. The praise they offered was often lavish. A discouraging word was not to be found:

The New York Times said (7):

> *"Certainly the most novel, if not the most successful, enterprise that New York has seen for many a day is the Pneumatic Tunnel under Broadway. A myth or humbug it has been called by every body who has been excluded from its interior: but hereafter, the incredulous public can have the opportunity of examining and judging its merits."*

The *Evening Mail* said (8):

> *"The problem of tunneling Broadway has been solved. There is no mistake about it. Even as we write, a comfortable passenger car is running smoothly and safely between Warren and Murray streets, demonstrating, beyond contradiction, that it is only a matter of time and money to give us rapid and comfortable transportation from the Battery to the Harlem River...It is truly most gratifying to see how admirably successful the affair has been carried out this far..."*

The *Herald* declared that the pneumatic was (9):

> *"... the opening day of the first underground railway in America."*

The *World* gushed (10):

> *"Any description of the pneumatic must necessarily be imperfect: the work must be inspected to be thoroughly appreciated."*

Not to be outdone, when *Scientific American* published its own description of the event and the pneumatic itself on March 5, 1870 (3), it quoted these and other complimentary reports.

The most surprising aspect of these reports is that the pneumatic did not actually run during the dignitaries' reception due to a minor mechanical problem. Beach had pulled off a major public relations coup, and his little subway became the talk of the town.

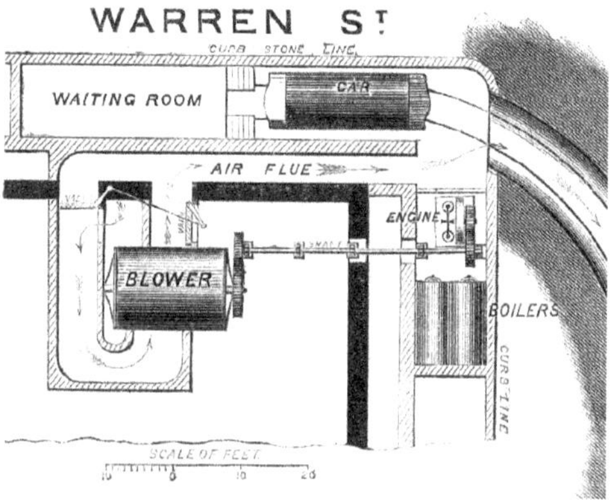

Fig. 7.8 The Beach Blower System at Warren Street: The Western Tornado

Fig. 7.9 Interior of Beach's Pneumatic Subway Car

Fig. 7.10 Boarding platform for the Beach Pneumatic Subway

At night, when the railcar was not running, Beach opened his tunnel as a promenade, which attracted significant numbers of people anxious to examine the waiting room and the tunnel.

At some point around November of 1870, it appears that Beach may have introduced a second car into operation. Several newspapers refer to "vehicles" in this period, although others still refer to a single car. Some descriptions refer to the second car as an "open" car.

7.5.3 An Expanded Plan for the Pneumatic Subway

After the surprising and popular opening of the pneumatic subway demonstration, an injunction was filed against its operation because it violated the charter, which only authorized a mail and packet service. While the charge was clearly correct on a legal basis, the popularity of Beach's experiment weighed on the court, and the suit was dismissed.

At this point, the story gets murky. It was generally assumed that it was Boss Tweed who had instigated the injunction. In fact, Beach himself, in numerous publications, indicated that Tweed was a major opponent of the pneumatic. Joseph Brennan, in his brilliantly-detailed history, *Beach Pneumatic* (2) provides a much more nuanced picture.

One week after the pneumatic opened to the general public, Beach and his superintendent (Joseph Dixon) were lobbying to get a broad charter to build a full transit system. It was none other than Boss Tweed, then a state senator, who introduced the bill in the legislature, and was a vocal proponent. The bill provided a charter for the Beach Pneumatic to build a transit system virtually anywhere in the city, giving the company broad powers to acquire necessary property through eminent domain.

Senator Henry Genet objected to the bill, and had it sent to the Committee on Municipal Affairs. Genet, however, was supporting a competing bill to charter the Arcade Railway, a many-faceted proposal that kept coming back in revised form. In fact, the Arcade Railway had failed in its attempts to get a charter in 1868 and 1869. The two proposals were in competition, both with the support of a major Tammany Hall politician. Both plans were opposed by a Broadway merchant, A.T. Stewart, who organized opposition to any plan for a subway or elevated on Broadway. The legislature subsequently defeated the Pneumatic bill overwhelmingly, but approved the Arcade Railway. With A.T. Stewart and several cohorts lobbying heavily against the Arcade Railway, the Governor eventually vetoed the bill.

Why did Tweed support the Pneumatic bill? One can only speculate, as he certainly opposed it in subsequent years. Perhaps, in writing a bill that provided broad powers for building a transit system, he saw an opportunity for massive graft. Perhaps he thought, in view of the popularity of Beach's subway, that he needed to jump on the bandwagon. Whatever the reason, Tweed was initially a supporter of the Pneumatic and its expansion; by the time later bills were introduced on behalf of the Pneumatic, Tweed had become an opponent in support of rival plans.

In 1871, Beach resubmitted his bill, with a great deal of lobbying in preparation. He had the support of a number of prominent newspapers, namely, the *World*, the *Evening Mail*, the *Journal of Commerce*, and (of course), the *Sun*. The 1871 bill was more specific than the 1870 bill in that it spelled out a specific route for a pneumatic subway: from Bowling Green to Central Park and Eighth Avenue, moving north under 4th Avenue to the Harlem River, with provisions for a tunnel under the river into the Bronx. The 1871 bill was introduced by Senator Genet, who had supported the Arcade Railway in 1870.

It was at this point in time that the "Viaduct Plan" was brought forth. The list of prominent directors in the New York Viaduct Railway Company included August Belmont (Sr.), John Jacob Astor, Charles Tiffany, and additionally, A.T. Stewart and William "Boss" Tweed. The Viaduct Plan called for the construction of a massive viaduct several blocks east of, but not on Broadway. At Houston Street, it would split into an eastside and a westside route. It would have essentially leveled many properties east of 3rd Avenue to accommodate the massive viaduct with new adjacent buildings. The Viaduct Railway was to consist of four tracks, the first time the concept of "local" and "express" service was formally proposed in New York.

While some historians claim that the plan was the brainchild of Tweed, Joseph Brennan points out that it was indeed A.T. Stewart who spearheaded the plan (2), with Tweed in support due to his general political alliance with Stewart. Stewart promoted the plan to get a transit plan in place that did not directly impact his properties on Broadway, and the properties of other businessmen.

Senator Genet introduced it just two days after introducing the Pneumatic bill. The Pneumatic bill passed the Senate on March 15, 1871, with Tweed in opposition; the Assembly passed the bill two days later.

The Viaduct Plan, however, also passed. Governor Hoffman signed the Viaduct bill and vetoed Beach's. While is it often portrayed as a political veto, with the Tammany Hall-connected Hoffman following Tweed's "advice," there were a number of other issues raised in the Governor's veto message. He questioned the broad authority given to the Pneumatic, and raised engineering issues involving the stability of Broadway and the buildings along it. In any event, the Tweed-backed Viaduct Plan was, for the moment, the plan that moved forward.

This was, however, Boss Tweed's last political hurrah. He was already under investigation for graft, having been exposed in July 1871 by the *New York Times* (11), and was thereafter convicted of a variety of offenses and jailed for life.

In 1873, Beach submitted his charter for the fourth time (an 1872 submission was also rejected). It was passed, and signed into law by a new Governor on April 9, 1873. The legislative victory was indeed pyrrhic, however. The financial Panic of 1873, brought on by the vast over-selling of railroad bonds, caused the collapse of over 5000 banks and commercial houses, and left the railroad industry in disarray. Moreover, the idea of pneumatic propulsion for a larger subway system had been generally rejected, and the age of steam-propelled elevateds had already begun. Gilbert, who had originally proposed an elevated atmospheric railway, was forced to change to steam propulsion. Eventually, even Beach realized that if he wanted to build a larger subway, that he would have to use steam power, but the change came too late. Indeed, by opening the door to the possibility of steam propulsion, Beach lost some of his supporters. The idea of steam locomotives in underground tunnels was still an unpopular one.

Beach continued to operate his demonstration line for four years. By the end of 1874, he could no longer afford to keep it running. After a brief period in which he rented the tunnel out as a shooting gallery, and later as a wine cellar, he sealed up the tunnel and abandoned it.

7.6 Last Remnants of the Pneumatic

After Beach sealed his tunnel, it was viewed twice more before disappearing into history. On December 4, 1898, a spectacular fire destroyed the building that had housed Devlin's Clothing Store. The fire destroyed everything in it, and the entire structure had to be leveled. In 1898 and 1899, as the foundation structures were being removed, workers came upon the tunnel, and several photos recorded the event.

(a) The Pneumatic Tunnel in 1899

(b) A dignitary poses in Beach's rediscovered railcar in 1899

Fig. 7.11 Beach's Pneumatic Subway after the fire in Devlin's

In February of 1912, workers building the Brooklyn Manhattan Transit subway (BMT) stumbled upon the tunnel for the last time. The station stood relatively intact, the car was still on the tracks, and the shield was hopefully poised against the lead end of the tunnel. The shield was given to Frederick Beach, Alfred's son, who in turn donated it to Cornell University. It was on display for several years in the University's Mechanical Engineering building, which was subsequently torn down. The fate of the tunneling shield thereafter is unknown.

The construction of the BMT City Hall station basically consumed what was left of the Pneumatic, where it was supposedly commemorated with a simple plaque. The plaque, however, is not there, nor is there any indication that it was there. Whether a plaque was ever placed remains another ongoing mystery.

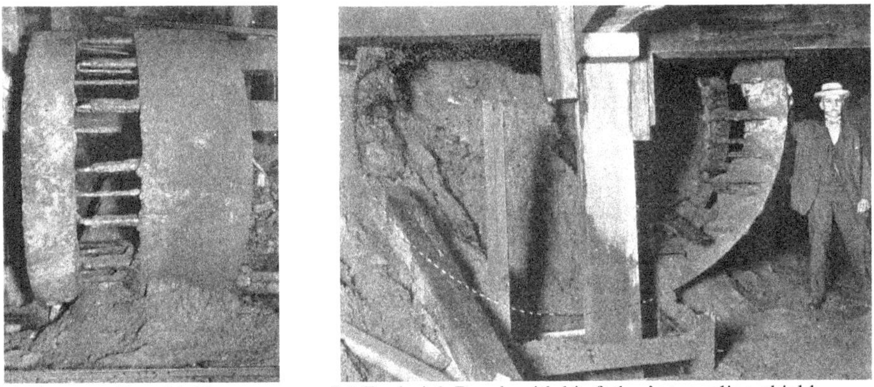

(a) The tunneling shield (b) Frederick Beach with his father's tunneling shield

Fig. 7.12 Last sighting of Beach's Tunneling Shield in 1912

7.7 A Footnote to History?

Alfred Beach's subway played a pivotal role in the development of New York City's rapid transit system. It demonstrated the viability of subway construction, and that people would in fact not fear, or be deterred, by underground travel. Both were key issues of the time.

His legacy as a publisher was perhaps even more important. His *Scientific American* was the principal publication for the presentation of myriad ideas and proposals for viable rapid transit systems. Without *Scientific American*, many of the details of early transit proposals and systems would have been lost. He was a critical part of the discussion of rapid transit in New York for a half-century. When he died, at the age of 69 in 1896 from pneumonia, he left a significant mark on the city behind. His dream of a subway, powered by a new source, electricity, would be a reality eight short years after he died.

References

1. Pneumatic Railways. Mechanic's Register and Journal of the Mechanics' Institute 6(7) (August 1835) (American Periodicals On-Line)
2. Brennan, J.: Alfred Beach's Pneumatic Subway and the Beginnings of Rapid Transit in New York (2003-2004) (an on-line web publication)
3. The Pneumatic Tunnel Under Broadway. Scientific American XXII(10) (1870) (American Periodicals On-Line)
4. The Pneumatic Railway. The New York Tribune (December 29, 1869) (American Periodicals On-Line)
5. The Pneumatic Railway: Statement of the Superintendant. New York Tribune (June 8, 1870) (American Periodicals On-Line)
6. The Broadway Tunnel Explored by a Tribune Reporter – First Bonifide Description. New York Tribune (June 11, 1870) (American Periodicals On-Line)

7. New York Times, New York (February 27, 1870)
8. The Evening Mail, New York (February 26, 1870)
9. The Herald, New York (February 27, 1870)
10. The World, New York (February 27, 1870)
11. The New York Times, New York (July 21, 1871)

Figure Sources

7.1: *Courtesy of* Wikimedia Commons
7.2: *Originally published in* Scientific American, Vol. XXII, No. 10, Munn & Co., New York, NY, March 5, 1870, pg 155.
7.3: *Originally published in* "Pneumatic Railways," *Mechanics' Magazine, and Journal of the Mechanics' Institute*," Vol. 6, No. 2, August 1835, Philadelphia PA, pg 97.
7.4: *Originally published in* Scientific American, October 5, 1861.
7.5: *Original source unknown, from* Brennan, J., "A Tube, A Car, A Revolving Fan," *Beach Pneumatic*, Columbia University, New York, NY, 2004–2005, pg 2–9 (On-Line).
7.6: *Originally published in* Beach, Alfred E., *Illustrated Description of the Broadway Pneumatic Underground Railway*, SLO Green, Printer, New York, NY, 1870, pg 7.
7.7: *Originally published in* Beach, Alfred E., *Illustrated Description of the Broadway Pneumatic Underground Railway*, SLO Green, Printer, New York, NY, 1870, pg 6.
7.8 *Originally published in* Beach, Alfred E., *Illustrated Description of the Broadway Pneumatic Underground Railway*, SLO Green, Printer, New York, NY, 1870, pg 9.
7.9: *Originally published in* Beach, Alfred E., *Illustrated Description of the Broadway Pneumatic Underground Railway*, SLO Green, Printer, New York, NY, 1870, pg 4.
7.10: *Originally published in* Beach, Alfred E., *Illustrated Description of the Broadway Pneumatic Underground Railway*, SLO Green, Printer, New York, NY, 1870. pg 11.
7.11: (a–b) *Originally published in* Scientific American, Munn & Company, New York, NY, April 15, 1899.
7.12: (a–b) *Originally published in* Scientific American, Munn & Company, New York, NY, April 15, 1899.

Chapter 8
The Interborough Rapid Transit System

The times surrounding the beginning of the 20th century were both turbulent and exploding with technological development. In 1892, Grover Cleveland was elected to the first and only nonconsecutive 2nd term as president; in 1902, William McKinley became the third president to be assassinated, making Theodore Roosevelt the youngest president to serve in the country's history. In February 1898, the USS Maine was blown up in Havana harbor, precipitating the Spanish–American War.

At the same time, men like Thomas Edison, Alexander Bell, Alfred Ely Beach, and others produced a seemingly endless stream of new inventions. Henry Ford developed the first assembly line for automobiles, and began making them at a price that many could afford. The Wright brothers flew at Kitty Hawk. In New York City, in 1902, architect Daniel Burnham built the Flatiron Building using a steel frame, generally recognized as the first true "skyscraper," ushering in a new era of vertical expansion.

Among the great achievements of the time was New York City's first full-scale operating subway.

8.1 Mayor Abram S. Hewitt and His Grand Plan for Transit

In 1886, Abram Hewitt was elected to a two-year term as Mayor of New York City after a difficult and bruising campaign. Already a congressman representing New York, he was called on to run by a coalition of factions in the Democratic Party, including Tammany Hall. Tammany Hall was still reeling from the revelations of Boss Tweed's indictment and conviction for massive fraud. He was opposed by two candidates, namely, Henry George for the Labor Party, and Theodore Roosevelt for the Republican Party. The final tally was Hewitt 90,199, George 67,581, and Roosevelt 60,332 (1), Roosevelt, finishing third, would become President in just six years, becoming the youngest man ever inaugurated when President McKinley was assassinated.

On January 1, 1888, Mayor Hewitt introduced a plan for a state-of-the-art subway that would operate at the outrageous speed of 40–50 mph (2). The startling nature of his proposal was its financing: he proposed that the city itself finance and own the subway, providing a franchise to a private company to build and operate it. This proposal was highly controversial. All previous transit construction

R.P. Roess and G. Sansone: The Wheels That Drove New York, STTT 1, pp. 155–178.
springerlink.com

and operation were completely private, including the highly successful elevateds. The public memory of Boss Tweed and the worst period of public corruption in the city's history were still fresh, and the idea of turning over the use of millions of dollars of public money to private companies was not well received.

Fig. 8.1 Portrait of Abram S. Hewitt (as a young man)

A little over a month after presenting his plan, the city was crippled by the Blizzard of 1888. Virtually all streetcar and elevated operations were stopped for a week, and the business in the city screeched to a costly halt. While the storm heightened public awareness of the need for a subway, it did not influence the Board of Aldermen, which promptly rejected Hewitt's plan.

In fact, the plan did not, originally, have a great deal of support. Most newspapers and organized groups opposed it. Owners of the elevateds also opposed it, despite the obvious overcrowding of their systems. The elevateds were making money on the crowded trains, and had no desire to dilute their investment by building additional lines.

Hewitt's concept would take hold almost a decade later, but Hewitt, sadly, would not live to see the city's first subway operate. *Scientific American*, in reporting on his death on January 18, 1903, (3) noted:

> *"Though his suggestions were not heeded at the time, he continued to actively promote the cause, and it was largely through his efforts that the present improvements were undertaken."* (Ref. 3, pg 7)

For his radical approach to subway development, he is often referred to as the "Father of the Subways."

Hewitt lost the support of Tammany Hall and served only one two-year term as Mayor. He was generally an austere man, and had angered the Hall by refusing to participate in a St. Patrick's Day celebration, and (more importantly) by refusing to hand out patronage to those designated by the Hall.

8.2 The Steinway Commission

On January 31, 1891, Governor David Hill signed the Rapid Transit Act of 1891 into law. Like the Husted Act before it, the law authorized formation of a new Rapid Transit Commission (RTC) to lay out routes for new subways and/or elevateds, and to assign a franchise for construction and operation to a private company. The act did not permit investment of public funds to do this. The chairman of the new RTC was William Steinway, and the group quickly became known as the Steinway Commission.

Fig. 8.2 William Steinway in front of his house in Astoria

Steinway's interest in transit came from an interesting angle. His father, Henry Steinway, was the founder of Steinway & Sons, which became one of the world's most famous makers of concert and other pianos. The company retains its reputation today, and is the holder of hundreds of patents related to piano design.

In 1866, the company opened the first Steinway Hall on 14th Street. It included a concert hall that seated 2000 people, and became the city's focal point for performances until Carnegie Hall opened in 1891. In 1925, the second Steinway Hall was opened on 57th street, its current location.

When Henry died, the company was turned over to two of his sons, that is, William and his older brother, C.F. Theodore. The latter was a concert pianist and was in charge of technical development. William was left in charge of the business aspects of the company.

In 1870, William purchased 400 mostly vacant acres in Queens, in the area now known as Astoria. He moved the company's main factory to the site, where it remains in operation today. Since the site of the new factory was remote and not readily accessible, William also built a company town around the factory, including housing for workers, a church, a library, a kindergarten, a recreation area called North Beach, and a trolley line. Some of the row houses he built are still in existence, and Astoria's main business district is located on the street honoring him, namely, Steinway Street. In 1939, North Beach became the site for LaGuardia Airport.

In the 1890s, Steinway also became heavily involved in building a railroad tunnel under the East River near 42nd Street. He hoped to use the double-tube tunnel to connect his trolley services to Manhattan, and/or for several streetcar lines he operated in Manhattan. Steinway, who died in 1896, did not live to see the tunnels, which are still known as the Steinway Tunnels, completed.

The tunnels went through a number of fits and starts, but were finished in 1907, and trolley services used the tunnel for a short time. A number of legal issues and one accident forced the tunnels to close shortly thereafter. The tunnels were not used again until 1915, when they were incorporated into the IRT rapid transit system.

After a period of study, the Steinway Commission issued its recommendations in October of 1891. The plan called for two subways to be built. A mainline would be built from South Ferry up Broadway and into the Bronx. A branch line would begin at Union Square, traveling up Madison Avenue to the Bronx. The Commission drafted a contract for construction and operation and invited bids. No one submitted a bid. The cost of building a subway was a major deterrent to private enterprise.

With the need for new rapid transit lines becoming clearer every day, the Steinway Commission, over Steinway's severe objections, began to negotiate with George Gould (Jay Gould's son, Jay Gould died in 1892) and the Manhattan Railway about building additional elevated lines. It is not clear whether Gould was serious about these discussions, or simply stalling the committee. In any event, the stock market crash of June 1893 intervened. It produced an almost decade-long depression, and the climate for any private investment in rapid transit became extremely poor. Despite its hard work and dramatic plans, the Steinway Commission failed to expand New York's rapid transit system.

8.3 The Chamber of Commerce and the Rapid Transit Act of 1894

At the behest of banker Richard T. Wilson, who himself wanted a franchise to build a subway, the NYC Chamber of Commerce appointed a special panel to study a plan for public financing of rapid transit. The panel, headed by Alexander Orr, citing the financing of the transcontinental railroads, recommended that the Chamber lobby in support of Wilson's idea. Wilson's plan, however, would have made both the construction and operation of rapid transit public. Former Mayor Hewitt and others persuaded the panel to drop Wilson's proposal in favor of Hewitt's original plan for public construction and ownership, but private management and operation.

The panel proceeded to draft legislation incorporating Hewitt's model. It further guaranteed its own control over the process by proposing a new Rapid Transit Commission in which the Chamber of Commerce would control five of the eight seats. The legislation passed, and was signed into law on May 22, 1894.

Unions, which had generally opposed the legislation, managed to get an amendment included that required a public referendum on the proposal in the November 1894 general election. The referendum was held, and it was approved in an overwhelming vote of 132,647 to 42,916. The legislation, however, had a few other hurdles to overcome. It limited the total cost of constructing the new line(s) to $50,000,000. Further, the specific plan had to be reviewed and approved by the mayor, the Board of Alderman, and either local affected property owners or the state courts.

On February 16, 1895, the third Rapid Transit Commission adopted a plan for a single subway, more or less following the path of the Steinway Commission's mainline. With construction cost and the $50,000,000 limit in mind, the Commission made one significant change, that is, the section between 92nd Street and 124th Street, crossing Manhattan Valley, would be built as an elevated, saving about $2,000,000. Real estate developers with land along the elevated portion strongly opposed this plan, which the commission eventually withdrew. It was, however, reinstated during the construction process.

The referendum called for in the Rapid Transit Act of 1894 required that property owners directly affected by the proposed line be polled, and that the votes be weighted by the market value of the properties each held. Property owners, fearful that years of cut-and-cover construction work would ruin their businesses and property values, defeated the proposal.

The Commission went to court, as an alternative to a successful referendum had been provided in the legislation. They petitioned the NYS Supreme Court in fall of 1895 to appoint a panel of three to determine if the RTC plan was viable. The panel endorsed the plan on March 6, 1896, but their recommendation had to be approved by the Appellate Court (In NYS, the Appellate Court is higher than the Supreme Court). The Appellate Court issued a stinging negative decision, questioning the veracity of the construction cost and other estimates.

The new RTC, not wanting to suffer the same fate as the Steinway Commission, once again turned to Gould and the Manhattan Railway seeking new elevated lines. William Steinway, still closely following the course of rapid transit development in New York, was aghast at this move. He produced a compromise plan for a scaled-down version of the subway. It would run from City Hall to Grand Central on the East Side, turn on 42nd Street to Longacre Square (now Times Square), then head up Broadway to 96th Street. At 96th Street, the subway would divide with one branch following Broadway through Harlem, Washington Heights, Fort George, and Riverdale, and the other running up Lenox Avenue and crossing the Harlem River to Bronx Park. The RTC formally accepted Steinway's plan on August 6, 1896. By March of 1898, the Mayor, the Board of Aldermen, the citizen's review panel, and the Appellate Court had approved the revised plan. A plan for New York's first subway was finally in place.

8.4 August Belmont Jr., William Barclay Parsons, John B. McDonald and the Interborough Rapid Transit System

By the middle of 1898, the RTC was ready to accept bids for the construction and operation of the new subway. Only two bids were received. One, from Andrew Onderdonk, proposed to construct the system for $39.9 million, which the city would provide. He offered to share a percentage of operating profits above $5,000,000 per year with the city. The other was submitted by John B. McDonald for a flat construction cost of $35,000,000. On January 16, 1900, the RTC awarded the franchise to John McDonald.

While the franchise was awarded to John McDonald, the subway would be built with two significant partners, August Belmont and William Parsons.

8.4.1 John B. McDonald

John McDonald was a 56-year old Irish immigrant who had come to the U.S. as a child. He developed a considerable reputation as a contractor identified with a number of large projects. He had been General Superintendent of the Croton Dam, directed several large railroad rehabilitation projects, but was most famous for his work in building the Baltimore Belt Line, a project that faced a number of physical difficulties.

(a) John B. McDonald (b) William Barclay Parsons (c) August Belmont Jr.

Fig. 8.3 Partners in building the Interborough Rapid Transit

The contract required that McDonald post a $7,000,000 bond with the city comptroller to proceed. McDonald did not have that kind money, and he was forced to look for support elsewhere. He turned to August Belmont Jr., who became the leading force in the subway's development.

8.4.2 *August Belmont Jr.*

August Schonberg was born in 1816 into a poor German Jewish family. At the age of 13, he landed a job in Frankfort with the Rothchild's bank, starting as a janitor. He worked his way up through the bank, proving to be skilled and trustworthy in managing money. In 1837, on his way to Cuba where the Rothchild's sugar holdings were at risk due to political turmoil, he arrived in New York one week after a panic that wiped out 250 banks, investment houses, and insurance companies. Seeing this as an opportunity, Schonberg used his financial knowledge and his Rothchild connections to acquire the remains of a number of bankrupt firms. At the age of 21, August Schonberg became a wealthy man, and a major financial player in New York.

Schonberg, like a number of other prominent Jewish financiers, experienced considerable antisemitism when he tried to join New York's commercial aristocracy. He changed his name from Schonberg to Belmont, the French equivalent to "beautiful mountain" (the meaning of "Schonberg"), and became an Episcopalian, the predominate religion among New York's patriciate. He married the daughter of Commodore Matthew Perry, and bolstered his social standing by throwing elaborate parties, and by patronizing a number of old-guard institutions.

When August Belmont Jr. was born, on February 18, 1853, he immediately entered a world of enormous privilege. When his father died in 1890, August Belmont Jr. became president of a major bank and a major player in American capitalism. He served on dozens of prominent corporate boards, and maintained close friendships with the Rothchilds, J. P. Morgan, John D. Rockefeller, and others.

Belmont was an avid sportsman, and was heavily involved in horse racing. In 1902, he bought a 650-acre tract of land on the Queens–Nassau border, and opened Belmont Park (named after his father) in 1905, complete with a grandstand, 1-1/2 mile track, and a luxurious clubhouse with bars, dining rooms, and bedrooms. Belmont was the breeder of the most famous racehorse in history, Man o' War. The horse had been sold as a yearling, however, and never raced under Belmont's colors.

Belmont agreed to finance McDonald's bond, and to undertake the operation of the new subway. To do so, Belmont formed and incorporated the Interborough Rapid Transit Company (IRT) with an initial capitalization of $25,000,000 (4, 5).

On February 1, 1900, he signed a contract to build, equip, and operate the subway for a period of 50 years, with an option for another 25 years. Under terms of the contract (historically referred to as Contract #1), the city provided $35,000,000 to build the subway and an additional $1,500,000 to buy land on which to build yards and maintenance facilities. Belmont agreed to purchase all rail cars, signals, and other equipment from his own resources. The contract provided for Belmont to pay an annual rent to the city an amount equal to the interest on the construction bonds, plus a small additional sum to be placed in a sinking fund.

Shortly thereafter, he signed another contract (Contract #2) for an extension of the subway under the East River and through Brooklyn. The Brooklyn Rapid Transit Company (BRT) had also bid on Contract #2, but Belmont submitted a low-ball estimate and was awarded the contract in the fall of 1902. He had bid

$2,000,000 for the job, $6,000,000 below the Rapid Transit Commission's own estimate (6, 7, 8). Belmont did not want any competition in the subway world.

Subsequently, the IRT leased all properties and operations of the Manhattan Railway, unifying the subway with the elevateds in Manhattan and the Bronx. Rumors concerning the consolidation of the planned subway with the Manhattan elevateds had been circulating for months, driving up the value of the Manhattan Elevated Railroad stock. When completed on April 3, 1903, the lease was for 999 years (9).

Belmont and McDonald also formed a separate company to build the subway, the Rapid Transit Construction Company.

8.4.3 William Barclay Parsons

William Barclay Parsons was the third member of the triumvirate responsible for the city's first subway. He was born on April 15, 1859, and could trace his lineage to colonial times. He was educated at a private English school, and later taught by private tutors while his family traveled through Europe for four years.

He graduated from Columbia College in 1879. Three years later, he earned a civil engineering degree from Columbia's School of Mines. After a short period during which he worked for the Erie Railroad, Parsons opened a consulting engineering practice, the forerunner of today's Parsons Brinkerhoff. He worked on a number of railroad projects, and developed a strong interest in the development of rapid transit systems.

One of the early proposals for a subway, the Arcade Railway Company, hired him as a staff engineer. Riddled with strife, Parsons and others broke away to form their own company, the New York District Railway, but both were bankrupt by 1886. Parsons continued his interest, however, and developed close relationships with Steinway, Hewitt, and other subway enthusiasts.

In 1891, the Steinway Commission hired him as its deputy chief engineer, and he became the principal architect of its plan. In 1894, the new RTC hired him as its chief engineer, a position he held when Contracts #1 and #2 were signed. While the subway would technically be "built" by John McDonald, the contracts called for McDonald to be supervised by the RTC's chief engineer, that is, Parsons.

8.4.4 Contracts #1 and #2

The first two subway construction contracts defined a good deal of the IRT system as it exists today. Its lines are all still in use as part of the present-day system.

The Contract #1 subway began at City Hall, up Broadway, Lafayette St., Park Avenue and Lexington Ave. to 42nd Street. At 42nd Street, the subway turned west to Longacre (now Times) Square, continuing north up Broadway, crossing into the Bronx at 215th Street, continuing north to Van Courtland Park. At 96th Street, the subway split, with one branch turning east, then proceeding north to 145th Street, where some trains would (and still do) terminate. The branch continued to a crossing into the Bronx, proceeding north to Wakefield.

The Contract #2 subway pushed further downtown from City Hall to Fulton and Wall Street, and Bowling Green stations. From there, the subway crossed the East River into Brooklyn to Borough Hall, terminating at Atlantic Avenue.

In Manhattan, south of 42nd Street, the subway is now the Lexington Avenue subway; north of 42nd Street, it is the 7th Avenue Subway. The crossover is now used as the Grand Central–Times Square Shuttle.

Figure 8.4 shows a map of the IRT system, including the Contract #1 and #2 subways, and the elevated lines under control of the IRT.

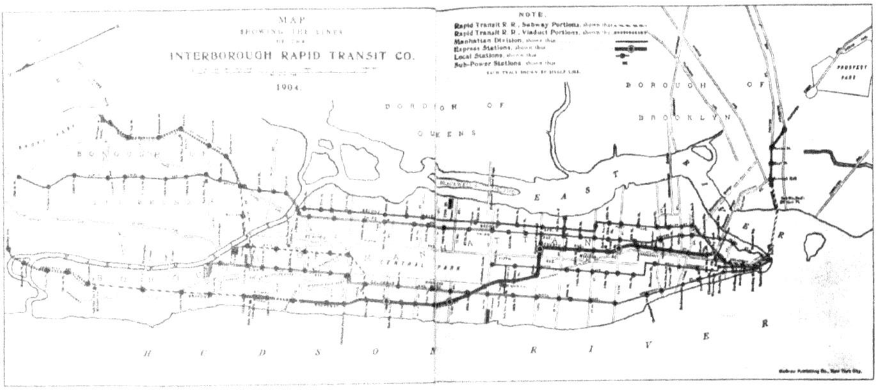

Fig. 8.4 Map of the IRT after completion of the Contract #1 and #2 subways in 1904

When Jay Gould had established full control over the Manhattan elevateds, he became a staunch and crafty opponent to further construction of rapid transit. Crowded trains produced enormous profits. August Belmont would soon adopt a similar stance, and would oppose additional construction of either subways or elevateds for the same reason.

8.5 Construction of the IRT

Two key technological decisions for the subway had been made under the auspices of the Steinway Commission, and the subsequent Rapid Transit Commission, and both were made at the behest of Parsons.

While Frank Sprague had made dramatic contributions to electric traction operations, and had opened the electrified Richmond trolley line of 12 miles to great success, it was not immediately clear that a New York subway would use electric traction power. Parson's had doubts that a system demonstrated on a single 12-mile line in a small city could be applied in a more massive system in a much larger city. In the summer of 1894, Parsons undertook a tour of European rail systems, paying particular attention to those that had converted to electric traction power. He was immediately impressed with the cleanliness and efficient operation of the electric traction system, and even more so with its economics, namely, conversion from steam to electric traction produced significant fuel savings, as well as fewer breakdowns and maintenance costs. On his return, he submitted a detailed report on his findings, and recommended the adoption of electric traction as the powering system for the subway.

His recommendation, made in 1894, was a good one. By 1902, electric-traction street railway operations in the country had increased from 35 miles in 1888 to 21,920 miles, and there were virtually no horse-streetcar operations left.

The geology of Manhattan created another problem. Most of Manhattan is underpinned by hard bedrock. While this provides an excellent material for supporting load, tunneling through it was very difficult and very expensive. Little, however, was known about the exact nature of the subsurface in the 1890s when the subway was being planned. Before the age of skyscrapers, few deep foundations had to be built, and very little exploration of Manhattan's subsurface had been necessary. Aside from those associated with the Croton Reservoir and the towers for the Brooklyn Bridge, virtually no subsurface investigations had taken place.

Because of this, Parsons undertook a series of test borings while investigating the originally-planned subway route for the Steinway Commission. They revealed a startling fact: the depth of Manhattan's bedrock was highly variable. In some places, bedrock could be found very close to the surface, as close as 16 ft at 30th Street and Broadway, and 20 ft at Whitehall Street on the Battery. In places between Whitehall and 30th Street, bedrock was located as much as 160 or more feet below the surface. This unusual feature had a significant influence on Manhattan's development. When the age of skyscrapers was underway, construction centered on the downtown area and what would now be called "midtown" because the close-to-surface bedrock in these areas provided an excellent footing for tall buildings.

Tunneling through a subsurface with widely-variant characteristics was both dangerous and expensive. Parsons presented two options. A tunnel through solid bedrock would have to be built at a depth of 200 ft; the alternative would be to build as close to the surface as possible using a cut-and-cover technique. As the latter could be accomplished in far less time, and at considerably lower cost, Parson strongly recommended the latter. The Rapid Transit Commission readily accepted his advice.

The mainline portion of the subway, from City Hall to point of divergence at 96th Street was planned as a four-track line, with local and express service to be provided in both directions. Prior to the IRT, there were few examples of this type of service. A short section of the 9th Avenue El had four tracks, but did not feature express and local service over its entire route. While the plan evolved out of general consensus, it was extraordinary for its time, and had a significant impact on New York. With the increased speed of express service, the practical commuting distance was significantly extended. While a four-track system had been proposed as part of the failed Viaduct Plan, it was also discussed by Frank Sprague in a *Railway Age* article in 1891.

Parsons was in China when the formal announcement of a contract to build the subway took place. At the time, he was surveying a 1000-mile railroad from Hankow to the sea on behalf of a group of investors headed by J. P. Morgan. The expedition itself became the stuff of legend, as it required passing through vast areas of China that were both remote and considered highly dangerous to foreigners. When news of the subway reached him, he left the survey and returned to New York to direct the construction effort.

While Parson's concept was to build tunnels as close to the surface as possible, a relatively smooth profile for the subway was also needed. To accomplish this, a section across Manhattan Valley, from 122nd Street to 135th Street, was built as an elevated structure. To negotiate the "highland" areas of upper Manhattan, from 158th Street to the Harlem River, a deep rock tunnel had to be built. Three stations, 168th Street, 181st Street, and 191st Street had to be located between 150 ft and 180 ft underground, requiring elevators for access and egress. Another section of the tunnel would also be built through solid bedrock, but close to the surface, between 33rd and 42nd Streets. Parsons also used a unique technique to build the tunnel under the Harlem River. He had a trench built on the riverbed, and "sunk" preconstructed iron tunnel sections into it. This technique would be refined and used again almost 70 years later in the building of the Bay Area Rapid Transit System tunnel under San Francisco Bay through an earthquake fault. It was also used in building the 63rd Street tunnel now used by the F-train.

Cut-and-cover tunneling was not a simple process. Most of the work was done by hand, with picks and shovels. Before a complete tunnel could be constructed, water, sewer, and electric lines existing just below the affected streets had to be temporarily moved, and rebuilt on completion of the subway. Adjacent building foundations had to be secured. In many places, a temporary street surface over the tunnel trench had to be built to allow for traffic, including streetcar lines, to operate while the tunnel underneath was constructed. In most cases, however, the trench remained open for most of the construction process.

Unskilled workers were paid an average of \$2.00–\$2.25 per day, while the more skilled workers received \$2.50. The highest paid workers were miners recruited to help build the two-mile deep tunnel through bedrock in upper Manhattan. They received a whopping wage of \$3.75 per day. Most of the workers were Irish and Italian immigrants, but other groups participated as well.

(a) Cut-and-cover tunneling at 64th St. and Broadway

(b) Water pipes replaced over the subway at 125th Street and Lenox Ave.

Fig. 8.5 Construction of the IRT

The dangerous nature of the work is evident in the fact that there were 54 deaths during the construction of the subway. One subcontractor, Ira A. Shaler, was involved in several accidents, and was referred to in the press as the "voodoo contractor." On January 28, 1902, a wooden powder house, where dynamite was stored, caught fire and exploded, killing five people and injuring over 100. The blast was so severe that it wrecked the Murray Hill Hotel, and damaged the face of Grand Central Terminal (10). In reporting on the tragedy, *Scientific American* had a moderate reaction to the loss of life:

> *"Although the results were disastrous, they were not nearly as fatal as one might have expected from the ordinarily crowded condition of the streets at this hour. In all, five persons lost their lives, three of them in the Murray Hill Hotel, and two or three hundred persons, the exact number of whom will never be known, were more or less injured."* (Ref. 10, pg 90)

Less than two months later, a portion of a tunnel he was building collapsed. No one was killed, but over $100,000 in damage to adjacent buildings was caused, and subway construction in the area was halted for several weeks (11). Finally, Shaler's luck ran out. On July 14, 1902, while on an inspection tour of his section with Parsons and another engineer, Parsons pointed out a rock formation protruding from above. Shaler stepped out from under the timber bracing to inspect the boulder just as it dropped (12). While being carried out on a stretcher, he was quoted as saying:

> *"Let the work go right along."* (Ref. 12, pg 16)

The falling rock had broken Shaler's neck, and he died eleven days later at Presbyterian Hospital. Work was indeed stopped for some time by order of the New York State Supreme Court while the causes of the accident were investigated. The accident had a profound effect on Parsons, who had been a good friend of Shaler's. He announced that he would resign from the project and was absent for several weeks. Ultimately, however, he was convinced to continue.

The worst single accident occurred on October 24, 1903, near the tunnel's Fort George portal. After a blasting operation, a foreman inspected the tunnel's walls and ceiling, declaring them safe. He ordered his work crews back into the tunnel to continue work. The foreman was unaware that an underground spring was close to the tunnel's face and had seriously compromised its integrity. Shortly after workers entered the tunnel, a 300-ton boulder fell through the roof, killing six men instantly and seriously injuring another eight. A number of workers were trapped under the rubble, and a dramatic rescue effort took place. Several of the trapped men were rescued, but three Italian workers had sustained such serious injuries that they could not be helped. Doctors administered morphine to ease their ordeal, and a priest, Father Thomas Lynch, remained with them (despite the chance of another cave-in) until they died. All told, ten people died, including the foreman.

The new subway was expected to be ready for service on January 1, 1904. However, the major accidents noted previously and other difficulties delayed the work several times. With most of the tunneling work completed, Mayor Low drove the first spike into the first rail laid at the Columbus Circle station on March 14, 1903 (13).

When the subway opened on October 27, 1904, it was not complete. It was finished to 145th Street, but sections further uptown, and the Contract #2 connection to Brooklyn were still being built. Nevertheless, the main portion of the subway had been built in four years. While it was not the first subway, or the largest (in 1904, London's underground still held that distinction), or the first underground electric-traction railway, it was a technological wonder, incorporating the best technologies available.

8.6 The Subway Opens

The opening of the subway was a major media event in New York. For weeks, private and corporate groups had been holding "subway parties" in anticipation of its opening. The official opening ceremony convened at 1:00 PM in the City Council chambers, with about 600 invitees that included most of New York's elite. The ceremony, which was planned for one hour, ran a bit over due to the length of remarks by August Belmont, Alexander Orr (President of the RTC), and others (14, 15). Mayor George B. McClennon Jr. (son of the Civil War general of the same name) gave the introductory remarks which highlighted the formative role of transportation in the development of New York:

> *"Without rapid transit, Greater New York would be little more than a geographical expression. It is no exaggeration to say that without Interborough communication Greater New York would never have come into being."*
>
> *"The present boundaries of our city included ten years ago a multitude of independent and heterogeneous communities which would have continued in all human probability to work out their own destinies independently had not modern genius and modern enterprise afforded their population the possibility of movement."*
>
> *"When the Brooklyn Bridge was opened, Greater New York was born. Every addition to transit facilities has stimulated her growth, which can only reach its full development when a complete system of rapid transit shall be rapid in fact as well as in name..."*
>
> *"We have met here today for the purpose of turning over a page in the history of our city; for the purpose of marking the advent of a new epoch in her development. If this new underground railroad which we are about to open proves as popular and successful as I confidently expect it to be, it will only be the first of many more which ultimately result in giving us an almost perfect system of Interborough communication."* (Ref. 15, pg 1)

At 2:35 PM, the official inaugural run of the subway left City Hall station, with Mayor McClennon at the helm. The plan was for the Mayor to start the train and turn over the throttle to a trained motorman shortly thereafter. The Mayor, however, was having too much fun, and stayed at the helm for most of the uptown trip to 145th Street. At one point, he managed to engage the emergency brake, bringing the four-car train to an abrupt and unexpected halt. He started up again, finally turning over control towards the end of the run (16).

At 7 PM, the IRT was officially opened to the public after carrying 27,000 passengers for free between 2:30 and 6:00 PM. Long lines of people had been gathering at stations for hours, and a crush was almost immediate. On that first evening, over 110,000 people rode the city's latest rage, while 125,000 tickets were sold (16, 18). All receipts from the five-cent fares were donated to charity.

In its first week of operation, the new subway carried more than 2,000,000 passengers (17). In the first month, 5,838,235 people rode the subway, not including fireman and policemen, who rode for free (18). On October 30th, the first Sunday of operation, almost 1,000,000 people tried to ride the subway. Unfortunately, the subway could accommodate only 300,000 people per day, and the police had to be used to quell disturbances at a number of stations. The events of October 30th were prophetic. Like the elevateds before it, the subway became overcrowded almost immediately, and calls for its expansion grew rapidly. It would take almost a decade of constant bickering, however, before the next expansion would occur.

8.7 The Wonders of the Subway

The subway that opened in 1904 was a true wonder of the world. It had adopted and adapted the best urban railway practices of the age, and became the model for subsequent subway construction throughout the world. In 1904, the IRT, which had acquired the elevated railroads of the Manhattan Railway, was the largest urban rapid transit system in the world.

8.7.1 Rolling Stock

To inaugurate service on the subway, 500 subway cars were needed. These were ordered in December of 1902, with the orders distributed among four rail car firms. Many trolley and other systems were keeping most of the manufacturers busy, and no one company could deliver 500 cars in time. The IRT decided that rail cars for all of its systems should be the same, and thus maximum flexibility in operating the subway and the elevateds could be achieved. In fact, the first 200 cars to arrive were initially placed into operation on the 2nd Avenue El in advance of the subway's opening.

The cars were state-of-the-art, and incorporated the best features available. Although there was interest in an all-steel car, none had ever been built, and no manufacturer was willing to accept such an order. The first 500 cars, therefore, were referred to as "protected wooden cars," with steel frames and copper-clad sides. In a promotional brochure distributed by the IRT in 1904, *Rapid Transit – The New York Subway, Its Construction and Equipment*, the principal attributes of the new cars were listed as follows (19):

- The length is 51 ft and provides seating capacity for 52 passengers. This length is 4 ft more than those of the existing Manhattan Elevated Railroad cars.
- Enclosed vestibule platforms with sliding doors are used instead of the usual gates. The enclosed platforms will contribute greatly to the comfort and safety of the passengers under subway conditions.
- Anti-telescoping car bulkheads and platform posts are used. This construction is similar to that in use on Pullman cars, and has been demonstrated in steam railroad service to be an important safety feature.
- The steel underframing of the car provides a rigid and durable bed structure for transmitting the heavy motive power stresses.
- Numerous devices protect against defects in the electrical apparatus.
- Window arrangements permit circulation without draughts.
- Emergency brake valves on trucks are operated by track trip.
- Emergency brake valves are provided in connection with master-controllers.

The track-trip feature insured that a train could not pass a "red" signal without being forced to stop. The last item eventually became known as the "dead-man" feature. The motorman had to keep downward pressure on the throttle. If the throttle arm was raised, the emergency brake system would be engaged.

The side-framing of the protected wood cars was made of white ash, with double braces and heavily trussed. The roofing was shaped by seven wrought-iron beams, preformed into the appropriate shape. The car was thought to be extremely safe, even if an accident should occur.

The cars had some other unique features as well. The two wheel trucks, located at each end of the car, were different. The "front" truck carried two motors, each powering one axle. The "rear" truck was not motorized. The "platforms," that is, the end structures through which passengers entered and left the train, were also unique. They were completely enclosed, not open at the sides. Sliding doors that moved into a slot in the car frame were used. This also allowed that the front vestibule of the train could be used as the motorman's cab for train operation.

While its initial order of 500 rail cars was for protected wooden cars, the IRT continued to pursue an all-steel design. When none of the existing manufacturers would consider building such cars, the IRT secured an agreement with the Pennsylvania Railroad to use its Altoona facilities to build an all-metal car. A sample car was completed in December of 1903, but it was too heavy to operate either on the elevateds or in the subway. Design modifications were made, and a contract for 200 all-steel cars was issued in early 1904.

8.7.2 Stations and Entrances

The stations and entrances that were the riders' first view of the IRT were designed with considerable care and attention to detail. While not opulent, they were designed to aesthetically pleasing, with detailed tiling that was generally appealing. Each station was individually designed with respect to its unique location and needs, but there was considerable standardization for the majority of stations.

(a) Exterior of protected wooden car

(b) Interior of protected wooden car

(c) Steel underframe of protected wooden car

Fig. 8.6 Views of the first IRT composite subway cars

Fig. 8.7 The first all-steel IRT subway cars

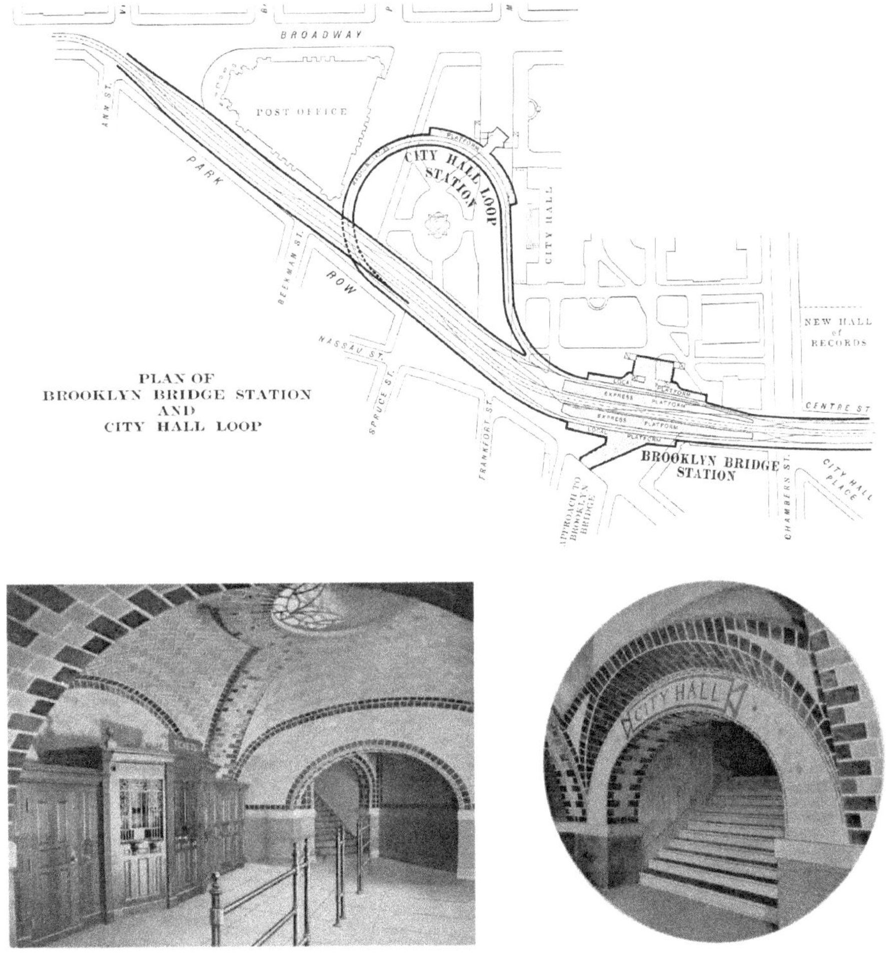

Fig. 8.8 Plan and views of the City Hall IRT Station

The most "luxurious" and unique station in the system was the City Hall station, from which the subway's inaugural run had begun. It was unique in that it consisted of a single side platform on a curve serving a single track that was on a loop. The loop serves as a turn-around for trains terminating at City Hall and returning uptown. After the addition of the Contract #2 subway to Brooklyn, only local trains used this turn-around. While the station itself has been closed, the turn-around is still used by No. 6 trains, and the station is often used for movie shoots and other promotions.

8.7.3 Four-Trackin'

The four-track mainline had both local and express stations. The typical local station consisted of separate side platforms adjacent to the two local tracks. In general, they were not connected by a mezzanine, and each side had its own entrances from the street. Express stations featured two island platforms located between the local and express tracks in each direction.

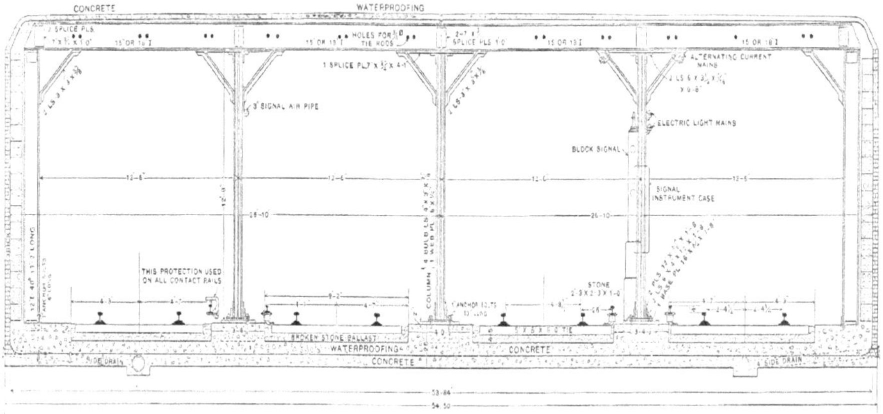

(a) Typical cross-section of a four-track tunnel

(b) Typical local side platform in a four-track tunnel

Fig. 8.9 Typical views of the four-track subway

Passenger platforms ranged from 200–350 ft. The shorter stations could accommodate a four-car train, while the longer platforms could handle six-car and seven-car trains. As time went by, lengthening of stations became a serious issue, as the IRT sought to run longer trains.

Stations were entered through street-level "kiosks" of standard design that provided an attractive portal, and were covered to provide some protection from the elements.

Fig. 8.10 Kiosks at 59th Street, Columbus Circle

8.7.4 Power Supply

The IRT built its own massive power generation facility between 58th and 59th Streets, and 11th and 12th Avenues. The plant generated three-phase alternating current (AC) at 11,000 volts for transmission to local substations, where it was transformed to direct current (DC) at 625 volts for the "third rail." In 1902, when the original rail cars were ordered, alternating current electric traction motors were in their infancy, leading the IRT to select the more tested DC technology.

Fig. 8.11 The IRT Powerhouse

New York City subways continue to run on direct current today, although the newest rail cars have AC traction motors (DC to AC traction inverters are on the cars). The transit system maintained its own power system until August 1, 1959, when it switched to purchasing power from Con Edison. This occurred a few years before the now-infamous East Coast blackout of November 9, 1965. Because the subways were tied into the national and regional power grid, the subways lost power for over 12 hours, leaving millions of commuters stranded in various parts of New York. Had the blackout occurred a few years before, subway operation would not have been affected.

8.8 The Dispute over Advertising

The public was generally impressed with the attractiveness of the stations, particularly the intricate tile work that decorated most of the walls, and the attractive station name mosaics.

When, a few short weeks after opening, advertising signs began to appear in subway stations, there was general outrage expressed, leading to a long set of legal proceedings over the issue. The dispute involved a number of entities, including the IRT, Ward & Gow (an advertising agency contracted by the IRT to sell and place advertising signs), the Rapid Transit Commission, the NYC Corporation Council, and the Architectural League of New York. The case moved through various hearings and jurisdictions over the course of two years. In the end, the right to place advertising in stations was deemed to be part of the IRT contract with the city. They remained, although some restrictions were agreed to, and have persisted to modern times. Later, similar disputes over flower and cigar stands in stations would arise, with similar ends. (20–30)

The IRT made about $500,000 per year from the advertising in the first several years. Subway advertising (both in stations and on trains) remains a significant source of revenue for transit system.

8.9 Impacts and the Call for Additional Subways

It would be hard to understate the impact of the subway on New York City. Longacre Square was a generally run-down area with many stables serving the horse trades. The new access provided by the subway would soon begin its transformation into Times Square, and with new theaters opening around it, creating the Great White Way. The name change had actually been approved by the Board of Alderman in April of 1904 in anticipation of the opening of the new subway (31).

Development on the upper west side exploded, with rapid movement of population and businesses to the area. The Bronx became accessible as never before, and the population of the borough rapidly increased. The speed of the subway allowed people to live further and further away from the city center. Between 1905 and 1920, the population in Manhattan above 125th Street increased by 265%, and the population of the Bronx increased by 150%.

Just as would be the case with highways in the late 20th century, the subways fueled new development, which in turn generated new demand for subway use. In 1904, the average New Yorker made 274 trips per year on public transportation. By 1914, this number had grown to 343 trips per year. The density of usage also climbed alarmingly. In 1904, the IRT served 3.6 million people per mile of track, that is, twice the intensity served by the elevateds. By 1914, the density of passenger traffic had become the highest in the world, 9.5 million per mile of route, more than the Paris Metro (7.2 million), the Berlin Railway (5.6 million), and the London Underground (4.4 million). The IRT responded by lengthening platforms and running longer trains, but could not keep up with demand.

With all of its success, overcrowding of the subway became an almost immediate issue, and with it, calls for additional subways. As Jay Gould before him, August Belmont Jr. steadfastly opposed subway expansion. The crowded subways represented healthy profits for the IRT.

The great success of the subway inspired the Rapid Transit Commission. In March 1905, less than one year after the subway's opening, the commission issued a mammoth $250,000,000 subway plan that literally dwarfed the IRT. Nineteen separate lines were proposed, totaling 165 route-miles (Contract #1 included only 22 route-miles). There is some doubt as to whether the commission ever intended to implement all of the lines, as many were included at the behest of local businessmen and real estate developers who were awed by the impact of the subway on adjacent land and businesses.

The RTC also sought to protect itself from the IRT exercising a monopoly. Their plan included lines that were clear extensions of the IRT system, and would clearly have to be built by the IRT. However, they also included subways running up First, Third, Eighth, and Ninth Avenues that would be unconnected to the IRT, and could, therefore, be theoretically built and operated by any company.

The RTC actively lobbied several companies to encourage bidding for its new lines, including the Hudson and Manhattan Railroad (forerunner to PATH), the BRT, and the Metropolitan Street Railway (who operated most of New York's trolley lines). Only the Metropolitan responded. Its standing was enhanced when John McDonald resigned his positions in the IRT and Rapid Transit Construction companies in December of 1904 and joined the Metropolitan (32).

August Belmont, however, did not want any competition. In late December of 1905, Belmont bought out the Metropolitan, creating a $220,000,000 holding company called the Interborough-Metropolitan. August Belmont now controlled all rapid transit and trolleys in Manhattan and the Bronx, and had a strangle-hold over the commission. The Rapid Transit Act of 1894 provided no leeway for the Commission to build or operate a rapid transit line. Lacking bidders, its lines were not going to be built. By late 1905 and 1906, the RTC became the focal point of public dissatisfaction concerning subway overcrowding and the lack of progress in getting new lines built. Considering the heroic role they had played in getting New York's first subway built, they might have expected a better standing with the public. Although it was August Belmont pulling the strings, the RTC became the "whipping boy" for everything that was wrong in New York City transit.

Belmont was aghast at the RTC plan for two reasons: it invited competitors into what Belmont felt was his (and his alone) market; and it included lines into sparsely-populated areas that would not produce sufficient usage to produce profits.

8.10 A Change in the Political Environment

Historian Clifton Hood explains the controversies over new subways as a transition from the era of leadership of the mercantile elite to a more populist model (33). In the mercantile model, the term "conflict of interest" was unknown. While graft was well understood, the RTC had no problem with August Belmont making a killing in real estate by buying up property along the subway he would be building. They valued his business acumen, and sought to work cooperatively with him. The evolving populist model viewed the line between private and public interests in a much sharper way, and understood them to be adversarial.

The populists were outraged by the operating lease granted to Belmont even before he took over the Metropolitan Street Railway. The Rapid Transit Act of 1894 offered no way to compel the construction of new subways if Belmont simply was opposed. Their outrage was inflamed further when he acquired the Metropolitan Street Railway.

The progressives were keenly interested in public transportation because in encompassed two of its major goals: urban population dispersal and prevention of monopolies.

In 1890, Jacob Riis published his landmark expose of Manhattan's Five Points tenements, *How the Other Half Lives*. The area was home to various ethnic conclaves at outrageous densities, that is, over 550,000 per square mile. Poor housing, poor sanitation, and enormous densities attracted infectious diseases such as tuberculosis, typhoid fever, and diphtheria. Progressives also believed that the tenements encouraged cultural separatism, criminal activity, and radical politics. They viewed public transportation as a way to disperse the population, and made it a central theme of their politics. Dr. Rufus Gilbert had thought that the elevateds would accomplish this a generation earlier, only to find that the poor, who remained poor despite transit, simply generated new tenements further from the city center.

In 1903, the Citizen's Union, a major progressive group, submitted the Elsberg Bill, designed to prevent a single company from dominating urban transit ever again. The bill would have separated contracts for construction and operation, limited the operating lease to 20 years, and provided ways to revoke contracts more easily. The RTC staunchly opposed the bill, and prevented it from being passed for several years.

The 1906 gubernatorial race was waged primarily around the issue of control of public utilities. William Randolph Hearst, the Democratic nominee, called for radical public control of utilities with little regard for private property rights. The Republican nominee, Charles Hughes, supported administrative regulation of public utilities, a middle course. Hughes won a narrow victory.

The new governor sponsored the Page-Merritt bill to create a pair of public utility regulatory commissions, one for New York City, and the other for the rest of the state. The commissions would have broad jurisdiction over electric, gas, railroad, and mass transit corporations. They would have authority to set rates, conduct investigations, and order schedule changes, and would have access to full-time professional staffs as needed. Despite strong opposition, the bill passed and was signed into law on June 6, 1907. On July 1st, the new Public Service Commission for the First District (NYC) replaced the RTC. A new era of subway expansion had dawned.

References

1. Hewitt Elected Mayor. New York Times, 1 (November 3, 1886)
2. The Mayor's Transit Plans. New York Tribune, 3 (February 1, 1888)
3. Abram S. Hewitt. Scientific American. LXXXVIII(5), 73 (January 31, 1903)
4. The Interborough Rapid Transit Railway Company Will Operate Subway Road. New York Tribune, 2 (April 25, 1902)
5. Subway Road Incorporated. New York Tribune, 6 (May 7, 1902)
6. Brooklyn Tunnel Construction Bids. New York Times, 1 (July 22, 1902)
7. Belmont to Build Tunnel. New York Times, 1 (July 25, 1902)
8. Tunnel Contract Signed. New York Tribune, 1 (September 12, 1902)
9. Subway and Elevated Unite. New York Tribune, 1 (November 27, 1902)
10. Explosion of Dynamite at the Rapid Transit Subway. Scientific American LXXXVI(6), 90 (February 8, 1902)
11. Cave-In at The Park Avenue Rapid Transit Tunnel. Scientific American LXXXVI(13), 221 (March 29, 1902)
12. To Stop Work on Park Avenue Tunnel. New York Times (June 19, 1902)
13. Mayor Drives Spike. New York Tribune, 1 (March 15, 1903)
14. Subway Opening Today with Simple Ceremony. New York Times, 1 (October 27, 1904)
15. Exercises in City Hall. New York Times (October 28, 1904)
16. Subway Travel On with Rush. New York Tribune (October 28, 1904)
17. Subway Carries 2,000,000 in a Week. New York Tribune (November 4, 1904)
18. Subway's First Month, 5,838,235 Passengers. New York Times, 1 (November 26, 1904)
19. Interborough Rapid Transit – The New York Subway, Its Construction and Equipment, Commemorative Brochure. Interborough Rapid Transit Company, New York (1904)
20. Prods Belmont on "Ads". New York Tribune, 4 (January 19, 1905)
21. Mayor Orders the Signs Out. New York Tribune, 1 (February 4, 1905)
22. Court Halts City in Subway Sign War. New York Times, 6 (February 7, 1905)
23. Argue Ward & Gow Case. New York Tribune, 4 (February 24, 1905)
24. "Ad" Injunction Still. New York Tribune, 2 (April 15, 1905)
25. Advertisements Mar It. New York Tribune, 1 (October 29, 1904)
26. Trample, Says Warner. New York Tribune, 1 (October 30, 1904)
27. Architectural League Condemns Subway "Ads". New York Times, 7 (November 2, 1904)
28. To Stand By Subway "Ad" Contract. New York Tribune, 6 (November 3, 1904)

29. The Shocking Disfigurement of the Subway. Scientific American XCI(19), 310 (November 4, 1905)
30. Subway Signs Barred, Then Go Up a Lot More. New York Times, 1 (November 25, 1904)
31. To Be Called Times Square. New York Times, 1 (April 6, 1904)
32. Contractor McDonald Quits Interborough. New York Times, 14 (December 2, 1904)
33. Hood, C.: 722 Miles –The Building of the Subways and How They Transformed New York. The Johns Hopkins University Press, Baltimore (1993)

Figure Sources

8.1: *Courtesy of* Library of Congress, Prints and Photographs Division, Brady-Handy Photo Collection, LC BH82 4763C.
8.2: *Courtesy of* Wikimedia Commons.
8.3: *Courtesy of* Library of Congress, Prints and Photographs Division, Pach Brothers Collection, (a) LC USZ62 75428, (b) LC USZ62 128453, (c) LC USZ62 128455.
8.4: *Originally published in* Interborough Rapid Transit – The New York Subway, its Construction and Equipment, *Commemorative Brochure*, Interborough Rapid Transit, New York, NY 1904, pg 24.
8.5: (a–b) *Originally published in* Interborough Rapid Transit – The New York Subway and its Equipment, *Commemorative Brochure*, Interborough Rapid Transit, New York, NY, 1904.
8.6: (a–c)*Originally published in* Interborough Rapid Transit – The New York Subway and its Equipment, *Commemorative Brochure*, Interborough Rapid Transit, New York, NY, 1904.
8.7: *Originally published in* Interborough Rapid Transit – The New York Subway and its Equipment, *Commemorative Brochure*, Interborough Rapid Transit, New York, NY, 1904.
8.8: *Originally published in* Interborough Rapid Transit – The New York Subway and its Equipment, *Commemorative Brochure*, Interborough Rapid Transit, New York, NY, 1904.
8.9: (a) *Originally published in* Interborough Rapid Transit – The New York Subway and its Equipment, *Commemorative Brochure*, Interborough Rapid Transit, New York, NY, 1904, (b) *Originally published in* Harper's Weekly, January 31, 1903.
8.10: *Originally published in* Interborough Rapid Transit – The New York Subway and its Equipment, *Commemorative Brochure*, Interborough Rapid Transit, New York, NY, 1904, pg 33.
8.11: *Originally published in* Interborough Rapid Transit – The New York Subway and its Equipment, *Commemorative Brochure*, Interborough Rapid Transit, New York, NY, 1904.

Chapter 9
The Dual Contracts and Expansion of the Subways

The need for additional subways to service New York City was evident within weeks of the opening of the IRT in 1904. The path to the next phase of rapid transit expansion was long delayed by the opposition of August Belmont and the IRT, and by shifting political and financial conditions. Ultimately, the great expansion known as the "Dual Contracts" was adopted and implemented, while at the same time breaking the IRT's monopoly over the system.

9.1 Introduction

The vast majority of the New York City rapid transit system was built in four ambitious phases. The four phases were more than just physically distinct; they represented four very different economies and approaches to the financing of rapid transit.

- The elevateds were built and operated entirely as private enterprises with private capital.
- The IRT was constructed with public funds, but private funds were used to equip it and operate it under a 50-year lease.
- The Dual Contracts, discussed in this chapter, were very much a joint private-public enterprise. The public role in financing and control was increased over previous efforts.
- The final stage of subway expansion, the Independent Subway (Ch. 10) was fully a public enterprise, both built and operated with public funds.

Indeed, the expansion of transit facilities, from 1904 on, has always been delayed by concerns over financing and, initially, over the proper roles of public and private interests.

The need for additional subways was evident virtually from the first day of the IRT's operation in 1904. Living conditions and densities within the tenements of lower Manhattan were fundamentally intolerable. One sixth of all New Yorkers lived in Manhattan below 14th Street, an area encompassing only 1/82 of the

R.P. Roess and G. Sansone: The Wheels That Drove New York, STTT 1, pp. 179–197.
springerlink.com

city's area. While "New York" had encompassed the "five boroughs" since 1898 (the Bronx had actually been annexed in 1895), the population remained heavily centered in Manhattan. Reformers and planners sought to disperse the population into the "outer boroughs" by building transportation connections over and under the rivers that confined people to Manhattan.

The first connection was the Brooklyn Bridge, which opened in 1883. Shortly after its opening, a shuttle cable-train operation across the bridge was initiated. In 1903, the second East River crossing, the Williamsburg Bridge, was opened.

Indeed, it had been hoped that the IRT would help disperse the population. The subway's enormous success, however, caused land values along its route to skyrocket. Land became so expensive that only large apartments and tenements could be built to make a profit. The reformers' dreams of quiet uptown residential communities based upon low-density housing were dashed. Many thought that the only way to avoid this impact was to open a significant number of new subways and/or elevateds at the same time, making large quantities of land available for development in the same market, thus holding prices down.

9.2 Where Are the Bidders?

Given the rush of the initial success of the IRT, the Rapid Transit Commission made several attempts to attract bids for new subways under the same rules used in building the IRT. Concerns over whether the city had enough money to fund additional subway construction, and the lack of interest from the IRT or competitors in submitting bids stalled any progress on new routes. When it appeared that the Metropolitan Street Railway, with John McDonald as its chief engineer, appeared ready to submit a bid, August Belmont bought them out. The RTC never obtained a proposal for subway expansion, and Belmont continued his opposition to any expansions.

In 1906, the Elsberg Act was finally passed. The bill required separate contracts for construction and operation, and limited the length of operating leases to 20 years.

When Republican Charles Hughes was elected Governor, he proposed a new structure for control of public utilities, including rapid transit, and signed legislation creating the Public Service Commission (1), with a separate district for New York City. The PSC replaced the Rapid Transit Commission as the principal agency charged with planning new subways.

The *New York Times* gave the new structure high marks, claiming that a number of benefits would be forthcoming under the PSC, assuming that the legislation was upheld by the courts (2):

> *"A five-cent fare to Coney Island."*
> *"Continuous service on surface lines from the Battery through the Bronx to the city line."*
> *"Universal transfers between all surface lines."*

> *"Improved service on all the surface, elevated, and Subway lines within the city."*
> *"Side-door cars in the Subway."*
> *"Additional platforms at the elevated stations in the Bronx or else where."*
> *"More and faster suburban trains on the New York Central, New Haven, and Long Island Railroads."*
> *"Improved gas supply."*
> *"Cheaper gas and electricity."*
> (Ref. 2, pg 3)

While the creation of the PSC helped achieve some of these goals, some were not achieved for many years. The Elsberg Act, on the other hand, was thought to be a strong deterrent to private involvement in any rapid transit expansion plans. The NYC controller explained the principal problem on April 29, 1907 (3):

> *"I have long been of the opinion," he said, "that it was a mistake to limit the life of a subway contract to twenty years, and that part of the Elsberg Law at least ought to be repealed immediately. Such a limit makes it impossible for anybody except the Interborough-Metropolitan Company to bid for new subways because a new independent bidder would have the additional expense of building and equipping a power house. No one can get back the money sunk in such an enterprise in 20 years."* (Ref. 3, pg 9)

Thus, after passage of the Elsberg Act, there was only one potential bidder, that is, August Belmont and the IRT, who was not interested in expanding the system.

The repeal and/or amendment of the Elsberg Act became an ongoing political issue, attracting a great deal of discussion. In May of 1907, State Senator McCarren (a Republican) introduced legislation repealing the Act. The *New York Tribune* (4) reported that the civic groups behind the Elsberg Act were descending upon Albany claiming the repeal, "is regarded as playing directly into the hands of the traction companies."

Letters to the editors of major newspapers were filled with commentary on the issue. Sidwell Randall, writing to the *New York Times* (5) favored the repeal, and complained about the legislature's lack of action with a humorous poem:

> *"At Albany in days gone by,*
> *A good old school was kept,*
> *Where mornings the boys did play,*
> *And afternoons they slept."*
> (Ref. 5, pg SM11)

9.3 A First Attempt: The Triborough Plan

Six months after it was established, the Public Service Commission of District One (PSC) introduced a proposal for a \$150,000,000 expansion of the rapid transit system that would double the length of the original IRT.

- A Broadway-Lexington Ave. line would travel from the Battery up the eastside to 138th Street in the Bronx, where it would divide into two branches, that is, one following Mott and Jerome Avenues to Woodlawn Cemetery, and the other following Southern Blvd. and Westchester Ave. to Pelham Bay Parkway.
- A loop line would circle through parts of Manhattan and Brooklyn, crossing both the Williamsburg and Manhattan Bridges (under construction, but not yet open at the time).
- A Fourth Avenue line would run from the Brooklyn Bridge area through sparsely-populated neighborhoods of Sunset Park and Bay Ridge, terminating at the Narrows.

These routes were not new, as all had, in one form or another, been included in RTC's earlier proposals. While the plan would have doubled the size of the original subway, it served primarily built-up areas, and did not provide the kind of coverage needed to significantly disperse the population. Figure 9.1 shows a map of the proposed Triborough routes.

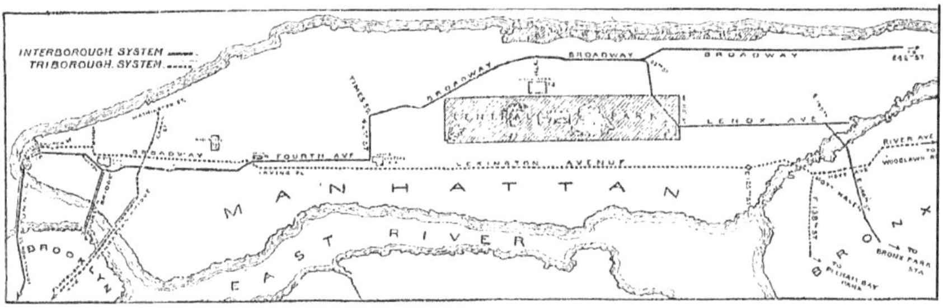

Fig. 9.1 The proposed Triborough Plan

The PSC was between a rock and a hard place with its dual intentions to spread development while at the same time preventing a monopoly from controlling the transit system. In order to attract bidders for new services, they had to serve areas sufficiently populated to generate profits from day one of operation. No single plan could have satisfied both objectives.

The Triborough Plan was endorsed by many progressive organizations, but was stiffly opposed by the Chamber of Commerce, who had been a major force behind the RTC. The Chamber criticized the routing of the Broadway-Lexington Ave. line, pointing out that it ran very close to the existing subway below 42nd Street. It also pointed out that there would still be no downtown westside service, more critical now, as the Pennsylvania Railroad was building Penn Station on 7th Avenue between 31st and 33rd Streets. Lacking general acceptance, the plan languished.

9.4 Revisions to the Elsberg Act and the Indeterminate Franchise

In 1909, the PSC successfully submitted legislation to amend some of the more restrictive aspects of the Elsberg Act. PSC Commissioner Bassett explained its significant changes (6):

> *"...The new rapid transit law was framed by the Public Service Commission, and it makes it possible for a private corporation to construct and operate, for the municipality to construct and let a private company operate, for the municipality to operate or to own the subway, and use private capital to build it, the private capital looking to an indeterminate operating contract for its remuneration."*
>
> *"Each of these different methods is safeguarded so that the blocking ability that goes with a long term contract or a perpetual franchise is prevented. This is sought to be accomplished by making it in every case possible for the city to buy, at any time after 10 years, at a price not exceeding the cost plus fifteen percent."* (Ref. 6, pg 7)

In effect, the length of operating leases was no longer restricted, and a new type of leasing arrangement was created, namely, the *indeterminate lease* or *indeterminate franchise*. In an indeterminate lease, the municipality could award a long-term franchise to a company that invested its own capital in rapid transit construction. However, the municipality could acquire the property after 10 years with appropriate compensation. The law set "appropriate compensation" as the total cost of the company's investment plus 15%, if the acquisition occurred exactly 10 years after opening. The municipality could, however, acquire the property at any time after 10 years, with compensation decreasing over time as the property depreciated. This set of provisions became known as the "recapture provisions."

The PSC, however, had yet another legal hurdle to face. In 1900, a new law gave the NYC Board of Estimate the power to appropriate municipal funds for rapid transit, and in 1905 added the right to approve the location of subway routes. Prior to 1909, the Board of Estimate was dominated by Tammany Hall Democrats, who staunchly opposed the PSC. In the election of 1909, however, a reformist-Republican fusion ticket won six out of eight positions on the Board of Estimate, including city comptroller, president of the Board of Alderman, and four out of five borough presidencies. The fusion ticket did not, however, capture the Mayor's seat.

The new Mayor, William J. Gaynor, after expressing support for the Triborough Plan during the campaign, wound up opposing it in favor of simple expansions of the IRT, that is, to be built and operated by the IRT. His position changed after a series of meetings with the President of the IRT, Theodore P. Shonts, who had become the public face of the IRT, with Belmont acting in the background. The Mayor's opposition did not alter the outcome, as the Board of Estimate approved release of the Triborough Plan contracts. The PSC released two sets of contracts for bids: one set for construction of the new lines with private capital under the "indeterminate franchise" arrangement; one set for public financing of the new lines. When the bids were opened, not one company had bid under the "indeterminate franchise" arrangement, but 23 company's had bid for municipal financing (7). Public reaction to public financing of the new lines was harshly negative, and the PSC did not move forward.

9.5 William McAdoo and a New Proposal

On November 18, 1910, the PSC received an offer from William G. McAdoo, President of the Hudson and Manhattan Railroad Company, to build a new subway. Along the lines of the Triborough Plan, the proposal included linking the new subway to the Hudson and Manhattan tubes, providing direct subway service to New Jersey. McAdoo was thought to have the financial backing of J. P. Morgan & Company (8). Mayor Gaynor was pleased to have received any interest in building new subways, and thought that with the appearance of McAdoo, other proposals would be forthcoming (9):

> *"I am glad that Mr. McAdoo has submitted his offer. It is certainly a good indication. I suppose there will be other offers, and the public should not come to a conclusion on any route until all offers are received."*
> (Ref. 9, pg 16)

Mayor Gaynor was indeed nervous that the IRT, despite claiming an interest in expanding its system, was not actually going to submit a bid. By the time of McAdoo's offer, the Mayor had had no contact with IRT officials for over five months (9). Perhaps attempting to spur new interest, the Mayor issued a statement that new subways could indeed be financed directly by the city (10) on November 30, 1910, which surprised both the Board of Estimate and the PSC, neither of which thought this to be a viable approach.

The entry of McAdoo into the fray, however, finally caused the IRT to react. On December 5, 1910, the IRT submitted its own expansion plan to the PSC (11). Its plan called for extending the IRT to form an H-pattern. The original route of the IRT started on the eastside, crossing over to the westside at 42nd Street. The new plan would continue the east side route to Upper Manhattan and the Bronx, and the west side route to Downtown Manhattan.

August Belmont was not about to allow competitors into what he considered his private domain. In an excellent article on the subject, Burton J. Hendrick details the virtual war waged by Belmont and his cronies on McAdoo (7), first trying to buy him out, and failing this, discouraging any potential financial backers from supporting him. On December 15, 1910, McAdoo was forced to withdraw his offer (12) after failing to attract sufficient financial support.

After withdrawal of McAdoo's proposal, the PSC sent two plans to the Board of Estimate on December 20, 1910. One was the IRT expansion plan, which would be privately financed, and for which leases would have to be negotiated; the other was the Triborough Plan, which would have to be publicly financed. Unable to agree on which plan should move forward, the PSC literally passed the buck to the Board of Estimate, which itself remained deadlocked over which plan to adopt.

9.6 George McAneny and the Dual Contracts

Both PSC and Board of Estimate continued to be deadlocked over a subway expansion plan through January of 1911. Before them were two plans, both of which represented risky choices. On one hand, August Belmont was finally ready to extend his subways, but he remained a monopoly, and his terms, which had to be negotiated, would likely be unacceptable. The only alternative was to release the Triborough Plan for bid with full public financing of construction. The economics of full public financing were difficult, at best.

On January 10, 1911, the game changed once again. The PSC received a proposal from Edwin Winter, President of the Brooklyn Rapid Transit Company to build the 4th Avenue and Manhattan loop routes of the Triborough Plan. The BRT owned and operated virtually all of Brooklyn's elevateds and surface seashore railroad lines, as well as most of the trolleys. It was a loosely coordinated system, and it lacked a means to conveniently circulate its passengers in Manhattan. The proposal also included converting the last surface steam railroads into rapid transit service, and linking them to the new loop subway. Belmont, of course, was outraged at the appearance of another competitor. The BRT route proposal was described in a *New York Times* article later in the year (13). Figure 9.2 shows the proposal.

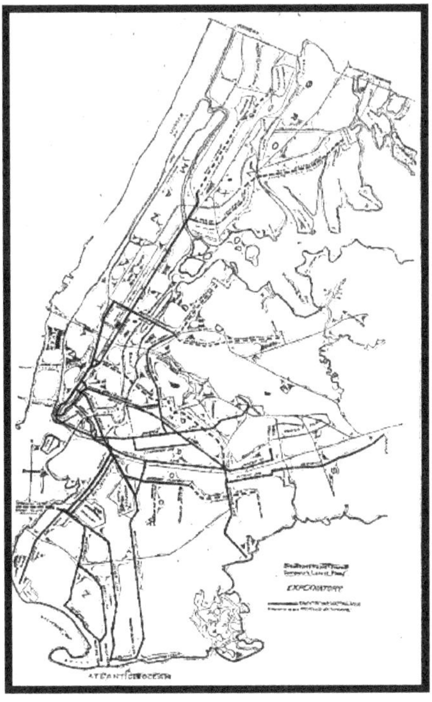

Fig. 9.2 Proposed Brooklyn Rapid Transit routes in 1911

On January 19, 1911, the Board of Estimate created a three-person transit committee to confer with the PSC, the IRT, and the BRT and recommend policies for the new subways. George McAneny, the Manhattan Borough President, was named to chair the committee. McAneny was clearly in charge of the committee, and its other two members had little impact on it. Through McAneny, the Board of Estimate became the driver of rapid transit decision-making in New York City.

George McAneny was a reformer with a keen interest in the emerging profession of city planning. He was first elected to public office as Manhattan Borough President on the fusion ticket of 1909. He was President of the Board of Aldermen between 1913 and 1916 and comptroller of New York City in 1933. He was an active member of such "good government" organizations as the City Club, the Regional Plan Association, and the Citizens Budget Commission. He was executive manager of the *New York Times* for five years, and later became chairman of the 1939-1940 World's Fair in Flushing Meadow.

Fig. 9.3 Photograph of George McAneny, circa 1910

McAneny believed that the best possible system would result from dealing with both the IRT and the BRT. Through the winter and spring of 1911, McAneny negotiated with both companies, and persuaded both to improve their plans. The IRT agreed to convert the Steinway Tunnel into a subway that would proceed (via elevated) to Astoria and Woodside. The BRT augmented its plan by including a Manhattan trunk line from the Battery to Central Park South.

On June 13, 1911, the committee issued its recommendations in a document that became known as the "McAneny Report." The report proposed that the city join with the IRT and the BRT to construct 87 route miles of subways and elevateds. The city would provide $123 million in financing for construction, while the IRT would contribute $75.8 million and the BRT, $50.4 million for construction and equipment. The plan became known as the "dual system." It proposed that the city own the dual contract lines, leasing them to the respective companies for 49 years. The contracts included a recapture provision after 10 years along the lines of the indeterminate lease provisions (14). Figure 9.4 shows the "dual system" recommended by the report.

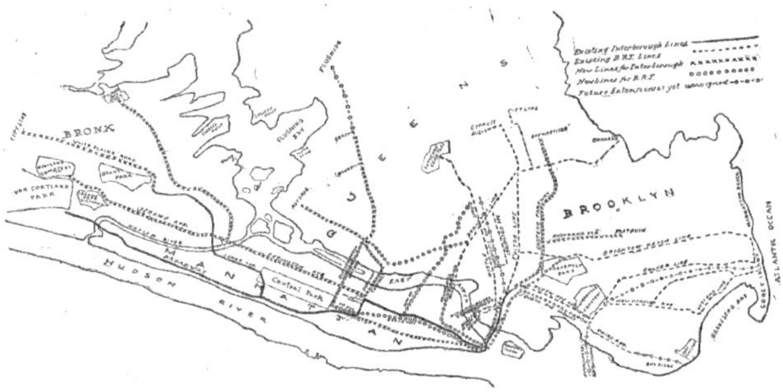

Fig. 9.4 The proposed dual system

The report also recommended a complex process for sharing revenues. Each company would first recover its interest and sinking fund expenses. The city would then receive payment for its fixed costs, after which the city and the companies would divide remaining profits equally. The BRT, however, was to be awarded an additional provision, called the "preferential." BRT agreed to pool its revenues from previously-owned elevateds and surface lines with the new subway revenues, essentially giving the city ownership of a large system at the price of a smaller subway construction. In exchange, the BRT would be guaranteed a sum equal to the net profit of the railway in the year prior to operating the new system. The "preferential" was also a reward to the BRT for agreeing to build lines into sparsely-populated areas.

The IRT was not offered a similar proposition, as its lines were making enormous profits, and because its elevated and subway lines were in separate corporate divisions. Further, the Contract #1 and Contract #2 subways were already publicly owned.

Belmont was, of course, outraged at what he considered to be a deal that deliberately favored the BRT. The plan provided some inducements, however. Both the BRT and the IRT were given a limited amount of time to positively respond to the plan; if either side rejected the plan, the routes would be transferred to the other. If both rejected the plan, the city would begin to build the system on its own. Nevertheless, the IRT vowed to reject the plan (15), labeling the terms "unjust discrimination."

The Board of Estimate approved the McAneny Report on June 21, 1911 (16). Within days, it was accepted by the BRT and rejected by the IRT (17). IRT President Shonts was most upset over allowing the BRT to have a route directly serving Manhattan, which he claimed "will take $5,000,000 from the Interborough earnings" (18).

For a while, Belmont, Shonts, and the IRT refused to negotiate with the Board of Estimate. By November of 1911, however, seeing that negotiations with the BRT were proceeding smoothly, he returned to the table. As described in the *New York Times*, negotiations were far from smooth (19):

> *"Various routes and methods of dividing the net earnings of the new roads were debated, but the discussion was general. No offer was made on the part of the Interborough, and no conditions were laid down by the city. As soon as one snag was overcome, another seemed to appear, and the conferees had almost agreed that a compromise was impossible."* (Ref. 19, pg 1)

In a key concession, McAneny agreed to give the IRT a preferential that, while similar to BRTs, was far more costly. The Board of Estimate vigorously debated this and other issues, but finally approved it on May 24, 1912 (20). Lawsuits and the passage of enabling legislation caused additional delays.

Finally, on March 19, 1913, the City of New York signed Contract #3 with the IRT, and Contract #4 with the BRT, the Dual Contracts. The two companies obtained 49-year leases, and agreed to pay for part of the construction costs and for all equipment costs. The contracts incorporated provisions for recapture, pooling, and preferentials. Nine years after the IRT first rumbled under New York's streets, the second and most dramatic expansion of the rapid transit system was finally underway.

Fig. 9.5 Signing of the Dual Contracts at PSC Quarters, March 19, 1913

9.7 The Dual System

The dual system would more than double the trackage of the rapid transit system, from 296 to 619 miles. If one tried to cover every route-mile of the system consecutively, the trip would take 21 hours. This is an interesting statistic that has become the subject of many contests. Periodically, a group tries to negotiate the entire present system to set a new record.

The dual system, however, incorporated one major flaw that has created difficulties over the years. While the IRT and the BRT used the standard track gauge, their cars were a different size. IRT cars were basically 51 ft by 9 ft because the IRT was built so that the same cars could be used on both the subway and the elevateds. The original BRT subway cars were larger, that is, 67 ft by 10 ft, because the BRT wanted to use fewer, but larger, cars to give an economic advantage. While all cars could technically operate over all tracks, the IRT tunnels were smaller than the BRTs and its curves were sharper; further, BRT cars could not operate through an IRT station platform. IRT cars could deadhead on BRT tracks, therefore, but not vice-versa. This fundamental difference remains today, requiring two different sets of railcars to equip the subway.

The dual system took almost a decade to build. When completed, it comprised a system that made cross-borough traffic (except to and from Richmond) easy, and opened areas of the Bronx, Brooklyn, and Queens to new development. By the mid-1920s, a New Yorker could start his/her day riding a roller coaster or taking a swim at Coney Island, watch Babe Ruth swat a home run at Yankee Stadium in the afternoon, and finish the day off with a movie or play at Times Square in the evening.

9.7.1 The Interborough Rapid Transit (Contract #3) Routes

Before the dual contracts, the IRT system (which had incorporated the Manhattan and Bronx elevated lines in 1902), had a major shortcoming: the line moved up the center of Manhattan to 42nd Street, then up the westside of Manhattan. Thus, above 42nd Street, there was no subway on the eastside, and below 42nd Street, there was no subway on the westside. The dual contracts rectified this by providing subways on both the east and westsides that traversed the length of Manhattan. When completed, the IRT would convert the 42nd Street crossover to a shuttle service; it is still operated as a shuttle.

Table 9.1 summarizes the lines added by the IRT as a result of the dual contracts (21).

Table 9.1 Interborough Rapid Transit lines covered by the dual contracts

Line	Section	No. of Tracks	ROW	Borough
New Subways/Elevateds				
Lexington Ave.	35th Street and Park Ave. to 135th Street (20.6 mi)	4	Subway	Man/Bronx
Jerome Ave.	135th Street to Woodlawn Road (18.2 mi)	3	Subway/El	Bronx
Westchester Ave.	135th Street to Pelham Bay Park (21.4 mi)	3	Subway/El	Bronx
White Plains Rd.	179th Street to Becker Ave. (13.7 mi)	3	Elevated	Bronx
Seventh Ave.	Times Square – Battery (15.2 mi)	4	Subway	Manhattan
Lower Manhattan	Park Place to Clark Street (4.5 mi)	2	Subway	Man/Brooklyn
Steinway Tunnel	Park Ave. & 42nd St. to Long Island City (3.2 mi)	2	Subway	Man/Queens
Steinway Tunnel	Times Square to junction (0.8 mi)	2	Subway	Manhattan
Steinway Tunnel	Junction to Queensborough Plaza (2.6 mi)	2	Subway/El	Queens
Astoria	Queensborough Plaza to Ditmars Ave. (7.5 mi)	3	Elevated	Queens
Flushing	Queensborough Plaza to Prime St. (16.5 mi)	3	Elevated	Queens
Eastern Parkway.	Atlantic/Flatbush to Buffalo Ave. (12.7 mi)	4	Subway	Brooklyn
Nostrand Ave.	Eastern Pkwy. to Flatbush Ave. (4.8 mi)	2	Subway	Brooklyn

Table 9.1 *(continued)*

Livonia Ave.	Eastern Pkwy. to New Lots Rd. (5.1 mi)	2	Elevated	Brooklyn
Extensions to Existing Elevated Lines				
Putnam Bridge	River Ave. to Jerome Ave. (1.3 mi)	2	Elevated	Bronx
West Farms	143rd St. and 3rd Ave to Harlem River Bridge (3.1 mi)	2/4	Elevated	Bronx
Queensborough Bridge	2nd Ave. El connection to Queensborough Plaza (0.5 mi)	2	Elevated	Man/Queens
Third Track Additions to Existing Elevated Lines				
2nd, 3rd Ave. Els	City Hall to 129th Street (8.1 mi)	3	Elevated	Manhattan
9th Ave. El	Rector Street to 155th Street (2.4 mi)	3	Elevated	Manhattan

In addition to the additional route-miles and track-miles, the IRT expansion under the dual contracts added 116 stations to the rapid transit system.

There were a number of unique features incorporated into the IRT expansions. At 42nd Street, the new subway would continue uptown under Lexington Avenue, as opposed to swinging west along 42nd Street to Broadway. At the same time, the 42nd Street connection was retained as a three-track shuttle service. At Broadway and 42nd Street, a new subway was built down the westside.

North of 42nd Street, the Lexington Avenue line, express tracks were laid in separate tubes below local tracks, as opposed to having four tracks at the same level. This allowed for slightly different alignments as needed for local and express tracks. For the higher speed of express trains, a straighter alignment was preferable to one that followed the alignment of the street. This type of arrangement was originally proposed by Frank Sprague. In later subways (the IND), express and local routings were different in several cases.

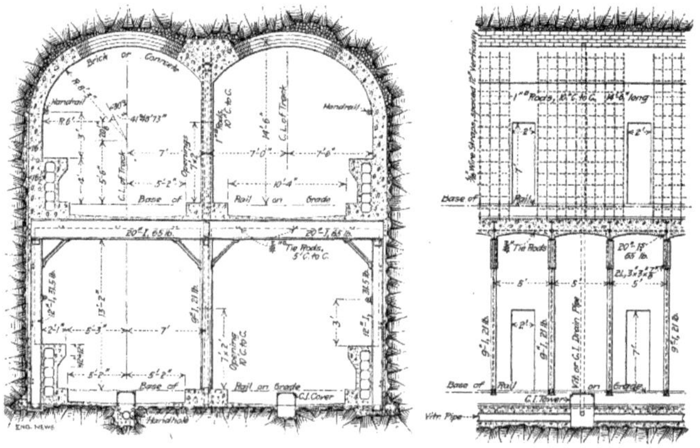

Fig. 9.6 Cross-section and plan view of double-deck tunnels on Lexington Avenue Line

In building new sections of elevated lines, particular attention was paid to those running over major avenues and streets. This was particularly true of a section of the Flushing Line (it only went as far as Woodside in the dual system) that traveled over Queens Blvd. This was the first place in the system that an elevated structure of reinforced concrete was used. The reinforced concrete structure was solid under the tracks, which were placed on rock ballast. This design minimized the noise that would emanate from the elevated. Further, reinforced concrete allowed for a more ornamental design to be used.

(a) Early depiction of the Queens Boulevard elevated section of the IRT

(b) Typical station on the Queens Boulevard elevated section

Fig. 9.7 Views of the Queens Boulevard elevated section of the IRT

Where steel elevated structures were erected, they were of simpler design and presented a less daunting image than older elevated structures. Tunneling for the IRT lines was still primarily cut-and-cover, but more use was made of temporary street coverings to minimize the impact on street operations and adjacent land uses.

(a) Cut-and-cover on Jerome Ave. through rock cut

(b) Temporary street cover on Lexington Avenue and 73rd Street

Fig. 9.8 Views of Lexington Avenue Subway construction

Fig. 9.9 Temporary street surface and relocated gas lines during Broadway Subway construction

9.7.2 The Brooklyn Rapid Transit (Contract #4) Routes

With the implementation of the dual system, the BRT created two additional subways in Manhattan. The first was basically the original Loop Line specified in the Triborough Plan. The second was a trunk line subway under Broadway to 59th Street. It was part of an open-loop system, starting in Brooklyn, and returning (via the Queensborough Bridge) to Queens at 59th Street.

The BRT also committed to significant expansion of its Brooklyn elevated system, including a major line in Queens. The expansion included a subway link from Eastern Parkway to Malbone Street. The entry to the Malbone Street tunnel would later become the site of the worst accident in the history of New York's rapid transit system.

Table 9.2 summarizes the Contract #4 additions to the rapid transit system. The BRT portions of the dual contract system were built by a separate corporate entity established for that purpose, that is, the New York Municipal Railway Corporation (22). The Brooklyn Rapid Transit Co. also added 81 stations to its routes.

Like the IRT, most of the tunneling for the BRT lines was constructed using the cut-and-cover technique. In many cases, temporary street surfaces had to be built, and utilities had to be temporarily located above ground.

Table 9.2 Brooklyn Rapid Transit Company Lines included in the dual contracts

Line	Section	No. of Tracks	ROW	Borough
Lines Built Jointly by the city and the company				
Broadway	Morris Street to 59th Street (17.6 mi)	4	Subway	Manhattan
59th Street	7th Avenue to Queensborough Plaza (4.5 mi)	2	Subway & Bridge	Man/Queens
Loop Line	Brooklyn Bridge to Whitehall Street (2.0 mi)	2	Subway	Manhattan
Loop Line	Broadway to Manhattan Bridge (1.2 mi)	2	Subway	Manhattan
Fulton St.	Flatbush Ave. to Malbone Street (2.4 mi)	2	Subway	Brooklyn
Canarsie	14th Street and Sixth Ave. to East New York (14.8 mi)	2	Subway/El	Man/Brooklyn
East River Tunnel	Whitehall St. to Monague St. (4.0 mi)	2	Subway	Man/Brooklyn
Fourth Ave.	Manhattan Bridge to 89th Street (23.9 mi)	4	Subway	Brooklyn
Loop Line	Park Row to Williamsburg Bridge (5.6 mi)	4	Subway	Manhattan
Connections	38 Street, 4th Ave., 9th Ave. (1.9 mi)	3	Subway Open Cut	Brooklyn
New Utrecht	39th St. to 86th St. to Coney Island (16.5 mi)	3	Elevated	Brooklyn
Culver	9th Ave. to Coney Island (16.1 mi)	3	Elevated	Brooklyn
Lines Built by the Company				
Jamaica	Cypress Hills to Jamaica (13.2 mi)	2	Elevated	Brooklyn/ Queens
Brighton Beach	Connection (2.0 mi)	2	Open Cut	Brooklyn
Sea Beach	65th Street and 4th Ave. to Coney Island (16.1 mi)	2	Open Cut Embankment	Brooklyn
Eastern Pkwy.	Malbone Street Tunnel (1.8 mi)	2	Subway	Brooklyn
Loop Line	Canal Street Spur (0.2 mi)	2	Subway	Manhattan
	Third Tracking of Fulton St., Broadway, and Myrtle Ave. Elevateds (9.3 mi)		Elevated	Brooklyn

9.8 Impacts of the Dual Contract Expansion

The impacts of the dual system lines on New York City cannot be understated. The new lines broke down the last of the physical barriers confining residences to Manhattan. The floodgates had been opened. With the rapid transit system expansions, a massive population shift from Manhattan to the now accessible outer boroughs began.

Between 1910 and 1940, the population of Manhattan decreased by 19% (about 333,000 people), while the city population as a whole increased by 56%. In the same period, population in the Bronx increased 309%, in Brooklyn 165%, and in Queens by 209%. Only Richmond remained relatively unaffected. Richmond, to this day, remains a more isolated borough with no direct rapid transit connection to the rest of the city.

The tenement districts of the lower eastside also underwent drastic changes, as many immigrants began to move to the less dense areas of Brooklyn, Queens, and the Bronx. Certainly, as the total population of New York continued to increase, higher densities migrated to areas of the outer boroughs as well. The population of Manhattan, however, has never been higher than it was in 1910, and its population density today remains lower than it was in 1910.

9.9 Changing Times and a Disaster on the BRT

By the time the dual contract lines were completed, the economics of the rapid transit business had been altered significantly. With the end of WWI, both the IRT and the BRT were financially stretched. A severe postwar inflationary spiral caused significant increases in their operating costs, while they were prohibited from raising the traditional fare, which had been set at five-cents as part of Contracts #1 through #4. By 1918, the BRT was on the edge of bankruptcy, while by 1921, the IRT had seen its profits decrease by 50%.

The automobile was also cutting into the rapid transit system's market. In 1908, Henry Ford had introduced the "Model T," which made the automobile accessible to large segments of the population. In 1918, the auto was well into the transition that would eventually make it the primary mode of transportation in the country.

November 1, 1918 occurred in the midst of turbulent times in the U.S. and in New York City. World War I, the "war to end all wars," was coming to a close, although the formal end would not come until 10 days later on November 11th. The influenza epidemic of 1918, which had killed over 320,000 Americans, was waning. The Brooklyn Dodgers had completed their first season in their new home at Ebbets Field. Women were about to vote in their first gubernatorial election since winning the franchise. At 6:22 PM that evening, a Brighton Beach BRT train crashed at the entrance to the Malbone Street Tunnel, killing between 93 and 103 people (23).

The BRT had been hit with a strike of motormen represented by the Brotherhood of Locomotive Engineers on the morning of November 1st. Not wishing to curtail its service, BRT officials pressed a number of employees into service as motormen. One of these was Edward Luciano, who was a dispatcher with minimal experience in operating trains in rail yards. He was given two hours of hurried training, and put into service as a motorman on the Culver Line. After completing a full shift, he was asked to make an additional, though fatal, run on the Brighton Line. In addition to being unqualified as a motorman, Luciano was not in the best of health, having recently recovered from influenza, which had claimed the life of one of his daughters shortly before.

The site of the accident was the portal into the Malbone Street Tunnel, which was operating at the time as a temporary bypass around a construction site. The "speed limit" for trains entering the portal on a sharp curve, was 6 mph. The tired Luciano was already having difficulties, having overshot several elevated stations approaching the tunnel. He was driving a five-car elevated train made of up wooden composite cars, two of which were light trailer units without motors. The two trailer cars were the second and third of the train, even though BRT policy prohibited two trailer cars from being linked consecutively. He entered the tunnel at a speed of 30 mph. The second car of the train derailed before entering the tunnel, crashing into the tunnel portal. Car three followed suit. The two trailer cars sustained most of the damage. The train did not stop immediately, and the fourth and fifth cars followed into the tunnel, running over scores of passengers that had been thrown to the tracks from the 2nd and 3rd cars. Despite dramatic rescue operations that continued through the night, 93 people were killed, although some reports list the number as high as 103.

Luciano left the scene of the accident and was not located until days later. Beginning in March of 1919, five criminal trials were held in Mineola, Long Island in which Luciano and various officials of the BRT were charged. No convictions were obtained.

On a modern map of Brooklyn, it is pretty hard to find Malbone Street. Shortly after the accident, Malbone Street was renamed Empire Blvd. in order to dissociate the street from the memories of the tragic accident. A short dead-end spur still bears the name "Malbone Street."

(a) Cars two and three after the accident

(b) Lead car one in the yard after the accident

Fig. 9.10 Subway cars after the Malbone Street wreck

While it is highly unlikely that the Malbone Street accident caused the bankruptcy of the BRT, it certainly did not help. On December 31, 1918, the BRT entered receivership. The bankruptcy stymied a large number of lawsuits against the company emanating from the wreck. It would be over four years before the company's tangled finances and all of the lawsuits could be sorted out. In 1923, the company was reorganized into the Brooklyn-Manhattan Transit Company, that is, the BMT.

There would be one more major expansion of the subway system. Despite this, however, the economics of rapid transit operations continued to worsen. As the great depression and World War II approached, the era of private involvement in rapid transit operation was coming to an end.

References

1. The Governor's Message. New York Times, 6 (January 3, 1907)
2. How Utilities Bill Aid the Public. New York Times, 3 (June 9, 1907)
3. For New Transit Laws. New York Tribune, 9 (April 30, 1907)
4. To Repeal Elsberg Act. New York Tribune, 2 (May 2, 1907)
5. Letter to the Editor from Sidwell S. Randall. New York Times, SM11 (April 5, 1908)
6. City Transportation Problem: Commissioner Basset's Views. Wall Street Journal, 7 (August 14, 1909)
7. Hendrick, B. J.: McAdoo and the Subway. McClure's Magazine XXXVI(5), 484 (March 11, 1911)
8. McAdoo Have Backing of Morgan. New York Tribune, 1 (November 20, 1910)
9. Gaynor Praises McAdoo. New York Tribune, 16 (November 22, 1910)
10. Subways Can be Built with the City's Money. New York Tribune, 1 (December 1, 1910)
11. The Interborough's Offer. New York Times, 12 (December 6, 1910)
12. Mr. McAdoo's Withdrawal. New York Times, 10 (December 16, 1910)
13. B.R.T. Puts in Subway Offer. New York Times, 1 (April 18, 1911)
14. Subway Plan Gives the City a Master Hand. New York Times, 1 (June 14, 1912)
15. Interborough Will Flatly Reject McAnany Subway Offer. Wall Street Journal, 7 (June 17, 1911)
16. McAnany Report Unanimously Adopted by Board of Estimate. Wall Street Journal, 2 (June 22, 1911)
17. B.R.T. Says 'Yes' Interboro 'No'. New York Times, 1 (June 26, 1911)
18. Interboroughs Refusal Final. Wall Street Journal, 7 (July 1, 1911)
19. Subway Ultimatum Goes To Interboro. New York Times, 1 (January 25, 1912)
20. Dual Subway Plan At Last Approved. New York Tribune, 1 (May 25, 1912)
21. Signing of the Dual System Contracts. PSC for the First District, ch. II, New York (September 1912)
22. Signing of the Dual System Contracts. PSC for the First District, ch. III, New York (September 1912)
23. Cudahy, B.: The Malbone Street Wreck. Fordham University Press, New York (1999)

Figure Sources

9.1: *Courtesy of* The Christian Science Monitor, December 3, 1910, pg 14.
9.2: *Originally published in* "BRT Puts in Subway Offer," *New York Times*, April 18, 1911, pg 1.
9.3: *Courtesy of* Library of Congress, Prints and Photographs Division, George G. Bain Collection.
9.4: *Originally published in* "Subway Plan Gives the City a Master Hand," *New York Times*, June 14, 1911, pg 1.
9.5: *Originally published in* "New Subways for New York: The Dual System of Rapid Transit," PSC for the First District, New York, NY, 1913.
9.6: *Originally published in* Lavis, F., "The New York City Rapid Transit Extensions," Engineering News, Vol. 72, No. 18. October 29, 1914, Figure 14.
9.7: *Originally published in* "The Dual System of Rapid Transit," PSC for the First District, New York, NY, September 12, 1913.
9.8: *Originally published in* "The Dual System of Rapid Transit," PSC for the First District, New York, NY, September 12, 1913.
9.9: *Originally published in* "The Dual System of Rapid Transit," PSC for the First District, New York, NY, September 12, 1913.
9.10: *Courtesy of* Wikimedia Commons.

Chapter 10
The Independent Subway

The third and last major stage of NYC Rapid Transit System was the Independent Subway, which opened in stages between 1933 and 1940. It completed the transition from private to public operation, as it was both built and operated by the city.

10.1 Introduction

The Dual Contracts forged by George McAneny not only provided for a massive and long-needed expansion of the subway system, they also contained two critical provisions that would affect the economics of rapid transit in New York City for decades:

- The contracts spelled out the financial relationship between the private operators of the consolidated subway and elevated lines and the city. The city would share in the net revenues generated by rapid transit after paying a "preferential" assuring both the IRT and BRT that their profits would remain at least at the levels established the year before the contracts were signed.
- The contracts required that the fare be kept at five cents per trip for the duration of the 49-year lease agreements.

In 1913, this seemed like a reasonable arrangement. Both the IRT and BRT were making huge profits, and the city expected to receive significant payments through the contract provisions.

The arrangements, however, were doomed in light of events that would change the national and local economy profoundly. The heart of the difficulty was the 5-cent fare. From the early days of the elevateds, fares had been strictly regulated. While some elevated fares, for a time, were higher than five cents, by the end of the 19th century, five cents had become a national norm for urban transit services. The low fare was intended to insure that the transit system would be accessible to virtually all citizens, and that few would be excluded from it because they couldn't afford it. While in the mid-1920s many transit systems across the U.S. were abandoning the five-cent fare, in New York it became a part of the political base, a fundamental right of the population that no politician would oppose without severe risks to their future.

For the IRT and BRT to make money, two things were necessary: (1) they had to be able to control costs, and (2) ridership had to continually increase. The five-cent fare was not based upon distance. With the dual contract expansions, many

R.P. Roess and G. Sansone: The Wheels That Drove New York, STTT 1, pp. 199–222.
springerlink.com

patrons were now traveling vastly greater distances than previously, as the population dispersed from the Manhattan tenements to the outer boroughs. Even without inflation, the cost of each trip had increased, but the fare had not.

The inflationary spiral of the post World War I period was the last "nail in the coffin" for the dual contract financial arrangements. Rapidly-increasing costs and the Malbone Street horror forced the BRT into receivership by the end of 1918. The financially-stronger IRT was able to forestall its bankruptcy until 1932, when it too went into receivership. In the interim, its profits dwindled, and in a number of years, deficits occurred. The result of this was that the city received far less than anticipated from the financial agreements. It never received a penny of the BRT's $207 million revenues during the 21 years of its profit-sharing agreement. It did receive some funds in some years from the IRT, but the total of $2.1 million was a small fraction of the $307 million the IRT earned from 1919–1940.

As the "Roaring '20s" dawned, the nation and city reached new cultural and economic heights. The rapid transit system, however, was already in deep financial trouble. By the end of the decade, the Great Depression would engulf the entire nation, and push the transit system into desperate times. The era would change the fundamental management and operation of rapid transit systems from private to public. As the age of public operation approached, the subways became increasingly politicized and a focal point of many municipal elections.

10.2 Mayor Hyland Enters the Fray

John F. Hylan came from a poor background in rural upstate Greene County. Forced to work to help support his family, he received little education in a one-room schoolhouse; he did not attend high school.

In 1887, at the age of 19, he moved to Brooklyn and got a construction job with the Brooklyn Union Elevated Railroad, a forerunner of the BRT. A hard worker, he was promoted several times, and eventually became a locomotive engineer, the top position for a manual laborer. The new job enabled him to pay off a $1500 mortgage on his parent's farm, get married, and support several of his siblings.

Unhappy with his lot, Hylan enrolled in evening classes, eventually earning a law degree from New York Law School. While studying, he juggled family life, his railroad job, and a clerkship in a Long Island City law firm.

About a month before he graduated, while driving a steam-powered elevated train near Navy Street, he almost ran over a railroad official named Barton. Some reports indicated that Hylan was studying a law text while driving the train, but this was never verified. The enraged Barton fired Hylan, though Hylan denied that he had done anything wrong, maintaining that position throughout his life.

Hylan opened a law office in Bushwick, and joined his local Democratic club in order to meet potential clients. An active member, he was able to rise in the party. In 1906, Mayor McClellen appointed him as a city magistrate. In 1914, Governor Martin Glynn appointed him to fill a vacancy on the Kings County Court. In 1916, he ran for a full term, strongly backed by William Randolf Hearst and his newspapers. He won easily, but attracted much attention, as he ran well ahead of the rest of the Brooklyn Democratic ticket.

As a judge, Hylan had been vocal and vehement in his opposition to the Dual Contracts for two reasons: he believed the contracts unduly favored the IRT and BRT; and the subways built did not serve enough older, built up areas, particularly in Brooklyn, and more particularly where his Bushwick constituency lived.

10.2.1 The 1917 Mayoral Election

The race for mayor in 1917 was, in many ways, a watershed event for the city and the nation. The sitting mayor, John P. Mitchel, had been elected in 1914 on the Fusion Ticket, a coalition of Republican progressives and other reformers. As mayor, he is credited with cleaning up corruption in the police and fire departments. He was elected at the age of 34, the youngest mayor in the city's history, and was frequently referred to as "the boy mayor."

Having lost power in the city government, Tammany Hall was determined to recover in 1917. It had great difficulty focusing on a candidate, and as late as August of 1917, was still considering a list of 19 potential candidates. The list included Judge John Hylan, William R. Hearst, and Alfred E. Smith (1). Hearst, the newspaper magnate, had an interesting relationship with Hylan, who had caught his attention while fighting the Dual Contracts. During the primary run, it was not clear whether he favored Hylan or was himself an active candidate. Finally, on August 14, Tammany Hall selected Hylan as its candidate for mayor. Al Smith, the future NYS governor, was nominated as President of the Board of Estimate (2, 3).

The nomination caused immediate controversy within Tammany Hall. It was well-known that Tammany's leader, Charles Murphy, was not particularly enthused by Hylan, who was a relative unknown. A week later, the Sullivan Club, a major component of Tammany Hall, announced it would support Mayor Mitchel, who was once more the Fusion candidate (4). Petitions were filed for Hearst to run in the Democratic primary for mayor, but it was thought that these were filed without his permission (5). Finally, on September 1, 1917, Hearst formally withdrew from the primary, and made clear his support for Hylan (6).

There was a third major candidate for mayor, Morris Hillquit, running on the Socialist Party ticket. The Socialist Party had gained a reasonably large base of support in the early 1900s, and was a significant political force in 1917.

The campaign was particularly nasty, with various charges hurled throughout. Three major issues arose, only one of which was directly related to the city:

- Mitchel was attacked for his failure to build schools to serve the rapidly increasing population of school-age children amongst immigrant groups. He had supported something called "the Gary Plan," which essentially relied on double-sessions (opponents called this "half-day" school) for most immigrant neighborhoods.
- WWI was approaching, and there was a great debate over whether the U.S. should enter the war. Mitchel was the only candidate to clearly favor U.S. participation in the war; Hillquit was strongly opposed, and after waffling for a while, Hylan also opposed U.S. participation.

- The campaign for universal suffrage was also underway. While the candidates accused each other of opposing extending the vote to women, all three candidates eventually expressed support. Hylan's support was unexpected, as Tammany Hall was against it (7).

Hylan attacked Mitchel for being the candidate of the wealthy and special interest groups. Mitchel attacked Hylan for his lack of patriotism and for his strong connections to the German community, which was staunchly against U.S. participation in WWI.

The messy campaign made it difficult to predict an outcome, even days before the election. The *New York Times* remarked that "The election of 1917 is a puzzle without parallel" (8). The election, however, provided Hylan with an unexpectedly easy victory. His vote more than doubled that of Mayor Mitchel, who ran second (9). While he did not win a majority, he carried the Democratic/Tammany Hall ticket to victory with him, yielding a decisive Democratic majority on the Board of Estimate.

Fig. 10.1 Portrait of Mayor John F. Hylan

John Hylan was a fiery personality with firm views. At the same time, he was not a very confident man, and many considered him to be unimposing, even embarrassing at times. Throughout his political career, and afterward, Hylan railed against what he considered to be the "invisible government," secretly running virtually every aspect of American life. In a speech given at the Knights of Columbus meeting in Chicago on March 26, 1922, his remarks included his most oft-quoted statement on the issue (10, 11). The quote appears in several different forms, as no one account contained the entirety of his comments:

> *"The real menace to the republic is the invisible government…a little coterie of international bankers that virtually run the United States Government for their own selfish interests. The octopus of invisible government seizes in its long and powerful tentacles our executive officers, our legislative bodies, our schools, our courts, our newspapers and every agency created for the public protection."* (Ref. 10, pg 3; Ref 11, pg 1 and 4)

Mayor Hylan, who was often referred to as "Red Mike," had a long list of public enemies. The traction companies were high on his list. He took on the IRT and BRT with missionary zeal, denouncing them at every possible turn. His attacks were further fueled when it became obvious that they improved his standing with the electorate.

10.2.2 Mayor Hylan and the Malbone Street Wreck

The Malbone Street wreck occurred almost one year after Hylan's election. He immediately involved himself in the investigation and prosecution of the incident (12). The *New York Times* reported:

> *"Mayor Hylan arrived at the Snyder Avenue Police Station shortly after midnight, and consulted with District Attorney Lewis and (Police) Commissioner Enright as to what steps should be taken in ordering the arrest of the officials of the B.R.T."* (Ref. 12, pg 1)

The Mayor immediately commenced a "John Doe" hearing which was intended to identify those responsible for the accident, including crew members and officials of the BRT (13). He invoked a rarely-used law which allowed him to personally sit as a committing magistrate of the hearing. In addition, he personally interviewed many of the surviving victims of the crash (14).

The motorman, who had left the scene of the accident, was arrested at his home. Originally identified in news accounts as "Anthony Lewis" and "William Lewis," it was eventually determined that his name was Edward Luciano. In one account (14), he was quoted as saying he had tried to slow the train before entering the tunnel, but after applying the brakes, "they didn't seem to work."

The hearings focused on the potential guilt of the motorman, and of various officials of the BRT who had authorized the "green motorman" to operate the train. Luciano, who was not a regular motorman, received only two hours of training before being allowed to operate trains during a strike.

The consist of the train, which included three wooden cars (two in succession in violation of BRT policy) was also questioned. Public Service Commissioner Whitney suggested that Hylan might "prove to be one of the John Doe's whom the inquiry was intended to identify." He asserted that the loss of life had been largely due to the use of the wooden cars, and noted that Mayor Hylan had been "for months the sole obstacle blocking action to compel the use of steel cars on this line" (14).

After a month of investigation and testimony, Hylan's hearings had not produced much. As the hearings closed, the *New York Tribune* (15) noted that:

> *"...the identity of John Doe and Richard Roe (is) still a mystery."*
> (Ref. 15, pg 16)

By the end of December 1918, a grand jury had indicted the motorman, Edward Luciano (his correct name was, by then, being used), and five officials of the BRT.

for manslaughter (16). The first to go to trial was Thomas Blewitt, Superintendent of the Southern District of the BRT. His trial introduced evidence that an object had been thrown on the tracks, causing the accident. After a 12-day trial, Blewitt was acquitted after only three and a half hours of deliberation (17).

The second to go to trial was the motorman (18). During the trial, only three witnesses testified, that is, Luciano, the motorman, and two others who testified on his behalf. After a one-week trial, the jury deliberated less than five hours before returning a verdict of "not guilty" (19).

After two "not guilty" verdicts, the district attorney began to question whether or not the trials of the other indicted officials should go on, or whether the charges should just be dropped. Mayor Hylan began to put pressure on the district attorney to continue the trials, prompting a terse reply (20):

> *"District Attorney Harry E. Lewis of Kings yesterday sent the Mayor a letter in which he virtually stated that he is capable of handling the affairs of his office without outside interference."* (Ref. 20, pg 6)

Two more trials were held in 1920, with both resulting in acquittals (21, 22). Finally, on January 17, 1921, all charges against remaining BRT officials were formally dropped (23).

In the end, despite public outcry and political pressure, the Malbone Street wreck resulted in not one criminal conviction of anyone. How much of this was due to Hylan's heavy-handed meddling, and how much was due to the inability to convince any jury that any of the indicted persons had acted with criminal negligence difficult to say. Guilt for an accident that need not have happened was never officially determined.

10.2.3 Hylan's First Term

Throughout his first term, Mayor Hylan fought against any transit proposals that he deemed would have benefited either the IRT or the BRT. He vigorously opposed various proposals from the PSC to reorganize the railways' finances, denouncing them as an attempt to protect investors and raise the five-cent fare. Defense of the five-cent fare became a major issue for Hylan, and highlighted his successful reelection campaign in 1921. In his second term, Mayor Hylan would focus his efforts on building a new city-owned and operated subway to compete with the existing privately-run traction companies.

10.3 A Plethora of Plans

10.3.1 The Office of the Transit Construction Commissioner (TCC)

After Mayor Hylan was elected in 1917, Tammany Hall and the Democrats captured the governor's seat with the election of Alfred E. Smith in 1918. In April of

1919, the state legislature created the Office of the Transit Construction Commissioner for New York City (24). The Office replaced the Public Service Commission and the Board of Estimate as the decision makers regarding transit in the city. It placed the responsibility for planning and implementation in the hands of a single entity as opposed to the divided responsibilities of the PSC and the Board of Estimate. One of its original aims was to speed the completion of the Dual Contract lines, as well as to plan for the next generation of the rapid transit system. The PSC, however, retained the responsibility of regulating transit operators. The creation of the office effectively removed any responsibility for rapid transit from the city, a situation that Mayor Hylan did not like.

The commissioner was to be appointed by the governor. On May 27, Governor Smith announced the appointment of John H. Delaney to head the office. Delaney was a Tammany Hall regular who had been serving as the Commissioner of Parks and Structures. The position had reportedly been offered to William Barclay Parsons, who declined to accept it (25).

By 1920, there was open discussion of "recapture" of the IRT and BRT under the Dual Contract provisions. While this did not happen for over a decade, there were almost constant discussions and threats of doing so throughout the 1920s (26). The potential for recapture was used by the PSC to compel the traction companies to make improvements in service to ease overcrowding.

The chief engineer of the Transit Construction Commissioner's Office (TCC) was Daniel L. Turner. In 1900, he had been a junior engineer on Parson's RTC engineering staff. He was chief engineer of the PSC in 1917, and moved to the same position in the TCC. In July of 1920, the TCC issued Turner's bold plan calling for 830 additional track-miles of subway and elevated lines in New York City (27, 28, 29). Turner believed that the dual contract system was not extensive enough, as it did not touch enough open areas, and did not provide for a 20–30 year master plan in which subways would continue to expand. In his vision of the future, subways would eventually reach the city's geographical limits in all of the boroughs. The system was not to be built at one time, but rather over a 25 to 30-year period. Commissioner Delaney stated:

> *"It is not intended that the comprehensive scheme as laid out in its entirety will be undertaken at once, but that it should be proceeded with graduality and continuously. For the 25-year program, based on current utilization of tracks and present type of equipment, the proposed additional trunk lines would be able to develop a capacity of 5,000,000 passengers per year."* (Ref. 27, pg 1)

The plan also anticipated the replacement of *all* surface public transportation in the city with elevated or subway lines. This massive plan included:

- An 8-track, double-deck trunk line under Eighth Ave. and Amsterdam Ave. Only the first four tracks would be constructed initially. Connecting branches would be built at both the northern and southern ends of the trunk line.

- A 6-track trunk line under Madison Ave., which would relieve congestion on the Lexington Avenue line.
- A crosstown subway between Brooklyn and Queens linking the Astoria Line directly with the Brighton elevated in downtown Brooklyn.
- An extension of the Brooklyn 4th Avenue line through a new tunnel under the Narrows to Staten Island, with potential continuation through Staten Island on a route to be determined. The potential for multiple connections to Staten Island was included in later stages of the plan.
- A set of river-to-river crosstown subways on 14th Street, 42nd Street, and 57th Street. Some of these would be in the form of underground "moving sidewalks," while some would connect with new rapid transit lines.
- New rapid transit lines in Brooklyn, Queens, and the Bronx and extension of existing lines to the city line(s).

It is hard to imagine the overall extent of the proposed plan, which would have added 830 miles of track to the system and cost an estimated $360,000,000. The *New York Times* published a system map (29), taken from the TCC brochure created for describing the plan.

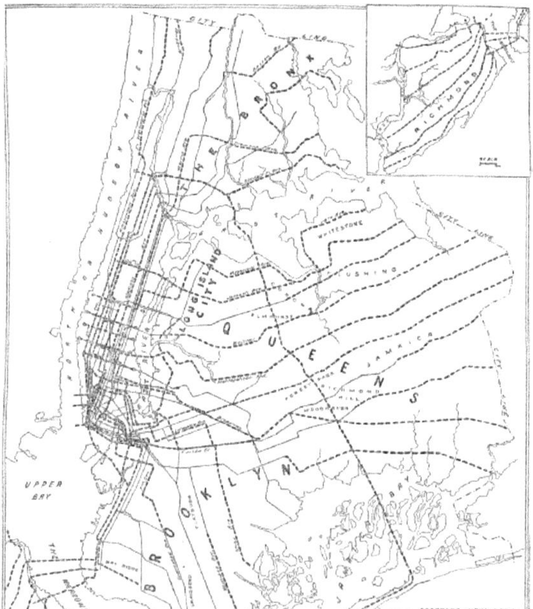

Fig. 10.2 Daniel Turner's TCC rapid transit expansion plan

10.3.2 The Transit Commission (TC)

Shortly after the TCC released Turner's massive rapid transit plan, the politics of the subway took another unexpected turn. In 1920, Alfred E. Smith was defeated in his bid for reelection as Governor to Republican Nathan L. Miller. Miller submitted legislation to once again change the power structure over NYC transit systems.

Soon after taking office, Governor Miller proposed a new commission to replace the TCC. The new Transit Commission would be appointed by the Governor, but members would have to be residents of NYC. The new commission would have broad powers, including the ability to set fares (30). The TC would seek city approval for any plan, but would be empowered to implement plans even in the case of city rejection (31).

The proposed changes infuriated Mayor Hylan and many Democrats. A heated debate raged for several months. The legislation was finally passed by the legislature on March 16, 1920 (32). Debate continued for several more weeks. The Governor finally signed the legislation into law on March 31, 1920 (33).

Finding candidates to fill the commission was another problem. The Governor wanted George McAneny to head the commission, but McAneny was initially cool to the idea, noting that his participation in the Dual Contracts was likely to be a significant political impediment (34). On April 17, the Transit Commission was introduced. McAneny did accept the chairmanship; the other two members were Leroy T. Harkness and Major General John F O'Ryan (35). In an attempt to mollify Mayor Hylan, two of the commissioners (McAneny and O'Ryan) were Democrats. The approach didn't work, however, as Hylan already detested McAneny, and took on opposition to the TC with his usual missionary zeal.

In an editorial, the *New York Times* lauded the Governor's selections (36):

> *"Governor Miller has lived up to his promise to choose, as Rapid Transit Commissioners for this city, men of special competence and unchallenged integrity. That they were also selected without regard to party considerations stands unquestioned. Two of the three, Mr. McAneny and Mr. Harkness, have the advantage through their previous connection with subway extensions and the dual contract between the companies and the municipality, of a wide knowledge of the details of the traction problem in New York. This fact will, no doubt, be flung in their faces by the Hylanites, in whose minds the possession of information is equivalent to treason, stratagems and spoils."* (Ref. 36, pg 9)

Mayor Hylan was having none of it, however. He ordered his corporation council, John P. O'Brien, to file for an injunction against the implementation of the Transit Commission. In court, the city argued that the legislation creating the TC was a violation of the NYS Constitution, particularly the "home rule" section under which the city exercised many powers normally undertaken by the state. The city specifically questioned the power of the TC to change or reverse contracts previously entered into by the city and the traction companies, and questioned the right of the TC to proceed with plans when the city formally indicates its disapproval (37). On May 17, 1921, the court denied the city's request for an injunction, and the TC was able to begin its work (38). The city continued to fight the TC through appeals, but did not prevail.

On May 12, 1922, the Transit Commission released its plan for new rapid transit lines (39, 40). The plan was more limited than that of the TCC, and was expected to cost $218,000,000. It included:

- Extending the IRT/BMT Corona Line to Flushing;
- Extension of the 42nd Street crosstown subway (the present shuttle) with moving platform equipment to meet all present and future north-south subways and elevated lines in Manhattan;
- A tunnel connecting the 4th Avenue Subway to Staten Island, where it would connect with existing steam and trolley lines.
- Extending the BMT Broadway Line from 57th Street and 7th Avenue to 155th Street and 7th Avenue;
- A crosstown Brooklyn Line from Queensborough Plaza to connect with the Brighton line at Franklin Ave. and Fulton Street in Brooklyn;
- A new tunnel under the East River to connect the Fulton Street El in Brooklyn with the Broadway-Fourth Ave. line at City Hall station in Manhattan.
- A new four-track trunk line from Washington Heights to lower Manhattan following Eighth Avenue and Amsterdam Avenue.

The system was expected to be built within six years. The TC plan was more like the dual contracts than the previous TCC plan in that it emphasized integration and connectivity with the existing rapid transit system.

Two days after release of the TC plan, Mayor Hylan announced his firm opposition to any or all of its aspects (41). The Mayor announced that he would fight the plan "through every court in the land," and that he would never allow any contract that included either the IRT or the BRT.

McAneny and Hylan were operating from a different set of objectives. McAneny viewed the subway as an integrated system that needed to be designed to best meet the transportation needs of the increasingly-dispersed New York population. Hylan wanted to build a separate system that would compete with the IRT and BRT, with the ultimate goal of having the city buy the two companies out (at a favorable price) and "unify" the system under full municipal control and operation. Hylan had some allies in his fight against the plan. The Queens Borough President was opposed to the plan due to its "neglect of Queens" (41), and the Manhattan Borough President didn't like the route of the proposed new trunk line.

10.3.3 The Hylan Plan

On August 28, 1922, Mayor Hylan released his own plan for a new municipal subway system, "independent" of the IRT and BRT. The event was somewhat unusual and understated. While Hylan had been publicly ranting over subway plans for a number of years, and pushing his concept of a city-owned and operated subway, he was not present when his own plan was released. His secretary, John F. Sinnott, also his son-in-law, presided over the official release. The Mayor was at Saratoga Springs with William Randolph Hearst and the elite of New York for the opening of the Saratoga horse-racing season. His son-in-law related the Mayor's regrets that he was unable to attend the release, after which one reporter (in an oft-quoted quip) said "I wish I couldn't be here too!" Despite the nature of the event, the plan that was released laid the groundwork for what would become the Independent Subway, and put the issue of "unification" formally on the table for the first time (42, 43). The plan included:

- "Recapture" of the existing IRT and BRT systems, under the provisions laid out in the Dual Contracts;
- A four-track west side trunk line in Manhattan between the uppermost part of Washington Heights under Broadway, Amsterdam Ave., and 10th Avenue to the Battery and Brooklyn where it would link with existing lines;
- A four-track eastside trunk line from 181st Street and Third Avenue in the Bronx through First Avenue and the lower East Side to Park Row and Brooklyn. At 10th Street two, two-track branches would split, one via 3rd Avenue and the Bowery into Brooklyn, and one to a loop near City Hall. Two branches would also emanate from the Harlem River, one to Fordham Rd. and Southern Blvd., and the other to Fordham Rd. and Webster Avenue, where it would join the current IRT White Plains line to 241st Street;
- A line from the west end of 125th Street crosstown, extending under the East River, connecting to the Astoria Line in Queens;
- A new line from the Hunter's Point area across Queens and Brooklyn to Lafayette Avenue, where it would split into two branches, one continuing along Lafayette Avenue that would eventually link with the First Avenue Subway, the other to Flatbush Avenue and Franklin Avenue;
- A new four-track line from Brooklyn Borough Hall via the Lafayette Avenue subway to Bedford Avenue, then to Broadway then to Jamaica. This line would have, for a portion of its route, run directly under the Broadway Elevated, competing directly with the BRT;
- A new IRT branch from the Eastern Parkway line onto Utica Avenue to Flatlands Avenue;
- A new four-track subway under Flatbush Avenue in Brooklyn to Nostrand Avenue, to Emmons Avenue in Coney Island, then to Surf Avenue. A branch would reach Floyd Bennett Field under Flatbush Avenue;
- Extension of the 14th Street Line to the Jamaica Elevated at 121st Street in Queens;
- A new line running from Prospect Avenue to Ft. Hamilton Parkway in Brooklyn, terminating at 90th Street;
- Extension of the Fourth Avenue line south to 96th Street;
- Extension of the Fourth Avenue line east to the West End Elevated;
- A new two-track line from Fourth Avenue and 67th Street to Staten Island;
- Extension of the IRT New Lots Line to Lefferts Avenue;
- Extension of the IRT Corona Line to Bell Blvd. via Main Street and Northern Blvd. (then Broadway);
- A branch from the IRT Corona Line to Jamaica.

The *New York Times* published a map of the Mayor's proposed system (43):

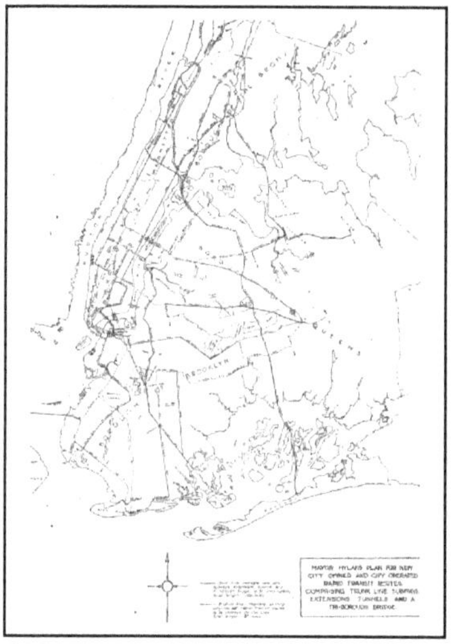

Fig. 10.3 The Hylan Plan for a city-operated rapid transit system

Even Hylan, however, realized that he did not have the power to implement his plan. As reported in the *New York Tribune* (44):

> *"The Mayor grudgingly concedes that the Transit Commission cannot be waved aside by the mere filing of his report. The Legislature must intervene to set up a Hylanized control of the transit situation. Therein lies the Mayor's safety and also his opportunity for political martyrdom. The Legislature has had some experience with him and he has had some experience with it. Each can say when they clash 'My spear knows no brother'. "* (Ref. 44, pg 8)

Mayor Hylan needed to have the Commission abolished in favor of a city-run group. In the meantime, the NYC Board of Estimate, which also had to approve expenditures for subway construction, made several attempts to bridge the gap between the two plans, and actually approved several common parts, including the 8th Avenue Subway, but did not release them for bid. It also began to consider a new 6th Avenue trunk subway to replace the 6th Avenue Elevated. This marked the beginning of the era in which the environmental problems of the elevateds would spell their eventual doom in Manhattan.

In 1922, Alfred E. Smith regained the governorship. With two Tammany men in place, it might be assumed that Hylan would have little difficulty in getting the required legislation. However, his relationship with Smith was tepid at best, and adversarial at worst. In the election of 1922, Smith did align himself with Hylan in opposing the Transit Commission and supporting the five-cent fare. While he won the race for governor, the legislature remained in Republican hands, further complicating the situation. Because of this, the discussion of the two transit plans continued into 1924 without resolution.

10.3.4 The Adler Act and the Board of Transportation

In 1924, Republican leaders sponsored a compromise bill, called the Adler Act (after Assembly Majority Leader Simon Adler). The legislation left the Transit Commission in place to monitor the existing rapid transit system of the IRT and BMT. It also transferred power to develop new rapid transit lines to a new city-based organization, namely, the Board of Transportation. The Board would have most of its members appointed by the Mayor, and would report to him. It gave the city the power to build its own subways (45). On May 2, 1924 the bill was signed into law by Governor Smith (46).

The bill also contained two aspects that were not to the city's (and Hylan's) liking. One prohibited the Board of Transportation from "recapturing" any parts of the existing system. The other set the fare at five cents for a period of only three years, after which the fare had to be set to insure the self-sufficiency of the system.

Mayor Hylan appointed John H. Delaney, the former chair of the TCC, to chair the new Board of Transportation (47). On December 9, 1924, the Board put forth its own plan for construction of new subways (48). Drawing on previous plans, it had several distinctive features. It would include two trunk lines in Manhattan, one under 8th Avenue, and another under 6th Avenue. It eliminated all mention of extending existing IRT and BMT lines to the City boundaries in Brooklyn and Queens. Their plan included:

- An 8th Avenue trunk line from 207th Street to Chambers Street, then under the East River to Brooklyn. A branch line would serve the Bronx under Grand Concourse to 205th Street;
- A 6 Avenue trunk line from 53rd Street to West 4th Street, where it would turn east to 2nd Avenue, continuing under the East River to Brooklyn, where it would connect with the 8th Avenue route;
- A line running along 53rd Street under the East River to Queens Plaza, then under Queens Blvd. and Hillside Ave. to Jamaica;
- A Brooklyn route running under Fulton Street to East New York.
- Another Brooklyn route going south from downtown Brooklyn to Red Hook, skirting the edge of Prospect Park.
- A Brooklyn-Queens crosstown route from Queens Plaza to Fulton Street.

With a few minor differences, this plan represents the Independent Subway as it was built by 1940.

There were additional hurdles before the Independent could be built. The 1924 law required that the Board of Estimate approve lines prior to construction. They were required to do a financial analysis of the plan. The plan would cost $674 million to build and equip; the city planned to issue bonds to raise that sum.

The results of the financial analysis were not encouraging: to comply with the self-sufficiency requirement of the law, the fare on the Independent would have to be set at 8 cents. To keep the fare at 5 cents for the first three years of operation (also required by the law), taxes would have to be raised by $187,000,000, and property assessments along the lines increased by $138,000,000. Moreover, in building the subway, the city would virtually exhaust its available credit, causing a significant delay in other capital projects, such as schools, docks, hospitals, sewers, bridges, and other needs. To build the system, the city needed to have a significant increase in its debt limit (49).

The Eighth Avenue trunk line, because it had been approved prior the 1924 law, could be started before the debt limit issue was resolved. Thus, on March 14, 1925, Mayor Hylan formally broke ground on the Washington Heights portion of the line. The ceremony was held in John Hancock Park to emphasize its role as a "people's subway." In tune with his fiery antitraction company positions, Hylan used the occasion to again blast the now BMT, IRT and Transit Commission (49).

In early 1926, the legislature approved a $300,000,000 increase in the city's debt limit, enabling other parts of the system to get underway.

10.4 The Independent Subway System (50, 51)

By the time construction was underway, Mayor Hylan would be serving his last few bitter months in office. In the summer of 1924, an accident occurred on the BMT, killing one person and injuring thirty more. Hylan seized upon this unfortunate event to mount yet another scathing attack on the private railway companies. In a communication to Governor Smith, he demanded the removal of the Transit Commissioners, namely, McAneny, Leroy Harkness, and John O'Ryan, accusing them of dereliction of duty and gross misconduct. He leveled other charges of corruption against Harkness as well. Governor Smith, who by that time had gotten his fill of Hylan, authorized a general investigation of New York's rapid transit situation. In reality, the investigation merely highlighted the unreasonable behavior of the Mayor in transit matters. His charges against Harkness proved to be baseless, and no one could connect any of the commissioners to the accident (49).

Tammany Hall decided to run another candidate against Hylan in the primary. Their choice, Jimmy Walker, would go on to make to the graft and dishonesty of Boss Tweed look controlled. At the time, however, Walker cut a dashing and fashionable figure, and was a gifted orator, none of the things that defined Hylan. Walker won an easy victory and went on become Mayor. At the moment of his greatest achievement, just as the Independent Subway was about to become a reality, Hylan left office a bitter man.

The Independent Subway was, in many ways, vastly different from its predecessors of the IRT and Dual Contracts. The politics and reality of the subway had changed. The Independent was not planned to cause further dispersion of New York's population, and it didn't. Because it was to be a separate entity competing with the IRT and BMT, it was built primarily into older areas that were already developed to insure enough ridership to meet its sufficiency requirement.

Indeed, the task of causing increased development of outer areas, both within and outside the city, had now passed to the automobile. Robert Moses was early in his career as a Master Builder as the Independent got underway. He would eventually become the dominant force in New York City and New York State planning. Just as Hylan had an almost-religious zeal for his "people's subway," Moses had a fundamental belief that the future of transportation was the automobile, and only the automobile. His career as an advocate for road construction and expansion was matched by his opposition to public transportation.

The Independent also included the first subways built primarily as replacements for aging elevated lines. It would lead directly to the removal of the 6th and 9th Avenue Els in Manhattan and the Fulton Street El in Brooklyn.

The Independent left unresolved a major bottleneck in the Brooklyn rapid transit system, namely, DeKalb Avenue. In 1924, the following BMT lines had 16 tracks converging at DeKalb Avenue, that is, Brighton, Culver, and West End Els, the Sea Beach line and the Fourth Avenue Subway. Unfortunately, only six tracks left DeKalb; two via tunnel to the Broadway Subway in Manhattan and four via the Manhattan Bridge. Although there have been some improvements over time, DeKalb Avenue remains a major system bottleneck today.

Nevertheless, the Independent represented a departure from previous rapid transit projects in a significant way:

- Construction of the Independent used all of the latest techniques, and was the first that did not depend almost completely on manual labor. Nevertheless, it did provide employment for almost 10,000 people at a time when the general economy was in trouble, particularly after the stock market crash in 1929.
- In building four-track lines with express and local service, it adopted an approach in which the express and local routings could be different. This is particularly clear in Queens, where the routing of express tunnels followed the straightest line between express stations, while local tunnels followed a more meandering route to service the greater number of local stations.
- Rolling stock for the Independent was state-of-the-art, incorporating the latest operational features, with the exception of a few features that had been patented by the BMT. Further, for almost 10 years, the fundamental design of Independent cars would remain standardized. The car orders started a numbering sequence that is still used. The first cars ordered are referred to as the "R-1" cars. Subsequent orders over a nine-year period were numbered as the R-2 through R-9 cars. The basic design of the car did not change until the R-10 models were ordered many years later.

One of the efficiencies adopted in the construction process was the provision of ramps connecting the street to the construction level. This allowed trucks to directly haul earth, rock, and other removed material from the site without requiring a separate system to move the material to the surface, where it would be loaded onto trucks. It also provided some direct access for material to be delivered.

(a) Car no. 100, the first R-1 delivered for the Independent Subway

(b) Interior of an R-1 car

Fig. 10.4 Standard subway car for the Independent

Fig. 10.5 Construction site with truck ramps for ease of materials movement

Like the IRT before it, the upper Manhattan portion of the 8th Avenue Subway involved four stations that were deep below the surface, requiring elevators to provide access. The 181st Station had a unique design. While the station was over 120 ft below the surface directly above it, the ground was sharply sloped, and a nearly-horizontal tunnel of 264 ft allowed direct access at street level as well. The station could be accessed by walking through the long tunnel from Overlook Terrace, or by elevator from Ft. Washington Avenue.

(a) Even though the 181st St. station is 122 ft under Fort Washington Ave., it is only 12 ft below Overlook Terrace at 184th Street and Bennett Ave., allowing a virtually horizontal access from that point

(b) 181st Street station entry from 184th Street and Bennett Ave

(c) Elevator Entrance to 181st Street Station

Fig. 10.6 A unique station at 181st Street on the Independent

Because of the complexity of the Independent Subway, and the fact that its lines were subject to several different approval sequences, they did not all start construction at the same time. The 8th Avenue Subway was the first to have work started, and was the first to open. It had gotten a head start because the NYC Board of Estimate had approved it prior to the 1924 act establishing the Board of Transportation.

The most difficult line to construct was the 6th Avenue trunk line between 8th and 53rd Streets. The major difficulty is that it would be built below the existing 6th Avenue Elevated, which would be operating throughout the construction period. It would pass under the BMT's 14th Street and Broadway lines, and over the Catskill Aqueduct, the IRT's Queens connection, Long Island Railroad tunnels and the Hudson and Manhattan Railroad's tunnel and stations under 6th Avenue. In addition, work was delayed while a complicated negotiation with a major streetcar operator on 6th Avenue to discontinue service was concluded. An agreement was reached whereby service was discontinued during construction, after which the city would restore the streetcar tracks, or give the company a bus franchise.

10.5 Opening the System

The 8th Avenue Subway was the first to open its doors to the public at midnight on September 10, 1932. The initial system included 57 route-miles and, at the time, the longest station platform in the world at 42nd Street and 8th Avenue, stretching 1200 ft. Unlike the fanfare that accompanied the opening of the IRT in 1904, there was virtually no pomp and circumstance accompanying the beginning of service. Mayor Hylan was long gone, and so was his successor, Jimmy Walker. Walker had resigned months before in the midst of an enormous scandal in which his personal participation in bribes and other illegal activities had been revealed.

The sitting mayor was Joseph V. McKee, who had been appointed until a special election could be held to serve the rest of Walker's term. He was investigating a deadly steamboat explosion that killed over 35 people, and did not have time to participate. There was no official "inaugural" run as trains had been running for weeks testing the system. At the stroke of midnight, all stations opened simultaneously, as a full schedule of trains was already in operation.

As there had been no bids for a franchise to operate the Independent, it opened under full city operation. There was some reluctance by the city to open, but a lawsuit forced the issue, and the opening went forward. The only ranking official present at the opening was John Delaney, chair of the Board of Transportation.

Fig. 10.7 Crowds awaiting the opening of the Independent on September 9, 1932

The Independent grew in fits and starts due to the financial difficulties facing the city during the depression. Most sections were completed by 1937, but the critical 6th Avenue Subway would not be opened (and then only to local service) until 1940. Table 10.1 shows the sequence in which various segments of the Independent Subway were opened.

Table 10.1 Opening of the Independent Subway by section

Date	Segment
September 10, 1932	Eight Avenue service inaugurated
February 1, 1933	Brooklyn service extended to Jay Street
March 20, 1933	Brooklyn service extended to Bergen Street
July 1, 1933	Bronx service extended to 205th Street and Bainbridge Avenue via Grand Concourse
August 19, 1933	Queens service extended to Roosevelt Avenue
October 7, 1933	Brooklyn service extended to Church Avenue
January 1, 1936	Connection between West 4th Street and East Broadway was opened
April 9, 1936	Brooklyn service extended from East Broadway to Jay Street; Fulton Street Line extended from Jay Street to Rockaway Avenue; Court Street Shuttle placed in service.
December 30, 1936	Queens service extended to Union Turnpike
April 24, 1937	Queens service extended to 169th Street
July 1, 1937	Brooklyn-Queens Crosstown service was opened between Nassau Avenue and Hoyt-Schermerhorn Street
April 30, 1939	World's Fair Service inaugurated; it ended on October 28, 1940
December 15, 1940	Sixth Avenue service initiated

Figure 10.8 shows a map of the Independent Subway as of mid-1936. It does not include the 6th Avenue Subway, which was not yet completed, or other sections which opened later.

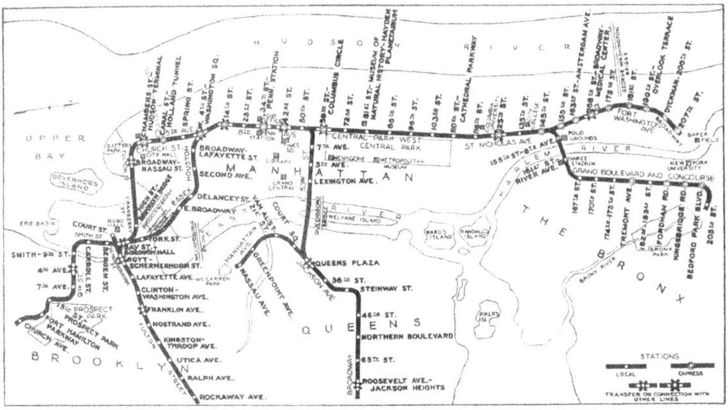

Fig. 10.8 The Independent Subway system in mid-1936

There would be other additions in the years that followed, but the bulk of the system was complete by 1940. The Independent would begin running a shuttle along the Dyre Avenue line, acquired from the NYW&B Railroad in 1941, but it would eventually become part of the IRT in 1957. Service to the Rockaways would begin in 1956 over former LIRR trackage.

Since 1940, there have been only a few notable additions to New York's massive rail rapid transit system. After completion of the Independent Subway in 1940, the era of massive rapid transit development in New York had ended.

10.6 The Train to the World's Fair (52)

In the darkest days of the Great Depression, in 1934, an engineer named Joseph Shadgen contacted Edward F. Roosevelt, a promoter and distant relative of Franklin Delano Roosevelt, with the idea of staging a World's Fair in New York City. The city had never hosted such an event, and Shadgen thought it might provide a boost to the city's economy and add luster to its reputation.

Roosevelt arranged a meeting with the newly-formed Regional Plan Association headed by none other than George McAneny. The idea gained support, and soon Robert Moses would be knee-deep in planning the event. Moses had to be involved, as he was the Parks Commissioner under New York's new mayor, Fiorello LaGuardia.

The Fair was to be held in Flushing Meadow Park, a veritable swamp at the time. It was chosen primarily because of its accessibility. The Flushing line, jointly operated by the IRT and BMT, already had a station in the middle of the park at Willet's Point Blvd. The new IND could provide direct access if a two-mile spur were built from the entrance to its yards to the Fair grounds. Robert Moses' Grand Central Parkway provided almost direct automobile access, and ferries could dock at a nearby marina.

Providing rapid transit access to the Fair was critical, and despite a number of physical and financial problems, all parties would wind up cooperating in the effort. Providing access involved three major projects:

- Complete renovation and expansion of the existing station at Willets Point Boulevard.
- Purchase of an additional 50 railcars to provide additional service on the IRT/BMT Flushing Line.
- Construction of a temporary two-mile spur connecting the Independent Subway to a new terminal platform at the World's Fair.

In the overall history of New York City transit, the World's Fair arrangements are but a footnote. They did, however, demonstrate what could be done when all of the private and public players agree to a goal and work in harmony towards achieving it. The IND spur would be temporary indeed, and would be removed almost immediately when World's Fair closed after the 1940 season. The improvements to Willets Point station, of course, remained, and are used today to provide access to Flushing Meadow Park, Shea Stadium (formerly) and Citifield (now), and the Tennis Center.

Fig. 10.9 Willets Point Station before the World's Fair

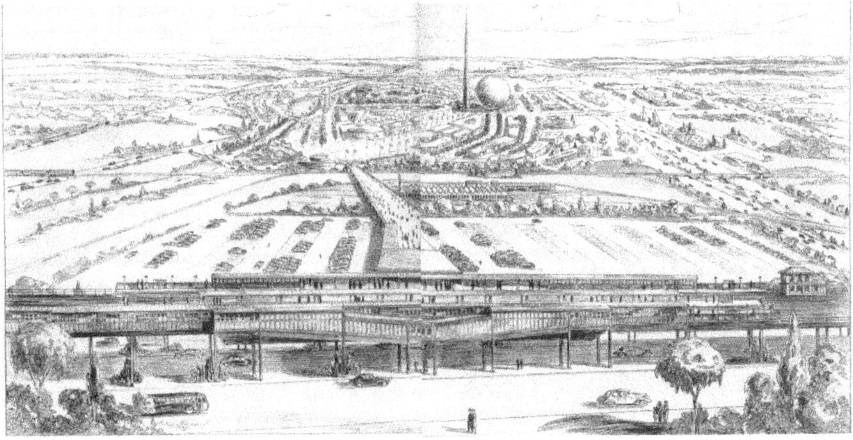

Fig. 10.10 Artists conception of the new Willets Point Station for the World's Fair

Fig. 10.11 New IRT cars for the 1939 World's Fair

The new World's Fair rolling stock was hardly futuristic. The IRT was in receivership, and bought cars that were as basic as possible, yet clearly new and attractive. It was the first time that the IRT abandoned its door layout, which featured two vestibules at the each end of the car, with one center door. It adopted a three-door layout within the car body, similar to recent BMT car designs. The IND bought no special cars for the World's Fair, as its own fleet was virtually brand new, and featured a four-door layout within the car body.

10.7 A New Day Dawns

In the late 1920s, the Regional Plan Association released its landmark Regional Plan of New York and Its Environs (1929–1931). It introduced a vast arterial and circumferential highway system, complete with bridges and tunnels running across the Hudson and East Rivers, and parkways going through the city and the suburbs. It was not silent on rapid transit, however. It included a transit plan put together by Daniel Turner, and it incorporated many elements of his earlier TCC plan. It called for 297 miles of new trackage, replacement of the elevateds, a new subway on the upper eastside, and extension of all rapid transit lines to the boundaries of the city.

The plan is best remembered for its prophetic vision of a vast highway system because that is only part of the plan that was fulfilled. Within a few short years, Robert Moses would be acquiring the powers to oversee and implement highways as the means of dispersing urban population. There would be little money left for investment in anything else, including the rapid transit system.

The age of using rapid transit to shape the city's physical development and future was over. Yet, that view would be incredibly short-sighted, as New York could not (and cannot) function without its rapid transit system. As the nation and the city went wild for the automobile, its rapid transit system was left to fend for itself, insuring a downward spiral that would extend through the 1970s and early 1980s.

References

1. Tammany Still Seeks a Man for Mayor. New York Times, 4 (August 14, 1917)
2. Pick Judge Hylan to Head Ticket Against Mitchel. New York Times, 1 (August 15, 1917)
3. Hylan Picked by Tammany for Mayor. New York Tribune, 1 (August 15, 1917)
4. Triangle Fight for N.Y. Mayor Splits Tammany. Chicago Daily Tribune, 7 (August 24, 1917)
5. For Mayor of New York. Outlook, 639 (August 29, 1917)
6. Hearst Declines to Run for Mayor; Supports Hylan. New York Times, 1 (September 1, 1917)
7. Hylan for Suffrage, Just Like Mitchel, He Asserts. New York Tribune, 9 (October 12, 1917)
8. City's Most Complex Election. New York Times (November 4, 1917)
9. Hylan Victor Over Mitchel by 145,000. New York Tribune, 1 (November 7, 1917)
10. Says Bankers Run Nation. The Washington Post, 3 (March 27, 1922)

11. Hylan Shies Hat Toward Next Presidential Ring. New York Tribune, 1 (March 27, 1922)
12. Scores Killed or Maimed in Brighton Tunnel Wreck. New York Times, 1 (November 2, 1918)
13. Two Inquiries to Fix Blame For "L" Wreck; 89 Are Dead. New York Tribune, 1 (November 3, 1918)
14. The Mayor to Begin B.R.T. Inquiry Today. New York Times, 24 (November 4, 1918)
15. Mayor Fails to Find 'Doe' or 'Roe' In Wreck Inquiry. New York Tribune, 16 (1918)
16. Wreck Jury Indicts Five B.R.T. Heads. New York Tribune, 16 (December 21, 1918)
17. Find Blewitt Not Guilty in B.R.T. Case. New York Tribune, 1 (March 19, 1919)
18. Trial of Motorman in B.R.T. Wreck Started. New York Tribune, 8
19. Lewis Acquitted of Manslaughter in B.R.T. Wreck. New York Tribune, 6 (April 5, 1919)
20. Lewis Hints That Hylan Mind His Own Affairs. New York Tribune, 6 (April 12, 1919)
21. Menden Acquitted in B.R.T. Trial. New York Times, 3 (January 27, 1920)
22. J. J. Dempsey Acquitted. New York Times, 22 (April 15, 1920)
23. Malbone Cases Are Dropped. New York Tribune, 7 (January 18, 1921)
24. Legislature Ready to Adjourn Today After Final Rush. New York Times, 1 (April 15, 1919)
25. Makes Delaney Transit Builder. New York Times, 17 (May 27, 1919)
26. City May Take Over Interborough Lines. The Christian Science Monitor, 6 (March 18, 1920)
27. $350,000,000 Plan For Subway Routes Has Been Completed. New York Times, 1 (September 26, 1920)
28. Plans to Double Subways for $350,000,000. New York Tribune, 16 (September 26, 1920)
29. City's Growth Discounted in Plans for Adding 830 Miles of Track to Rapid Transit Systems. New York Times, 114 (October 3, 1920)
30. Miller to Tell Plan on Traction Monday. New York Times, 4 (January 22, 1921)
31. Full Power Over Fares and Franchises Given to the New Transit Commission Bill Made Public by Gov. Miller. New York Times, 1 (February 15, 1921)
32. Miller's Transit Bill Put Through Senate, 33 to 18. New York Times, 1 (March 17, 1921)
33. Traction Bill Signed by Gov. Miller. New York Tribune, 1 (March 31, 1921)
34. Miller in Quandry Over Transit Heads. New York Times, 30 (April 4, 1921)
35. Transit Board Named, McAneny is Chairman, Prendergast Heads P.S.C. New York Tribune, 1 (April 17, 1921)
36. The New Transit Commission. New York Times, 9 (April 18, 1921)
37. Transit Injunction Argued in Court. Wall Street Journal, 6 (April 28, 1921)
38. Court Denies City Injunction Against Transit Commission. New York Tribune, 1 (May 18, 1921)
39. Seven More Lines in Transit Plan. New York Times, 1 (May 12, 1922)
40. Transit Board Announces $218,000,000 Plan. Wall Street Journal, 6 (May 12, 1922)
41. Hylan Out to Block New Subway Lines. New York Times, 12 (May 14, 1922)
42. Hylan Announces His $600,000,000 Plan for Transit. New York Times, 1 (August 28, 1922)
43. Mayor Explains His Plan. New York Times, 2 (August 28, 1922)
44. Mr. Hylan's Vision. New York Tribune, 8 (August 2, 1922)

45. Republicans Offer Home Rule; City to Pick Board. New York Times, 1 (February 24, 1924)
46. New Transit Plan Approved by Smith. New York Times, 2 (May 3, 1924)
47. Hylan Swears In His Subway Board. New York Times, 21 (July 22, 1924)
48. Hylan Subway Plan Links Four Boroughs at $450,000,000 Cost. New York Times, 1 (December 10, 1924)
49. Hood, C.: 722 Miles: The Building of the Subways and How They Transformed New York. The Johns Hopkins University Press, Baltimore (1993)
50. Kramer, F.: Building the Independent Subway. Quadrant Press Inc., New York (1990)
51. Feinman, M.: History of the Independent Subway, `http://www.nycsubway.org`
52. Kramer, F.: Subway to the World's Fair. Bells & Whistles, Westfield (1991)

Figure Sources

10.1 *Courtesy of* Wikimedia Commons.
10.2 *Originally published in* "City's Growth Discounted in Plans for Adding 830 Miles of Track to Rapid Transit Systems," *New York Times*, October 3, 1920, pg 114.
10.3 *Originally published in* "Hylan Announces Plan for Transit," *New York Times*, August 28, 1922, pg 3.
10.4 *Courtesy of* New York Transit Museum.
10.5 *Courtesy of* Frederick A. Kramer, *from* Kramer, *Building the Independent Subway*, Quadrant Press Inc., New York, NY, 1990, pg 20.
10.6 *Courtesy of* Frederick A. Kramer, *from* Kramer, *Building the Independent Subway*, Quadrant Press Inc., New York, NY, 1990, pg 28.
10.7 *Courtesy of* Frederick A. Kramer, *from* Kramer, *Building the Independent Subway*, Quadrant Press Inc., New York, NY, 1990, pg 43.
10.8 *Courtesy of* Frederick A. Kramer, *from* Kramer, *Building the Independent Subway*, Quadrant Press Inc., New York, NY, 1990, pg 25.
10.9 *Courtesy of* Frederick A. Kramer, *from* Kramer, *Subway to the World's Fair*, Bells & Whistles, Westfield NJ, 1991, pg 9.
10.10 *Courtesy of* Frederick A. Kramer, *from* Kramer, *Subway to the World's Fair*, Bells & Whistles, Westfield NJ, 1991, pg 4–5.
10.11 *Courtesy of* Frederick A. Kramer, *from* Kramer, *Subway to the World's Fair*, Bells & Whistles, Westfield NJ, 1991, pg 17.

Chapter 11
The Staten Island Rapid Transit

Staten Island is unique among the five boroughs of New York City. It is more isolated than the other five boroughs, and until the opening of the Verrazano Bridge on November 21, 1964, it was only directly connected to Manhattan and the rest of New York City by ferry services.

While Staten Island, despite many plans to do so, remains unconnected to the rest of New York via rapid transit (express bus services do exist), it has had its own "railroad" for many years. This chapter explores the background and history of the Staten Island Rapid Transit, and its impact on the island itself.

11.1 Introduction

The name is generally credited to Hendrik Hudson, who named the island "Staten Eylant," Dutch for "the island of Staten" (1). When it was sold to the Duke of York in 1670, it was referred to "as the commodious state and richest land in America" (1). A June 15, 1860 article in the *New York Daily Tribune* (2) put Staten Island into a group of very well-known recreational havens:

> *"The fine summer weather we are now enjoying tells favorably upon the various watering places in this vicinity. Jonce's Woods, Elm Park, Central Park, Elysian Fields, and Staten Island are visited by many thousands of New Yorkers daily."* (Ref. 2, pg 8)

The island became a magnet for recreation, and for summer and permanent residences of some of Manhattan's wealthiest families. To handle the desire for travel from Manhattan to Staten Island, ferry services existed from as early as 1755 when Martin Duckett introduced a semiregular service (1). Cornelius Vanderbilt operated one of the earliest ferries, using vessels named "Dread" and "Nautilus." Initiated in 1810, his Staten Island ferry service was his first financial venture, and marked the beginning of his financial empire which would eventually make him one of the richest men in the world. Vanderbilt became one of the island's earliest and most influential residents.

Staten Island's relative isolation has had a dramatic affect on its development, and it remains to this day a very different place than other four boroughs of New York. Residential densities are far lower than in Manhattan, Brooklyn, the Bronx, or Queens. Significant parts of the borough are best described as "suburban,"

R.P. Roess and G. Sansone: The Wheels That Drove New York, STTT 1, pp. 223–248.
springerlink.com

looking more like Nassau and Suffolk Counties on Long Island than a part of New York City. More remote portions of the borough can still be thought of as "rural."

Figure 11.1 shows a map of New York City and its five boroughs. Looking at the map (minus the lines designating NYC), one would be led to believe that Staten Island is part of New Jersey, as it hugs the NJ shoreline, separated only by the narrow Arthur Kill.

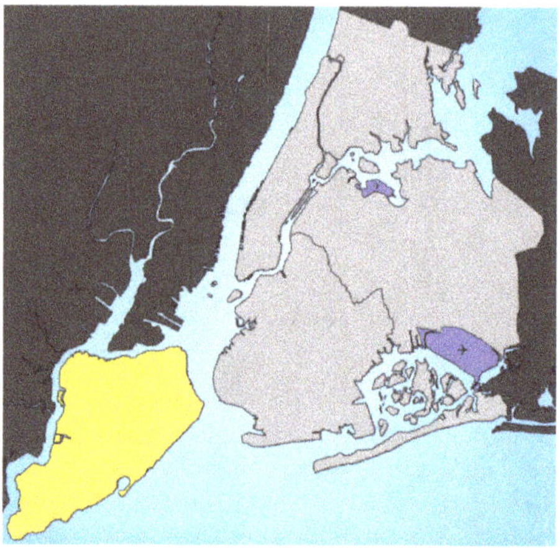

Fig. 11.1 Staten Island is closer to New Jersey than to the rest of New York City

The attachment of Staten Island to New York City is itself an accident of its history. When English rule occurred, the colonies of New York and New Jersey both laid claim to Staten Island. The Duke of York, appointed by the Crown to oversee both areas, declared that whoever could circumnavigate the island in a boat in less than 24 hours would be given jurisdiction. Captain Christopher Billop accomplished this on behalf of the New York colony in 1687, and New York was duly awarded jurisdiction. Shortly thereafter, the Duke named it "Richmond." To this day, both names survive: Staten Island as a borough of New York City; Richmond as a county in New York State.

In previous chapters, the history of omnibuses, horse streetcars, trolleys, the elevateds, and subways has been presented with very few references to Staten Island. In fact, Staten Island had several horse streetcar lines and a small system of trolleys. None, however, provided a direct connection to the other boroughs.

In the years leading up to the Dual Contracts, plans for a subway tunnel connecting the 4th Avenue Subway (BRT/BMT) in Brooklyn to St. George, Staten Island, were drawn up and included in many proposals for expansion of the rapid transit system. The cost, however, always resulted in this portion of the plan being dropped.

Staten Island, however, was not without rapid transit service. In fact, all of the plans for a connection to Brooklyn joined with the service that had existed independently on Staten Island for many years, that is, the Staten Island Rapid Transit system. The system began as a railroad, slowly developing into an operation that more resembled "rapid transit" over the years. Because the effectiveness of the rail service was strongly dependent upon its ability to link up with various ferry services to Manhattan (and elsewhere), the fortunes of the rail system and the ferries have been historically intertwined. Over its history, the railroad has had several operators, some of whom also operated ferries, and some of whom did not. The desirability of controlling both, however, has always been understood.

The Staten Island Rapid Transit system has had an impressive pedigree of owners and participants, including the Vanderbilts (Cornelius and Jacob), George Law (for whom the town of St. George is named), Erastus Wiman (who was known as the Duke of Staten Island), and Robert Garrett of the B&O Railroad. Like New York City's elevated and subway empire, the leading financiers and businessmen of the day were involved from the very beginning in the development of the railroad, the ferry system, and the island itself.

11.2 Early Days

The first charter to build a railroad along the east shore of Staten Island from Tottenville to the area around the future site of St. George was issued in 1836. Financing for the venture, however, could not be achieved, and the railroad was never started (3).

Cornelius Vanderbilt provided the financial backing for the first railroad on Staten Island, which opened on April 23, 1860 (4).

> *"The Staten Island Railroad was opened to the public, yesterday, as far as Ettingville, about eight miles from Vanderbilt's Landing. It is expected that it will be completed, in a month, to Tottenville – sixteen miles – where it will connect by ferry boat with the trains to Philadelphia, by the Camden and Amboy line."* (Ref. 4, pg 8)

The Tottenville extension went into service on June 2, 1860 (5). While ferry connections from Tottenville to New Jersey provided a link to the railroad to Philadelphia, Vanderbilt's Landing was nearby ferry slips providing service to Manhattan.

Initially, the service operated with only two locomotives. When the New Jersey Locomotive Works threatened to foreclose on the locomotives in 1861, Vanderbilt had his son, William, declared the receiver, allowing the service to continue (3).

The financial success of the railroad depended heavily on the coordination of the railroad and ferry operations. It was imperative that the few daily runs of the trains had to meet the infrequent ferries to and from Manhattan. This proved very difficult during the first few years of operation. The ferries serving Vanderbilt's

Landing were owned by George Law. An article in the *New York Times* in August of 1860 described Vanderbilt's attempt to begin a competing service (6):

> *"It is also in contemplation to establish a new ferry line to New York, to run in connection with the cars, and afford the public better accommodations than are now furnished by the miserable boats which ply between this city and the Island under the auspices of Mr. Law."* (Ref. 6, pg 5)

A franchise rights battle ensued, and Vanderbilt eventually had to sell his ferry service to Law (7):

> *"It appears that Commodore Vanderbilt has sold his ferry to George Law, who has secured a monopoly of the ferry rights to the eastern shore of the Island."* (Ref. 7, pg 8)

The article went on to indicate that the Staten Island Railroad timetable would be adjusted to link up with the ferries. The ferry sailings, however, were not very reliable, and establishing a regular linkage between the railroad and the ferries remained a problem. In March of 1864, Vanderbilt bought the ferries from George Law, bringing both the railroad and the ferries under common control for the first time (8).

11.3 The *Westfield* Disaster

On July 30, 1871, the boiler of the ferryboat *Westfield* exploded while loading passengers at the Whitehall Street terminal in Manhattan. The horrific accident, which occurred at approximately 1:30 PM, just before the loaded ferry was to leave the dock, caused extensive damage and many deaths and serious injuries (9). The ferry was owned by the Staten Island Railroad Company, and had been built in 1862 at a Brooklyn shipbuilding yard. The boiler that exploded was described as a "low-pressure boiler 24 feet in length, 10 feet diameter, with 12 feet front..." and was designed for a pressure of 25 psi (10).

Given the records of the day, the number of people on the vessel when the boiler exploded was variously reported as 800 (11), from 200 to 300 (12), and from 300 to 400 (13). Similarly, the number of killed and injured varied widely from report to report. An article in the *New York Tribune* on August 1, 1871 summarized some of the published numbers and commented on their variety (12):

> *"The Tribune, therefore, took great care to publish as correct a list of the sufferers and their number as could possibly be obtained. The names and addresses of all those identified were published, and although the sad list was longer and more numerous than that of other leading journals, it appears, unfortunately to be more correct. The numbers given in the Herald were: 45 killed...and 78 wounded....; The Times: 82 killed...and 103 wounded; The World: 91 killed...and 208 wounded; The Tribune: 67 killed...and 121 wounded."* (Ref 12, pg 1)

Investigations and inquests were immediately launched to try and discover the cause of the accident, and who was to blame. The results of initial inspections were presented in the *New York Tribune* on August 2, 1871 (14):

"*1. The boiler was unsound, though it had only been recently inspected and pronounced perfectly safe.*
2. The safety valve was out of order.
3. The pressure of steam was in excess of the amount allowed by law, and
4. The engineer was not at his post." (Ref. 14, pg 4)

Testimony and published opinions, however, varied widely, depending upon who was giving the answers and who was asking them. An article in the *New York Tribune* on August 8, 1871, seemed to illustrate the frustration of the public (15):

> *"Thus far, the evidence elicited by the Board of Steamboat Inspectors has determined that the boiler of the ferry-boat Westfield was cracked – and that it wasn't cracked; that the iron was good – and that it was very bad; that the water was low – and that it wasn't low; that the engineer was neglecting his duty – and that he was giving usual and sufficient attention to his boiler; that the initial fracture at the moment of explosion occurred at a point opposite the lower range of the flues – and that it was at a point near the 'man-hole' on top; that steam was escaping through the safety valve – and that it wasn't; that the 'gas' did it – and that the 'gas' didn't.'* (Ref. 15, pg 4)

The article goes on to say that there was agreement only that the explosion was caused by "over-pressure," but points out that this conclusion meant very little, as the cause of the over-pressure was not detected. It decried the use of expert witnesses with connections to the steamboat industry, and finally concludes that the investigation would arrive at the usual verdict of "Death by the visitation of God."

Fig. 11.2 The Westfield Ferry Explosion

An article appearing in the *San Francisco Chronicle* on August 10, 1871, attacked the Vanderbilt family directly (16). The article points out that the ferry boats had long been a subject of complaint of Staten Islanders because of their "niggardly accommodations and unfitness for the service," and that the Vanderbilt family enterprise is a "closed corporation and monopoly." In fact, however, virtually all ferry services from Staten Island had been subjected to similar complaints over the years. The Vanderbilts were in no sense unique in this regard.

On August 17, the results of a coroner's inquest into the death of passenger Andrew Coyle and others was released (18):

> *"We find that the deceased, Andrew Coyle, and others, died through the rupture or explosion of the boiler of the ferry boat Westfield on July 10, 1871; that said explosion or rupture was caused by a flaw in the iron and by the negligence of Engineer Robinson in carrying an over-pressure of steam; that the company is responsible for the disaster, as the defect could have been detected if the Staten Island Ferry Company had had a competent superintendent, engineer, and mechanics in their employ, and is therefore, criminally negligent."*
>
> *"We recommend the District Attorney to take prompt action in the matter. We also think that the government inspection is very imperfect as now conducted."* (Ref. 18, pg 1)

Pursuant to this verdict, Jacob Vanderbilt (President of the Staten Island Ferry Company), James Braisted (the Superintendent), and Harry Robinson (the Engineer on duty at the time of the explosion) were arrested (19).

While several trials were conducted, none of the company executives were convicted. The *Westfield* itself was eventually renovated and returned to service for a number of years. The *Westfield* disaster, however, had a dramatic effect on the finances of both the Staten Island Ferry and the Staten Island Railroad (which owned the Ferry company). By July of 1872, the railroad (and with it, the ferry) were in receivership (20). Later that year, George Law purchased the railroad, and with it, the ferries from receiver L.H. Meyer (21).

Law renamed the company the Staten Island Railway. He pursued the idea of a joint railroad-ferry terminal with Vanderbilt, and a location at St. George was selected. Vanderbilt began construction, but abandoned the project due to a number of lawsuits and a storm, which washed away much of the structure (3). The Company, however, remained in receivership, with Jacob Vanderbilt still occupying the nominal position of President.

11.4 The "Duke of Staten Island"

In 1867, Erastus Wiman, a Canadian businessman, arrived in New York to oversee the main office of Dun, Barlow, and Company in New York. He acquired the interest of Barlow on his death in 1880. Wiman had extensive interests in the Canadian telegraph industry, and was often involved in issues regarding the U.S. telegraph industry.

Wiman located his residence in a mansion on Staten Island, and quickly became one of Staten Island's most prominent citizens. Dubbed the "Duke of Staten Island," he was interested in developing Staten Island, and recognized that to do so successfully required a coordinated transportation hub, and connections to both New York City and New Jersey.

By mid-1880, the Staten Island Railroad Company was barely operational. In May of 1880, the Attorney General sued on behalf of New York State to have the Company dissolved (21). The suit claimed that the Company had become "insolvent in September 1872, to have then surrendered its rights to others, and to have failed to exercise those rights." An injunction was obtained, and the legal proceeding continued for some time. It highlighted, however, the dire plight of the both the railroad and the ferries.

Wiman entered the fray to this difficult background. To help advance a transportation plan for Staten Island, Wiman incorporated the Staten Island Rapid Transit Company in April of 1880 (22, 23). This was the first use of the word "rapid transit" in connection with the Staten Island Railroad. Wiman's vision included the expansion of the Staten Island Railroad into a belt system, and the coordination of all ferries from a single terminal, located near St. George, which was the closest point on Staten Island to Manhattan. Such a system would eliminate the need for ferries, which at the time arrived at different locations in Staten Island, to run expensive and time-consuming connections to various shore points. In Wiman's plan, distributing passengers along the shores of Staten Island would be the job of the proposed rapid transit system.

Wiman's company began to seek legislation to acquire various rights-of-way that would be needed to implement the plan. These actions were started at a time when the company, in fact, did not own or control a railroad. Had the company gained charters to build connections, it would have had nothing to which to connect. Nevertheless, the company proceeded, and even began surveying for portions of the proposed routes (24, 25, 26). In April of 1881, the company acquired 1.5 miles of critical right-of-way directly from George Law (27). In October of 1882, Wiman made application for a wharf to land passengers from the company's planned new ferry service to Manhattan (28).

On April 3, 1883, Wiman gained control of the railroad (29). At the annual meeting of the Staten Island Railroad Company, Erastus Wiman and two of his cronies were elected to the Board of Directors of the railroad. The move was the result of much behind-the-scenes discussion, and surprised many of the railroad's directors. This step was the first of several that would lead to a consolidation of the Staten Island Railroad and the Staten Island Rapid Transit Company. By the end of the month, Wiman resigned from the Rapid Transit Company to avoid any conflict of interest (30). On June 27, 1883, a meeting of the directors of both companies formally ratified the merger of the two companies under the leadership of Erastus Wiman, who was named president (31). Finally, Wiman had control of the railroad and the ferry and could proceed to implement his ambitious plan.

While the control of the railroad included control of Vanderbilt's ferry, the North Shore Ferry was at this time separately leased and operated by John. H. Starin. His lease, however, was to expire on May 1, 1884. On July 18, 1884, the Staten Island Rapid Transit Company was awarded a new lease on the North Shore Ferry operation (32). Mr. Starin continued to fight the lease in the courts, in a battle that would go on for several years.

The most difficult part of the implementation of Wiman's plan involved obtaining rights-of-way through both government and privately-held land allowing the extension of the original railroad from its terminus at Vanderbilt's Landing in the vicinity of Clifton to the location of the planned ferry terminus at St. George. After a number of legal proceedings, however, construction on this extension began on March 17, 1884 (33):

> *"The construction of the new Staten Island Rapid Transit Railway along the shore of the Narrows from Clifton to New Brighton was begun yesterday, when a party of surveyors marked out the grade and broke ground for the road-bed. Next week, a large force of men will be put to work by the contractors."* (Ref. 33, pg 8)

On August 1, 1884, Wiman opened his service on the East Shore line through Tompkinsville (34). On November 16, 1884, Wiman and others incorporated the St. George Improvement Company (35); the new company was formed to handle the land and waterfront purchase from the estate of George Law. The *New York Times* indicated the following concerning the new company:

> *"The powers conveyed...are of varied scope, including not only the building and renting on Staten Island of docks, warehouses, and residences, but also the building of a casino or theater, skating rink, and the operation of play grounds for cricket, baseball, lawn tennis, lacrosse etc. The principal place of business named is New-Brighton, and the special locality stated for the operations of the company is at St. George, the terminus of the Staten Island Rapid Transit Railroad."* (Ref. 35, pg 13)

The new company was to handle the building of a new terminal for ferries at St. George as part of an overall development plan for multiple attractions at the location.

11.5 Robert Garrett and a New Partnership with the B&O Railroad

On November 21, 1885, the Staten Island Rapid Transit Company was leased to the Baltimore and Ohio (B&O) Railroad for a period of 99 years (36). B&O provided a guarantee of principal and interest on $2,500,000 worth of 40-year bonds at 5%. Wiman needed the financing to complete the railroad and terminal facilities at St. George. Robert Garrett, the President of the B&O, needed a right-of-way into New York. With the lease, another planned facility was added to the overall plan, that is, a bridge across the Kill von Kull at Elizabethport, with supporting "coal piers, round houses, and other terminal facilities."

Less than one month later, Robert Garrett was introduced to Staten Island movers and shakers by Wiman at a lavish dinner at the Pavilion Hotel in New-Brighton (37). Garrett, Wiman, and selected dignitaries began their evening from the Whitehall Street ferry terminal, boating to Staten Island via steamship in 20 minutes. Garrett acquired a residence on Staten Island, and began to take an active part in its politics and social activities.

On March 7, 1886, the fundamental elements of Wiman's transportation plan went into full operation (38, 39, 40). The *New York Times* of that day reported:

> *"The Staten Island Rapid Transit Company will open on the north and east shores today, and all of the old ferries will be closed at once. Passengers will get to the new ferry from the foot of Whitehall Street, New York to St. George. Trains will be run along each shore between St. George and Elm Park, St. George and Clifton for local travel, and St. George and Perth Amboy."* (Ref. 38, pg 7)

On March 9, the *New York Times* noted the improvements that the new service enabled (40):

> *"The great change which has been made by the introduction of this system can be realized by the fact that while five years ago the number of daily (ferry) trips between this city and Staten Island was only 12, on Sunday and yesterday, the number was increased to 84 trips each way. Port Richmond has hitherto been reached in one hour from this city, and the boats ran once an hour. Today it can be reached in 36 minutes, and three boats start from the Battery every hour. On the east shore, the time to Clifton, which under the one ferryboat system was 47 minutes, has been reduced to half an hour, with trains every 25 minutes. In the interior of the island, the facilities for communication are even more noticeably improved. A year ago, the time to Tottenville, opposite Perth Amboy, was 1 hour 45 minutes; yesterday, the trip was made regularly in 1 hour, 10 minutes. During the summer months, it is intended to run the trains much more frequently than under the present time table."* (Ref. 40, pg 8)

Within a few days, service on the South Shore Beach line was initiated (3). The final two major improvements took somewhat longer. The B&O freight connection over Kill von Kull, after a lengthy legal battle with competing railroads, finally opened on June 13, 1889. The joint railroad-ferry terminal in St. George was not completed and opened until 1897 (3).

Unfortunately, neither Robert Garrett nor Erastus Wiman would be involved in the latter event, both falling on hard times in the late 1880s and 1890s.

Figure 11.3 shows a map of the Staten Island Rapid Transit with all passenger services and the Cranford freight connection. The map represents the Staten Island Rapid Transit system at its peak.

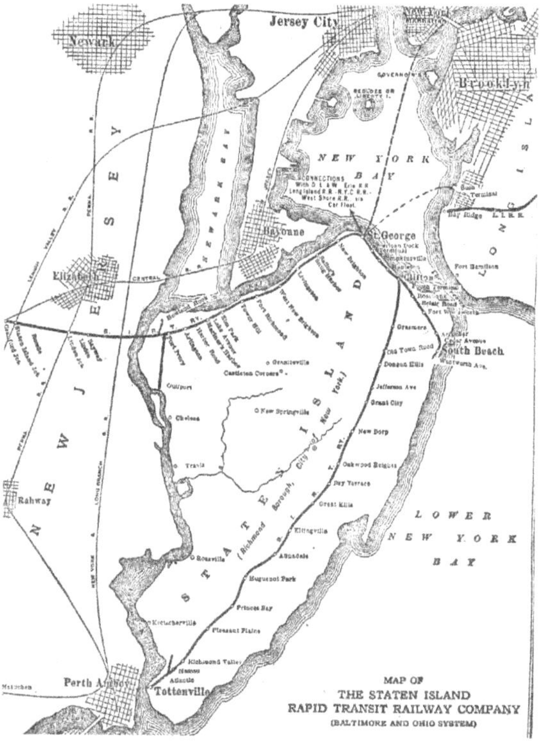

Fig. 11.3 A map of the Staten Island Rapid Transit at its peak; map is from 1952, before the closure of some service

11.6 Robert Garrett and the Sale of the B&O

The B&O Railroad was, in and of itself, a valuable asset. It owned and operated important trackage along the East Coast, and was considered by other railroads to be predatory, often charging lower freight rates than its competitors to drive them out of the market. The B&O, founded by Garrett's father, was therefore, frequently a target of competing railroads and other business groups, who wanted to acquire it. The B&O also had a subsidiary telegraph company that was considered to be desirable. As Robert Garrett held the single largest block of stock in the B&O, no purchase could take place without his blessing.

On March 8, 1887, the *Chicago Daily Tribune* announced that a sale of the B&O had occurred (41). It stated that after a lengthy negotiation with Garrett, control of the company had been acquired by Alfred Sully of New York. In an unusual statement, it indicated that "Tomorrow, Mr. Sully will be able to show in Wall Street the written contract by virtue of which Mr. Garrett put him in possession." If this suggested that a deal was not quite done, it was a gross understatement.

On March 14, 1887, the *New York Times* reported rumors that there was another potential buyer for the B&O: Jay Gould (42). Garrett, however, despised Gould, and was not about to sell to him. On the same day, the *San Francisco Chronicle* (43) reported that Robert Garrett was going to retire, quoting him as saying:

> *"I already have more money than I can spend; why be harassed to death with this enormous business, when I might just as well be getting out of life the most there is to be had."* (Ref. 43, pg 1

On March 22, *The New York Times* confirmed that a deal for sale of the B&O was, in fact, awaiting finalization (44). However, it also said that "...the Baltimore and Ohio Road will not pass out of the management of President Robert Garrett and that Jay Gould will have no hand or voice in the new arrangement." On July 20, 1887, Robert Garrett published a letter in the *Chicago Daily Times* unequivocally stating that the deal was off (45). On September 3rd, the *New York Times* reported that an agreement had been reached with "a huge syndicate" for purchase of the B&O, and that Robert Garrett would step down as President (46).

Finally, on October 13, 1887, the *Washington Post* reported that Robert Garrett had resigned as president at the monthly directors meeting, but had been elected as a director (47). At the same meeting, the directors voted no dividend on the main railroad, causing a steep loss in its stock value.

On October 24, 1887, the *New York Times* reported that Robert Garrett was "afflicted mentally" (48). On October 26th, the *Washington Post* suggested that reports of Garrett's mental problems were planted by Jay Gould, who was still trying to purchase the B&O telegraph company (49). Reports of Garrett's illness continued, however, for some time. Robert Garrett continued to block any transfer of B&O companies from his position as a major stockholder.

On June 8, 1888, Harrison Garrett, Robert's brother, drowned in a yachting accident on the Patapsco River (50). Reports of Robert Garrett's mental illness reappeared, and he commenced an extended tour of Europe to recover. While he had periods of better health, he was in and out of hospitals for the rest of his life, and died on July 19, 1896 (51).

By early 1896, the B&O itself was in receivership. It was, at the time, expected to remain in receivership for approximately five years (52). The Staten Island Rapid Transit Company, however, was excluded from receivership, even though the B&O held 51% of its stock.

11.7 Problems for Wiman and Staten Island Rapid Transit

On June 14, 1888, Erastus Wiman celebrated the completion of the Arthur Kill Bridge, and officially started the machinery to operate the swivel bridge (53). At the same time, the arrival of the new steel ferry boat, the *Robert Garrett*, at St. George was celebrated (53). Though the bridge would not be connected to the Staten Island Rapid Transit or the B&O Railroad for almost a year, the event marked the last high point for Wiman in the implementation of his original vision.

A number of events created a difficult environment for the Staten Island Rapid Transit over the next decade:

- With the resignation of Robert Garrett, the new leadership of the B&O Railroad was not as committed to the Staten Island Rapid Transit as was Garrett.
- Beginning in 1895, the interior of Staten Island began to be served by a variety of trolley services who were often at war with each other. Wiman himself actually acquired a trolley franchise (54). The trolleys, however, mostly competed with the SIRT, taking passengers directly to St. George and the ferries as opposed to serving as a feeder to the SIRT.
- Wiman suffered a significant loss in the U.S. financial crisis of 1893. By 1895, some of Wiman's property was seized and sold at auction for the benefit of his creditor's, who were concerned about his ability to pay his debts (55).

On June 15, 1894, Erastus Wiman was convicted of forgery, and was dismissed from his position at R.G. Dun & Company. He appealed the conviction and was awarded a new trial (56, 57, 58). Indictments for a new trial were ordered dismissed, clearing Wiman, on February 11, 1896 (59). While he had been cleared, the damage to his reputation and position were substantial. He had spent time in the Tombs between his conviction and the appeal, and his position with Dun was gone. This, on top of his financial losses in 1893, left him in a very weakened financial position.

When a contract for the construction of the new ferry terminal at St. George was signed on October 25, 1895, Wiman was not involved. He suffered a stroke in 1901, and died 1904 (3).

Competition from trolleys, the travails of the parent B&O Railroad, and Wiman's personal decline eventually affected the solvency of the Staten Island Rapid Transit Company. In 1897, the SIRT experienced an operating deficit of $72,455 (60). On October 2, 1898, acting on the directions of B&O receivers, the SIRT formally defaulted on its monthly interest charges on its bonds (61), throwing the SIRT into receivership. On April 21, the SIRT was sold at auction for $2,000,000 to the B&O Railroad, itself still in receivership (62 63, 64), giving it absolute ownership of the property, and the ferries held by SIRT for the first time.

The SIRT was formally reorganized on July 30, 1899 (65). While Erastus Wiman was no longer involved, the management of the SIRT remained intact, and the reorganized company's capitalization was reported to be $500,000.

11.8 The Northfield Wreck and City Control

The B&O Railroad continued to operate the SIRT and ferries for several years. On June 14, 1901, ferryboat *Northfield* was involved in a collision. As described by the *New York Tribune* (66):

> *"The Staten Island ferryboat Northfield, loaded with about one thousand passengers, was rammed a few minutes after six o'clock last night by the steel-hulled propeller Mauch Chunk, used as a ferryboat by the Central Railroad of New Jersey. The Northfield, with her hull badly stove, but under her own steam, headed a few minutes later for Pier No. 10, East River, near Old Slip, and in less than twenty minutes she went down in thirty feet of water close to the end of Pier 10."*
>
> *"...Over a hundred of the passengers on the sunken Northfield were dragged out of the water by people along shore and the crews of the fleet of river tugs which promptly responded to the ferryboat's call for help."* (Ref 66, pg 1)

The description gave differing views of the number of lives lost in the accident. While reporting "no loss of life known," it quoted a tugboat captain as saying "many were drowned." By June 16th, however, divers had found only one body, with only six passengers missing (67).

The accident caused growing concern on the part of New York City that the ferries were not being operated in a safe way. By 1905, the B&O was successfully pressured to relinquish control of the ferries to the city. On February 1, 1905, the city approved a plan to acquire the ferry terminal at St. George (68). Five new boats were ordered, and were named Richmond, Manhattan, Brooklyn, Bronx, and Queens (69). The city did not take actual ownership of the ferry until May 29, 1905, ending a negotiation that had taken place over three years (70).

All was not clear sailing with city operation. With five new boats ready for service, they could not be used until new ferry slips to accommodate them were built in Staten Island (71). The new service, using the new boats and new slips was finally kicked off on October 26, 1905 (72, 73) with the first formal sailing of ferryboat *Manhattan*, with Mayor McClellan and other dignitaries onboard.

What we now know as the city's Staten Island Ferry operation began in 1905. The B&O Railroad would, however, continue to own and operate the Staten Island Rapid Transit for another 65 years.

11.9 The Staten Island Rapid Transit between 1905 and 1970

The B&O operated the Staten Island Railroad through 1970, with periods of prosperity and periods of decline. Through WWI, there was little controversy, and the railroad generally prospered. In 1909, a minor controversy arose when the SIRT eliminated its discount fares for school children (74). In the same year, the SIRT posted its highest profits ever (75).

11.9.1 World War I and Operating Problems

With the start of WWI in 1917, the operation of the SIRT was severely impacted by several conditions, the most serious of which was the national rationing of coal. On August 2, the railroad petitioned the Public Service Commission to eliminate 89 trains from its daily schedule (76). While the cost and rationing of coal was a major reason, the railroad also cited requirements for greater freight capacity and reductions in manpower as significant causes. The reductions eliminated all night runs and many off-peak trains. While various civic groups opposed the reductions (77), they were eventually upheld. It was said that there was a 90,000 ton shortage of coal on Staten Island alone due to the war effort (78).

While the railroad did begin to replace some of the discontinued trains, it continued to have problems. Because of the coal situation, the railroad (and other users) was forced to use more soft coal, which resulted in a significant increase in pollution. The Staten Island Civic League lodged a complaint with the Health Commissioner (79), saying:

> *"Throughout the war period, the people of Staten Island have patiently endured the nuisance of having the whole island smudged with soot and cinders, but we feel that the time has come for relief from this plague of choking and fouling smoke."* (Ref 79, pg E6)

In April of 1920, a strike crippled the Staten Island trolley system, causing the city to declare all trolley leases forfeited. The city initiated action that would lead to a direct city takeover of the Staten Island trolleys (80).

Despite the difficulties of wartime operation, the SIRT survived and was able to return to full operation after its end.

11.9.2 Electrification

In previous chapters, it is noted that there was much discussion of connecting the SIRT to new subways to be built in Brooklyn by the BRT as part of the Dual Contracts. While a tunnel was indeed mapped and planned as part of the subway expansion efforts, the tunnel was never funded. The direct connection to the SIRT was also raised as part of some of the early plans that led to the Independent Subway.

A direct connection to the NYC rapid transit system, through, posed two problems: (1) Steam operation in tunnels was prohibited in NYC, and (2) the B&O, as an interstate railroad, was subject to federal railroad regulations concerning its rolling stock, which were larger and heavier than that of the NYC rapid transit system.

Yet another problem was the B&O's desire to have any tunnel connecting Staten Island with New York City accommodate freight, as well as passengers.

Of these problems, only one could be eliminated independently by the railroad: electrification. On December 29, 1923, the B&O announced its plan to electrify all of its passenger routes, beginning with the East Shore Line (81). The project was estimated to cost $15,000,000. On May 23, 1924, the railroad sought bids for 80 electric cars. They were to be of a size comparable to BMT rapid transit cars (67 ft long, 10 ft wide) and would be powered from a third rail. This would make them compatible with the Fourth Avenue Subway, to which a planned tunnel would connect (82). On August 25, 1924, the SIRT confirmed that work was underway to electrify 17 miles of double-track passenger routes, and that an order for 80 cars had been placed with the Standard Steel Car Company (83). General Electric would provide the motors for the cars.

With great fanfare, the SIRT inaugurated its electric service on July 1, 1925 (84). The *New York Times* declared:

> *"Staten Island took a half-holiday yesterday, in accordance with a proclamation by Borough President John A. Lynch of Richmond County, to celebrate the electrification of twenty miles of the Staten Island Rapid Transit tracks over which steam locomotives have operated for the last forty years."* (Ref. 84, pg 11)

The trackage electrified included the East Shore Line and the South Shore Beach Line. Later that evening, however, the Vice-President of the B&O sounded a cautionary note:

> *"Competing bus systems constitute the chief danger to the financial stability of the new roads. They have already taken 19,000 passengers from us at Port Richmond and New Brighton. We cannot hope to complete the electrification of the railroads of Staten Island unless you are willing to take advantage that will result from the use of them."* (Ref. 84, pg 11)

While the electrification of the North Shore Line was completed later in 1925, the comment indicated approaching financial difficulty for the Staten Island Rapid Transit.

Fig. 11.4 Electrified ME-1 Car for the SIRT, built between 1925–1928

11.9.3 Grade-Crossing Elimination

With electrification, the SIRT had to convert to high-level platforms. It also eliminated several grade-crossings, but many were left in operation after electrification. These became the source of safety concerns and disputes over the years.

In 1926, the state legislature passed the State Grade Crossings Act which called for the elimination of grade-crossings on many railroads in populated areas. The law required the removal of 73 grade-crossings on the Staten Island Rapid Transit (85). While the state would provide a portion of the funding, the cost to SIRT was estimated to be $6,000,000. The railroad claimed it could not afford this cost, and threatened to challenge the constitutionality of the Act (85).

On June 25, 1926, the Transit Commission also entered the fray, demanding the elimination of four specific at-grade crossings at Bay Street, Clifton, Hope Avenue and Tompkins Avenue (86). The cost was estimated to be $1,000,000, of which the railroad would pay half, with the state and city picking up 25% each. The railroad sued, claiming that the TC did not have the power to order such projects. On July 23, 1926, the Court of Appeals ruled in favor of the Transit Commission (87). The railroad, however, carried its case all the way to United States Supreme Court. On March 20, 1928, however, the Supreme Court declined to hear the case, dismissing it "for lack of jurisdiction" (88).

Because the railroad continued to balk at proceeding, the Transit Commission ordered legal action to take place against both the Staten Island Rapid Transit and the Long Island Railroad to compel them to submit plans for the ordered grade-crossing eliminations (89). On December 4, 1929, the Transit Commission asked the state legislature to provide an additional appropriation of $26,000,000 for grade-crossing removals in Staten Island, Queens, and Manhattan (90).

With additional state funding, removal programs began in earnest in the early 1930s. On February 25, 1937, the Staten Island Rapid Transit opened its single largest grade-crossing elimination project, the Port Richmond-Tower Hill Viaduct (91). The viaduct was more than a mile long and spanned eight grade crossings. The viaduct was the last piece of the $6,000,000 project which resulted in the elimination of a total of 34 grade-crossings.

On July 14, 1938, the SIRT announced an additional $2,000,000 project that would eliminate seven grade-crossings between Great Kills and Amboy Road (92). On December 13, 1948, Robert Moses announced that the state-aided program for elimination of grade-crossings in NYC was complete, except for "two major and seven minor" projects (93). One of the two "major" projects involved elimination of another 13 grade-crossings on the SIRT, a project that was not completed until the 1960s.

The last grade-crossing on the Staten Island Rapid Transit, however, at New Dorp, was not eliminated until 1966. On November 8, an SIRT train struck a school bus, killing an eight-year old girl and injuring 30 (94). Borough President Albert Maniscalco ordered the "closing" of the grade-crossing at New Dorp.

The Governor and New York City officials each blamed the other for the failure to proceed on the elimination of the 13 remaining grade-crossings (the same one's noted by Robert Moses in 1948). Finally, the State Public Service Commission authorized the project to eliminate those grade-crossings on August 29, 1964 (95). New Dorp, the site of the bus accident, was the absolute last crossing eliminated in 1966.

Fig. 11.5 The last grade-crossing on the SIRT at New Dorp, still operating in 1963

11.9.4 St. George Terminal Fire

On June 25, 1946, a fire destroyed the St. George Ferry Terminal, and with it, the contiguous Staten Island Rapid Transit line station (96). Three people were killed and 22 were hospitalized. Luckily, the fire broke out just after 400 to 500 passengers boarded the ferryboat *Miss New York*, and left the dock. The fire was found to have started in the railroad terminal, and spread to the ferry terminal and its four ferry slips.

The fire caused massive disruptions in service for Staten Island commuters. The ferry did not operate for several days (97), and commuters were forced to use a variety of alternative means to get to work, most involving indirect routes through New Jersey. On June 29th, partial ferry service was restored using a slip usually used by a ferry to 39th Street in Brooklyn (98). Fire inspectors eventually declared that some of the St. George ferry slips to be safe for use, and service resumed, albeit, only for foot passengers. The Staten Island Rapid Transit resumed service to the St. George Station on June 29th as well.

Normal service, including service for vehicles, was not restored until July 13, 1946. Ferries could use only slips 4 and 5, which had previously been used as layup piers for out-of-service vessels (99). Slip 4 had to be modified to allow ferry access by vehicles. Bids to build a temporary terminal were formally opened on August 21, 1946, and a contract was let (100).

A permanent replacement for the terminal would take a good deal longer. On February 10, 1948, Mayor William O'Dwyer promised to build a new $15,000,000 ferry terminal at the annual dinner of the Staten Island Chamber of Commerce (101). The promise was fulfilled three years later with the opening of the new terminal on June 8, 1951 (102). The new terminal was more efficient, coordinating railroad, bus, and ferry transportation in a single structure.

11.9.5 The Road to Public Ownership and Operation

Like transit in the other four boroughs, the financial condition of the SIRT declined gradually as a result of many unfavorable conditions. Eventually, it would become necessary for the public sector to own and operate the property, as it did in the other boroughs.

In February of 1947, twenty-four bus routes on Staten Island serving 60,000 daily passengers were threatened with closure. The NYC Board of Estimate sought a one-week delay in the closure to consider alternatives (103). The buses were operated by the Isle Transportation Corporation, and were losing about $4000 per month. The company was also $30,000 in arrears to the city on its franchise taxes. Their financial condition was worsened by an 11% wage increase for employees, agreed to after a one-week strike (103). On February 22, 1947, the Board of Estimate approved a city take-over of the property. A loss of $600,000 per year was expected (104).

The city take-over of the buses led to an unanticipated consequence. Fares on the buses were five cents *per zone*. The bus lines each ran through up to four zones, resulting in fare as high as twenty cents per passenger. With city operation, Staten Island riders demanded a change to a flat five-cent fare, as was the case for other city transit services (105). The city was forced to agree to the reduction, which led to a significant increase in ridership on the bus lines, which competed with the Staten Island Rapid Transit. Within a year, SIRT revenues were down by 60% (3).

Unable to compete with the lower bus fares, the SIRT started the process of reducing the number of trains operated to save money. As is usually the case, however, reductions in service led to further erosion of ridership and revenues, creating a destructive cycle.

On June 4, 1952, the SIRT informed the Public Service Commission that it would discontinue all passenger service on all three of its lines as of July 7th (106). The PSC, on June 17th, ordered SIRT to continue all of its passenger services while the commission conducted a study of its request (107). On December 19, 1952, the PSC authorized the SIRT to end passenger service on its North Shore and South Shore Beach lines (108, 109) as of March 31, 1953. The decision, however, also agreed to reorganize bus lines to provide alternative service for Staten Islanders, and to alter many to provide feeder service to the SIRT.

A little more than one year later, on September 8, 1954, the SIRT renewed its application to discontinue remaining rail service on the East Shore line as of October 5th (110). An agreement to reduce the number of East Shore trains run and new efforts to create bus feeder lines allowed the line to remain open.

By 1955, however, it was clear that the SIRT would not survive on its own, and discussions regarding a direct take-over by New York City began in earnest. On August 20, 1955, the *New York Times* reported that a deal had been reached (111):

> *"The city has worked out a tentative agreement that would assure continued operation of the Staten Island Rapid Transit Railway."*
>
> *"Under the arrangement, the line would lease its property to the city for an annual rental equal exactly to its yearly minimum tax bill. Management of the road would remain in the hands of the company."* (Ref. 111, pg 19)

On December 13, 1956, the PSC approved the agreement. It called for a 20-year lease under which the company continued to provide the service, with the city repaying the company all of the taxes and charges paid, including franchise taxes, water and sewer rents, real estate taxes and city sales and use taxes (111).

The agreement was essentially one that guaranteed a city subsidy to the SIRT for up to 30 years. While the lease could be renegotiated after two years if the railroad's revenue varied substantially from its estimated 1955 deficit of $655,000, the lease lasted until June 30, 1966, with automatic yearly renewals for ten additional years (111). The deal would not last that long.

In 1958, the PSC approved the raising of commuter fares on the SIRT by 28 to 43% (112). In doing so, the commission noted that the operating deficit of the SIRT had risen to over $1.5 million in fiscal year 1958. By October of the same year, the city and SIRT announced that they would try to renegotiate a deal "under which long-term commuter service will be guaranteed by the railroad in exchange for additional tax relief and other subsidies from the city" (113). In August of 1960, an agreement was reached that would double the city subsidy to SIRT through the end of the decade (114).

By the end of the decade, the Metropolitan Transportation Authority had been formed, and was ready to take on the New York City Transit Authority as a subsidiary. William Ronan, the Authority's chairman began to openly discuss direct acquisition of the SIRT. The 1960 agreement had included option for the city to acquire title to the SIRT for $1.00 on January 1, 1970 (115). Ronan also called for a $25,000,000 modernization program, which would be 75% financed out of the state transportation bond issue that had been approved in November 1969. At the time, there was not a single car on the railroad that was made after 1923. In May 1970, the City Planning Commission released the Staten Island volume of its proposed New York City plan (116). Among many proposals was the reopening of the North Shore Line of the SIRT for passenger service.

In a comprehensive interview with William Ronan published in the *New York Times* on May 17, 1970, entitled "The Wholly Ronan Empire," Ronan indicated his belief that the MTA would acquire the SIRT "by the end of the summer"

(117). It took an additional year. The SIRT was formally taken over by the MTA on July 1, 1971. The Staten Island Rapid Transit Operating Authority (SIRTOA) was established as a subsidiary to operate the system (3).

11.10 The Era of Public Ownership

The MTA immediately began to address the issue of aging and outdated rolling stock. In 1972, as a temporary measure, a number of electric cars from the Long Island Railroad were transferred for use on the Staten Island Rapid Transit. They negotiated a deal with the Federal Railroad Administration to allow some modifications to the normal intercity railroad standards for passenger cars. This allowed MTA to purchase cars based upon the then new R-44 cars for the NYCTA. The cars had additional structural reinforcement and some other modifications incorporated. In early 1973, the first modified R-44 cars began service, and by the end of 1974, SIRT was operating with a fleet of 52 new cars. An additional 12 cars were added as part of the NYCTA General Overhaul Program.

Fig. 11.6 Modified R-44 Cars for the SIRT

Freight service over the Arthur Kill Bridge continued through 1991, when it was abandoned due to lack of sufficient business. The bridge itself underwent reconditioning between 2004 and 2006, and new freight trackage was built along the historic route to Fresh Kills, known as the Travis Branch, and along a short portion of the old North Shore Line right of way. Regular freight service resumed on April 2, 2007, and continues today. A deal was also negotiated with CSX Transportation (the successor to the B&O), Norfolk Southern Company, and Conrail to haul waste from the Staten Island Transfer Station and container freight from the Howland Hook Marine Terminal and other industries. The waste arrangement was critical due to the closing of the Staten Island Landfill, which was the only operational landfill for NYC waste. Today, all of the City's waste must be transported to dump sites across the nation for disposal.

A small passenger spur and station was added to the system in 2001, providing game-day service to the Staten Island Yankees minor league ballpark. This service has since been abandoned.

11.11 The Staten Island Rapid Transit System Today

Today, the SIRT continues to operate as a rapid transit line along the original East Shore Line from St. George to Tottenville. Modern passengers now receive the "best buy" in transit in the entire city: A single fare allows a passenger to ride the SIRT, take the Staten Island Ferry, and a connecting subway trip in Manhattan.

11.12 Into the Future

The SIRT still operates with its fleet of now-aging R-44 Cars. These will soon need to be replaced. They could be replaced with subway cars, commuter railcars, or some hybrid. The Federal Railroad Administration has granted additional waivers of it rules that would allow the use of standard rapid transit cars on the SIRT. Use of anything other than subway cars would eliminate any possibility of a connection with the NYCTA transit system. On the other hand, it is difficult to imagine such a connection today, with all present capital projects (the 2nd Avenue Subway, East Side LIRR access, and No. 7 train extensions) all running over budget and behind schedule. It is not even clear that modern Staten Islanders would want or approve such a connection. After witnessing the rapid growth that occurred after the opening of the Verrazano Bridge, many Staten Island residents are concerned with preserving what is left of the suburban/rural character of the borough.

The South Shore Line has been demapped, and all tracks have been removed. Most of the land has been used for new residential developments and other purposes. The North Shore R.O.W., and some of the trackage, however, still exists. There have been serious discussions and proposed plans that would lead to some form of service being reinstated. These plans range from a full restoration of the North Shore Line, to use of the R.O.W. for light rail service as part of a more comprehensive light rail system on Staten Island.

11.13 A Historic Footnote

Despite the lack of a direct connection between the NYC subway system and the SIRT, rolling stock from the railroad did eventually reach Brooklyn. When the North Shore and South Beach Lines closed, SIRT sold 30 of its cars to the NYCTA, where they operated on the Brighton Beach Line for several years. Economic convenience accomplished what McAneny, Hylan, and several generations of transit planners in New York could not.

(a) Steam work train on the Staten Island Railroad, circa1880

(b) Steam train in the maintenance yard, circa 1875

(c) South Beach Station in 1952

(d) St. George Station of the Staten Island Railroad in 1914

(e) Arlington Terminal on the North Shore Line

(f) East Shore Line Station

Fig. 11.7 Views of the Staten Island Rapid Transit through the years

References

1. Staten Island's Beauties. New York Tribune, 8 (August 26, 1895)
2. City Items. New York Daily Tribune, 7 (January 16, 1860)
3. Leigh, I., Matus, P.: Staten Island Rapid Transit: The Essential History (January 2002), `http://www.thethirdrail.net`
4. Staten Island; Staten Island Railroad. New York Times, 8 (April 24, 1860)
5. City Intelligence; Railroad in Staten Island. New York Times, 8 (June 2, 1860)

6. Staten Island News; The Staten Island Railroad. New York Times, 5 (August 25, 1860)
7. City Items; Change in Ownership of the Staten Island Ferry. New York Daily Tribune, 8 (June 18, 1862)
8. Domestic. New York Observer and Chronicle 42(9), 70 (1864)
9. New York; Explosion of the Boiler of the Ferryboat Westfield. Chicago Tribune, 1 (July 31, 1871)
10. Appalling Disaster. New York Times, 1 (July 31, 1871)
11. Yesterday's Boiler Explosion. New York Times, 4 (July 31, 1871)
12. The Westfield Disaster. New York Tribune, 1 (August 1, 1871)
13. Awful Calamity at New York. Massachusetts Ploughman and New England Journal of Agriculture 30(45), 3 (August 5, 1871)
14. Who Was to Blame? New York Tribune, 4 (August 2, 1871)
15. Where Are the Savants? New York Tribune, 4 (August 8, 1871)
16. New York; The Vanderbilt Family to Blame for the 'Westfield' Disaster. San Francisco Chronicle, 1 (August 10, 1871)
17. Verdict in the Westfield Disaster. Chicago Tribune, 1 (August 17, 1871)
18. The Westfield Disaster. New York Times, 2 (August 18, 1871)
19. Metropolitan Suburbs; The Staten Island Railroad. New York Tribune, 3 (July 11, 1872)
20. Transit to the Suburbs; The Staten Island Railroad. New York Tribune, 4 (March 29, 1873)
21. A Railroad Charter in Peril. New York Times, 3 (May 14, 1880)
22. Rapid Transit on Staten Island. New York Times, 8 (April 20, 1880)
23. Local Miscellany; Rapid Transit in Staten Island. New York Tribune, 8 (August 20, 1880)
24. Staten Island Rapid Transit. New York Times, 2 (June 1, 1880)
25. Railroad Projects; Rapid Transit on Staten Island. New York Tribune, 1 (August 12, 1880)
26. Railway News; Staten Island's Rapid Transit Scheme. New York Tribune, 1 (April 12, 1881)
27. Railway Interests; Rapid Transit on Staten Island. New York Tribune, 2 (April 17, 1881)
28. The New Staten Island Ferry. New York Times, 8 (October 5, 1882)
29. To Staten Island; The Rapid Transit Men Secure Possession of the Railroad and Ferry. New York Times, 5 (April 4, 1883)
30. Railroad Interests; Staten Island Rapid Transit. New York Tribune, 1 (April 29, 1883)
31. Railroad News; Staten Island Rapid Transit. New York Tribune, 5 (June 27, 1883)
32. The North Shore Ferry. New York Times, 8 (July 18, 1884)
33. City and Suburban News; Staten Island. New York Times, 6 (March 18, 1884)
34. City and Suburban News; Staten Island. New York Times, 6 (August 13, 1884)
35. Improvements on Staten Island. New York Times, 13 (November 16, 1884)
36. Affairs of Railroads; The Baltimore and Ohio Railroad to Enter New York. New York Times, 2 (November 22, 1885)
37. Robert Garrett Dined. New York Times, 1 (December 12, 1885)
38. City and Suburban News; Staten Island. New York Times, 7 (March 7, 1886)
39. Staten Island's Rapid Transit. New York Times, 8 (March 9, 1886)

40. A Red-Letter Day for Staten Island. New York Tribune, 8 (March 9, 1886)
41. A Great Railroad Sold. Chicago Daily Tribune, 1 (March 9, 1887)
42. Betting on Jay Gould. New York Times, 1 (March 14, 1887)
43. Garrett to Retire. San Francisco Chronicle, 1 (March 14, 1887)
44. The Deal Accomplished. New York Times, 1 (March 22 ,1887)
45. The B&O Deal Off. Chicago Daily Tribune, 1 (July 21, 1887)
46. Mr. Garrett Stepping Out. New York Times, 1 (September 3, 1887)
47. Garrett Resigns. The Washington Post, 1 (October 13, 1887)
48. Mr. Garrett's Queer Ways. New York Times, 1 (October 24, 1887)
49. Did Gould Euchre Garrett? The Washington Post, 1 (October 26, 1887)
50. A Yacht Sunk; A Brother of Robert Garrett Loses His Life. Los Angeles Times, 1 (June 9, 1888)
51. Robert Garrett Dead; His Long Mental and Physical Illness Ended At Last. New York Times (July 30, 1896)
52. The Baltimore and Ohio. San Francisco Chronicle, 4 (March 2, 1896)
53. Swinging the Arthur Kill Draw; Erastus Wiman Views the Result of His Seven Years' Effort with Satisfaction. New York Tribune, 2 (June14, 1888)
54. Erastus Wiman Gets a Franchise. New York Times, 10 (March 12, 1895)
55. Large Sale of Lots at Rosebank. New York Tribune, 8 (July 17, 1895)
56. Wiman's Appeal Argued. New York Tribune, 4 (June 9, 1895)
57. He Gets a New Trial. Chicago Daily Tribune, 1 (March 16, 1895)
58. In Favor of Erastus Wiman. New York Tribune, 12 (February 11, 1896)
59. Erastus Wiman Free. New York Tribune, 12 (February 11, 1896)
60. New Ferry Boats on Staten Island. New York Tribune, 9 (October 25, 1895)
61. Staten Island Rapid Transit. Wall Street Journal, 1 (September 13, 1897)
62. Railroad Interests; Staten Island Rapid Transit Bonds. New York Tribune, 10 (October 2, 1898)
63. Staten Island Rapid Transit. New York Times, 2 (April 21, 1899)
64. Staten Island Rapid Transit Sold. New York Tribune, 2 (April 21, 1899)
65. Added to the B&O System. The Washington Post, 9 (April 21, 1899)
66. Staten Island Rapid Transit. New York Times, 4 (July 30, 1899)
67. Ferry Boats in Collision. New York Tribune, 1 (June 15, 1901)
68. Divers Find One Body. The Washington Post, 1 (June 16, 1901)
69. Staten Island Ferry Sites Bought By City. New York Times, 10 (February 1, 1905)
70. Names for the City's New Ferryboats. New York Tribune, 16 (February 4, 1905)
71. City Takes S. I. Ferry. New York Tribune, 7 (May 30, 1905)
72. Anger at Ferry Delay. New York Tribune, 9 (September 1, 1905)
73. City's Ferry Fleet Ready. New York Times, 18 (October 24, 1905)
74. Twenty-Minute Ferry to Staten Island Now. New York Times, 18 (October 26, 1905)
75. Staten Island Fares Jump. New York Times, 1 (June 4, 1909)
76. Staten Island's Growth. New York Times, XX2 (December 12, 1909)
77. Staten Island Road Asks Permission to Cut Off 89 Trains. New York Tribune, 12 (August 2, 1917)
78. Staten Islanders Want More Trains. New York Times, 74 (September 9, 1917)
79. Coal Shortage Acute. New York Times, 8 (October 15, 1917)
80. Complaints on Soft Coal. New York Times, E6 (November 30, 1919)
81. Asks City to Seize Staten Island Cars and Lighting Plant. New York Times, 1 (April 19, 1920)

82. B&O to Electrify Staten Island Line. New York Times, 7 (December 29, 1923)
83. 80 Electric Cars for Staten Island. New York Times, 29 (May 24, 1924)
84. Staten Island Rail Electrification Begun. Wall Street Journal, 5 (August 25, 1924)
85. Opens New Service on Electrified Line. New York Times, 11 (July 2, 1925)
86. Railroad Fight Grade Crossing Act. New York Times, 40 (May 20, 1926)
87. Four Grade Crossings Will be Eliminated. New York Times, 15 (June 26, 1926)
88. Power to Abolish Crossings Proved. New York Times, 2 (July 23, 1927)
89. Crossing Appeal Lost by Staten Island Line. New York Times, 16 (March 20, 1928)
90. Plans Suits to End 44 Grade Crossings. New York Times, 2 (November 22, 1928)
91. Asks $26,000,000 for Crossing Work. New York Times, 19 (December 4, 1929)
92. Staten Island Opens Mile-Long Viaduct; Thirty Four Grade Crossings Are Eliminated. New York Times, 3 (February 26, 1937)
93. Crossing Job Begun on Staten Island. New York Times, 7 (July 14, 1938)
94. Grade Crossings Nearly All Gone. New York Times, 38 (December 13, 1948)
95. Fatal Crossing Closed on S.I. New York Times, 33 (November 11, 1960)
96. $10,923,000 Job Authorized to End S.I. Grade Crossings. New York Times, 45 (August 29, 1964)
97. Staten Island Fire Wrecks Ferry Terminal, Kills 3; Damage Put at $2,000.000. New York Times, 1 (June 26, 1946)
98. Ferryboat Travel is Nearing Normal. New York Times, 33 (June 28, 1946)
99. Staten Island Ferries Normal, Take Vehicles. New York Times, 12 (July 13, 1946)
100. Ferry Terminal Bids Opened. New York Times, 48 (August 21, 1946)
101. Aid for Staten Island. New York Times, 48 (February 10, 1948)
102. Terminal Opening June 8. New York Times, 18 (May 29, 1951)
103. Bus Halt Put Off on Staten Island. New York Times, 3 (February 15, 1947)
104. City to Run Buses on Staten Island. New York Times, 1 (February 21, 1947)
105. Staten Islanders Seek 5c Bus Fare. New York Times, 23 (May 10, 1947)
106. Staten Island Road, Scoring Buses, Files to End Passenger Service Runs July 7. New York Times, 1 (June 4, 1952)
107. Staten Island Road Ordered to Continue Service Pending Study. Wall Street Journal, 5 (June 17, 1952)
108. Train Service Cut on Staten Island. New York Times, 19 (December 20, 1952)
109. Staten Island Line Would Cease Runs. New York Times, 33 (September 8, 1954)
110. Transit Deal Due on Staten Island. New York Times, 19 (August 20, 1955)
111. City Pact Approved on Tottenville Line. New York Times, 52 (December 14, 1956)
112. Rail Fare to Rise on Staten Island. New York Times, 28 (June 19, 1958)
113. Transit Pact Sought. New York Times, 34 (October 9, 1958)
114. Staten Island Rail Line Subsidy Doubled by City for Next Decade. New York Times, 13 (August 26, 1960)
115. City Urged to Take Over Staten Island Railway. New York Times, 43 (May 8, 1968)
116. Expansion Plans for Staten Island. New York Amsterdam News, 32 (March 7, 1970)
117. Shapiro, F.: The Wholly Ronan Empire. New York Times, 220 (May 17, 1970)

Figure Sources

11.1: *Courtesy of Wikimedia Commons*, Map by Julius Schorzman, used under Creative Commons Share-Alike 2.5 Generic.
11.2: *Originally published in* Frank Leslie's Illustration Newspaper, New York, NY, August 1871.
11.3: *Courtesy of* Wikimedia Commons.
11.4: *Courtesy of* New York Transit Museum.
11.5: *Courtesy of* New York Transit Museum.
11.6: *Courtesy of* New York Transit Museum.
11.7: *Courtesy of* New York Transit Museum (a–f).

Chapter 12
When the Rubber Hits the Road

The motor bus is the historic successor to all previous forms of surface public transportation systems. It proceeds from the horse-drawn omnibus and horse-streetcar technologies, and trolleys.

The bus provides public transportation service over public streets and highways, although separated rights-of-way (busways) are provided in some cases for faster service. While the capacity of a single bus varies widely depending upon its specific design, bus services can accommodate as few as 50 passengers an hour to as much as 7000 to 8000 passengers per hour. Because they use local streets, buses can pick-up and drop-off passengers closer to their desired origins and destinations than rapid transit. For many parts of the country, buses are the only, or primary, form of public transportation.

New York City has, and has had, one of the largest bus systems in the world, despite the fact that it operates in a region dominated by one of the largest rapid transit systems in the world. On a typical weekday, more than 2.4 million passengers ride the buses of New York City.

12.1 The First Motor Bus

One of the first inventors to develop a motor bus was Sir Goldsworthy Gurney of Great Britain. A professional surgeon and inventor, he is best known for the invention of the "gurney" used throughout the world's hospitals and clinics. Between 1825 and 1829, Gurney developed plans for a large stagecoach/wagon powered by a steam engine. In July of 1829, a prototype was driven from London to Bath without incident, except for an unruly crowd in Bath that tried to demolish the vehicle. A regular service between Cheltenham and Gloucester was operated by Sir Charles Dance using three Gurney vehicles. Because the new service was viewed as a threat to competing omnibus services, a series of taxes and fees on steam-carriages, known as the Turnpike Acts, were passed effectively putting an end to the service.

Once of the difficulties faced by these early steam-carriages was the real and perceived instability of steam engines. Many people feared riding in a vehicle on top of a steam engine that might explode. In an attempt to combat this, Gurney actually produced two steam-carriages with the steam engine on a separate articulated cart that "pulled" the carriage along. Neither of these was ever used in service. Unfortunately, during a test-charging of one such engine, it exploded, seriously injuring two people.

R.P. Roess and G. Sansone: The Wheels That Drove New York, STTT 1, pp. 249–279.
springerlink.com

(a) Gurney's steam-carriage in 1827 (b) Gurney's steam-drag vehicle, circa 1830

Fig. 12.1 Gurney's early attempts at a steam-powered carriage

Gurney was not the first to develop a steam-powered road vehicle. Richard Trevithick had built a steam-carriage as early as 1805, but it was not a commercial success. Others, both before and during the time of Gurney, developed parallel designs. Unlike the others, however, Gurney is the only one to have accomplished the implementation of a regular omnibus operation using the vehicle.

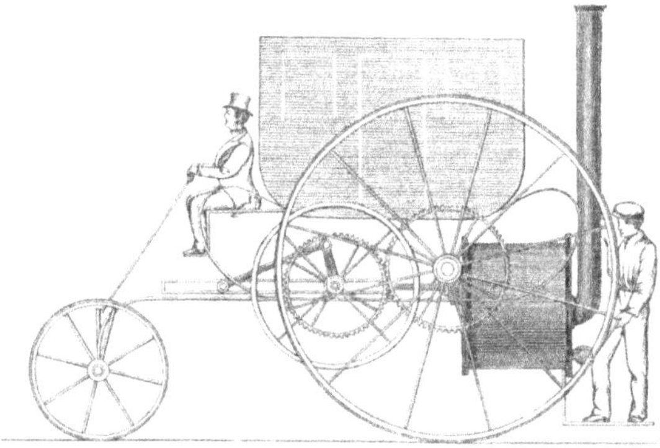

Fig. 12.2 Trevithick's Steam Carriage, circa 1805

12.2 Buses Come to Fifth Avenue

New York was the first city to make extensive use of motor buses, and it was indeed the test bed for a wide variety of vehicles throughout the early years of the 20th century.

The first "bus" was placed in operation along Manhattan's Fifth Avenue. Fifth Avenue never had either horse-railways or trolleys. Residents and business owners located along this elite thoroughfare had long opposed the laying of rails. Because of this, the horse-drawn omnibus was the only form of public transit on Fifth Avenue until the introduction of buses.

The first "bus" was placed in service in 1902. It was a battery-powered vehicle, and did not prove very reliable. In 1905, the Fifth Avenue Coach Company purchased a fleet of 15 French DeDion Bouton double-deck buses and put them in service on Fifth Avenue. The buses were a huge success, and within a few years, a fleet of 132 DeDion buses replaced all horse-drawn omnibuses. Similar buses were purchased from Daimler (English) to complete the fleet.

The first buses were quite interesting. They were not truly "motor buses," but were hybrids. A small gasoline engine turned an electric generator that supplied power to the axles. The second deck was open, and the wheels were solid rubber, making for a somewhat bumpy ride, but smoother than that of the omnibuses, which used wooden wagon-wheels.

Fig. 12.3 DeDion Bus on Fifth Avenue

Fig. 12.4 Early Daimler Buses in operation along Fifth Avenue

The early success of the Fifth Avenue operation was the spark that ignited a nationwide movement to replace horse-drawn vehicles with motor buses. The movement led to the development of several significant U.S. manufacturers of motor buses, and to a great deal of technical innovation. In New York, virtually all surface transit operations, including trolleys, would be replaced by buses by 1940, with the exception of a few trolley services in the outer boroughs.

In Manhattan, Mayor Fiorello LaGuardia hastened the conversion of trolley lines to buses. In the early 1930s, he established a policy to remove elevateds and trolleys from Manhattan. LaGuardia, along with Robert Moses, whom he appointed to a number of positions of power and influence, believed that the future of the city was more tied to growth in automobile use. The construction of the 8th and 6th Avenue subways led to the elimination of the 6th and 9th Avenue Els. The 2nd and 3rd Avenue Els were eliminated without having replacement subways. The same zeal was applied to eliminating trolleys and replacing them with motor buses.

12.3 Building a Better Bus

The first bus operations in the U.S. used foreign manufactured vehicles. At the turn of the century, there were no manufacturers of motor buses in the U.S. that could produce transit vehicles in any quantity. That would change in the early 1920s, as a number significant companies began to focus on the development and manufacture of buses.

12.3.1 Fageol Motors, Fageol Truck & Coach, and Twin Coach Companies

The first motor vehicles specifically designed and manufactured for intercity or urban transit use were designed and manufactured by the Fageol brothers. Originally formed to make luxury cars and trucks, their company found its greatest success in the motor bus business. Earlier bus models were basically a passenger body placed on a light truck or touring car chassis. Neither of these produced very comfortable or efficient vehicles.

The first vehicle designed for passenger transit was the Fageol Safety Coach, which was first manufactured in 1920. It featured a wider track and lower body, (only two feet above the street) than previous models, and could accommodate 22 passengers. It was powered by a four-cylinder gasoline engine. Shortly after its introduction, a six-cylinder model accommodating 29 passengers was introduced. In 1922, Fageol introduced air-brakes in yet another technological improvement. The Fageol Safety Coach was immediately successful. While most of these vehicles were used in intercity bus service, many of them were used in urban transit service.

In 1925, the American Car and Foundry Company (ACF) purchased both the J.G. Brill Co. and the Fageol manufacturing facilities. The Fageol brothers became vice-presidents of ACF. Both left in 1927 when it became clear that ACF was not interested in some of the innovations the Fageols were proposing. In 1926, the brothers formed the Twin Coach Company.

Fig. 12.5 A Fageol Safety Coach operated by Public Service of New Jersey in 1923

The first Twin Coach was the Model 40, which became one of the most successful bus designs in history. It was widely used throughout the U.S. The Model 40, so called because it had a capacity of 40 passengers, introduced a number of novel features:

- The bus was powered by two gasoline engines mounted on each side of the bus, each with its own driveline and differential.
- The entrance doors and driver's position were moved ahead of the front wheels.
- Because of the change in engine location, passengers could be seated throughout the bus, including over the engine compartments.

With the Twin Coach Model 40, the Fageols had created a template for future bus designs over the next fifty years. The Twin Coach proved to be a very reliable vehicle, with many remaining in service for 20 years or more.

The Fageol brothers had originally formed the Fageol Motors Company in 1916. In 1932, after a bankruptcy, they formed the Fageol Truck and Coach Company. Even after merging with ACF in 1925, the Fageols kept one manufacturing site in Oakland independent. The Twin Coach Company continued through the mid-1950s, after which it was sold to Flxible. Their truck business was sold to T.A. Peterman in 1939. The truck company was renamed "Peterbuilt" at that time.

Fig. 12.6 Two examples of the Twin Coach Model 40, one operated by the Brooklyn Bus Corporation in 1930, the other by Jamaica Buses Inc. in 1933

12.3.2 Yellow Coach Manufacturing Company

Yellow Coach was also an important innovator in the design and manufacture of early motor buses. The company had its roots in the Yellow Cab Company of Chicago, formed by John D. Hertz (of car rental fame) and an associate in 1915. In 1920, Hertz formed the Yellow Cab Manufacturing Company to build vehicles for the rapidly-growing taxi industry in U.S. and abroad. In 1922, Hertz and his associates acquired several smaller companies involved in bus building in Chicago. In 1923, the bus-building operation was established as the Yellow Coach Manufacturing Company.

Shortly after its formation, Yellow Coach became quite successful in selling its buses. A large order from Philadelphia attracted the attention of Alfred Sloan, the President of General Motors. In 1925, Sloan proposed a merger with Yellow Coach, establishing the Yellow Truck and Coach Manufacturing Company, with GM Truck Company as a subsidiary. The company continued through 1943, when a controlling interest was acquired by General Motors. The company continued under the name of General Motors Truck and Coach Division.

The first Yellow Coaches were built and operated in Chicago. The first buses were gasoline-electric hybrids, and were similar to the early Fageol models. The first buses to appear in New York were a fleet of double-decker buses for the Fifth Avenue Coach Company. The double-deckers seated 62 passengers, and the second level was fully enclosed for comfort. While Yellow Coach provided the chassis, the body was manufactured by Fifth Avenue in its own body-building facility.

Fig. 12.7 Yellow Coach Double-Decker, Fifth Avenue Coach, 1932

Yellow Coach was responsible for many significant technological innovations in bus design and manufacturing. In 1932, they were one of the first companies to move the engine to the rear of the bus. The original rear-engine drives were more efficient than the prior front-engine designs, but the engine intruded on the passenger compartment, creating both heat and noise. The engine, which could be removed from the rear, had to be serviced from inside the passenger cabin.

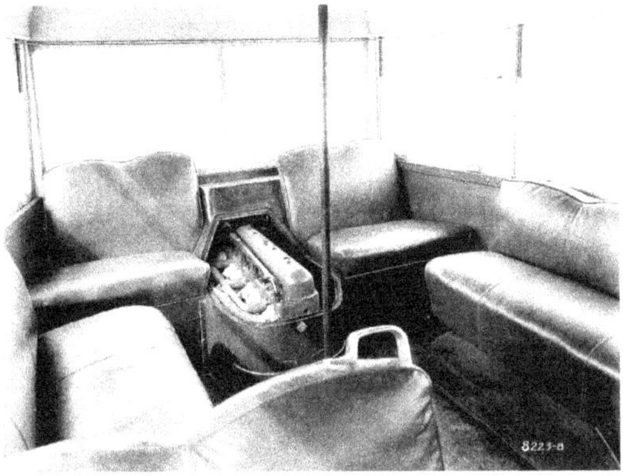

Fig. 12.8 Yellow Coach Company rear-mounted engine, 1933

In 1934, Yellow Coach hired inventor Dwight Austin. He had developed and patented a new rear engine angle drive for use in buses. This invention allowed both the rear engine and transmission to be mounted transversely, with the angle drive connecting the transmission to the drive shaft. In this configuration, the engine and transmission could be isolated from the passenger compartment.

Fig. 12.9 A Yellow Coach with a transverse gasoline engine

Three hundred and sixty-six of the angle-drive buses were ordered for New York City operation. One of the first purchasers was the Madison Avenue Coach Company, which put 76 into service in late 1934. The Madison Avenue Coach Company was a short-lived operation. It operated a single route using the franchise of the former New York and Harlem RR streetcar line, which had been abandoned. It was a subsidiary of the New York Railways Company until 1936, when it was merged into the New York City Omnibus Corporation.

Fig. 12.10 A Yellow Coach with a transverse rear engine, operated by Madison Avenue Coach Inc. in 1934

Yellow Coach was also a leader in the development of bus transmissions. Until 1936, all bus transmissions were manual. The process of manual shifting by the driver often resulted in a "jerky" acceleration motion which was uncomfortable for passengers. In 1936, Yellow Coach introduced a semiautomatic transmission which was not particularly successful. In 1938, however, it began to produce fully automatic hydraulic transmissions, which quickly became a mainstay in transit buses throughout the U.S.

Yellow was also heavily involved in the development of diesel engines for buses. The first Yellow buses produced with diesels were built in 1938, using two-cycle diesel engines made by GM.

All of the early buses were bodies mounted on separate chasses. In 1938, Yellow introduced the "monocoque" design. In this integrated design, the entire body frame supported all bus components. No separate chassis was required.

Yellow Coach was merged into General Motors in 1943, after which all buses were sold under the General Motors brand.

Fig. 12.11 A Yellow Coach with diesel power and monocoque design, operated by Triboro Coach Corporation in 1939

12.3.3 ACF–Brill

The American Car & Foundry Company was a major builder and refurbisher of railcars. It was formed in 1899 through a merger of 13 railroad manufacturing companies. During the same period, the J. G. Brill Corporation manufactured trolleys and interurban electric railcars, as well as cars for railroads. Through a series of complicated mergers and take-overs, by 1925, ACF had obtained control of Brill, the Hall–Scott Motor Company (an engine manufacturer), and Fageol Motors. The association with the Fageol brothers was a brief one, with the brothers leaving to form Twin Coach in 1927.

ACF was far more successful in building railroad cars than it ever was in building buses. ACF actually built the first all-steel transit cars for the IRT. While it managed a profitable bus business for a number of years, ACF was never considered to be a leader or innovator in the field. It sold many more intercity buses than transit buses. It didn't compare well with the Twin Coach in the early years. Later, it lost significant business to GM when ACF failed to follow the trend and place diesel engines in its transit vehicles. In 1953, ACF abandoned the bus and trolley field.

ACF never sold a significant number of buses to New York operators. In 1935 and 1936, it provided a fleet of 54 twenty-two passenger gasoline-powered buses to the Staten Island Coach Company. Triboro Coach, a private operator in Queens, bought 24 ACF buses during WWII, when overall bus production was hampered by wartime rationing of supplies.

(a) ACF Bus in Staten Island, 1934 (b) ACF–Brill Bus in Queens operation, 1943

Fig. 12.12 Examples of ACT–Brill Buses in NYC use

When ACF left the bus industry, production of ACF–Brill postwar designs continued in Canada under a previous agreement with the company. Buses manufactured in Canada bore the name CCF (Canadian Car Foundry)–Brill, and continued into the 1960s. No CCF–Brill buses were ever used in U.S. transit service.

12.3.4 Mack Buses

Mack is most known as a major producer of trucks, a business in which it still plays a significant role. At various times in its history, however, Mack also manufactured railcars and locomotives, fire engines, and buses.

An early Mack advertisement states that "The first Mack was a bus and the first bus was a Mack." The second part of this slogan exaggerates, as several foreign manufacturers produced bus-type vehicles before the turn of the century. The first Mack bus was sold to Isaac Harris for use as a sightseeing vehicle in 1900. After an initial trial period, additional vehicles were sold to Harris.

Fig. 12.13 First advertisement of a Mack Bus, Sightseeing Coach, 1903

Production facilities were originally located in Brooklyn, NY, but moved to Allentown in 1905, where truck and bus production facilities were combined. Bus and truck chassis were the same until the early 1920s.

During WWII, the Navy took over the Allentown bus plant to make torpedo bombers. Bus manufacturing restarted in 1945, and continued until Mack left the bus business in 1960.

Over the years, Mack earned a reputation for durability and reliability, and Mack buses were prominent in New York City through the 1960s. They were used by the Fifth Avenue Coach Company in Manhattan and extensively by the Surface Transit System in the Bronx. Several Queens operators also used Macks.

Mack exited the transit bus business in 1960. It continued to make intercity buses for several years, but eventually abandoned that business as well. In 1985, it attempted to reenter the intercity bus business in cooperation with a French manufacturer, but that experiment was short-lived.

Fig. 12.14 Mack Bus operating in Brooklyn

12.3.5 Flxible Transit Buses

Flxible was a very old business, but did not enter the transit bus business until 1951, when the company acquired Twin Coach. The company was formed in 1913 as a manufacturer of motorcycle side-cars. The company name results from dropping the letter "e" from the more common word. Because "flexible" was a word in common usage, the name could not be copyrighted and used as a trademark.

The side-car business dried up after Ford introduced his Model T at a cost of $360, which was less than the cost of a motorcycle and side-car. Flxible decided to enter the intercity bus market in early 1920s. Unlike most other bus manufacturers, Flxible focused on manufacturing bus bodies. They were mounted on chasses purchased from a variety of vehicle manufacturers, including Studebaker, Buick, and Chevrolet.

After several cooperative ventures with Fageol Twin Coach, Flxible bought the bus-building operation from Twin Coach in 1951, when the Fageols left the bus business to concentrate on trucks and aircraft parts. The Chicago Transit Authority was the first major client for the initial buses, usually referred to as "Flxible Twins," purchasing over 300 buses. The CTA buses were powered by propane. The Flxible Twin buses did not see extensive use in New York, although the Surface Transportation System acquired a used diesel-powered Flxible Twin in 1956 for use in the Bronx.

Fig. 12.15 Flxible Twin Coach in New York City operation, circa 1956

In 1961, Flxible discontinued production of the Flxible Twin, and introduced its version of the "new look" buses that had been introduced by General Motors. Through the 1960s and 1970s, the "new look" design were the only buses purchased for transit use in New York City. While General Motors snagged the lion's share of the sales, Flxibles were used by a number of private operators, and by the New York Transit Authority.

Fig. 12.16 Flxible "New Look Buses" operated by Queens Surface Co. and Triboro Coach Corp. circa 1980

In 1969, Flxible was taken over by Rohr Industries. Rohr continued to make "new look" buses until 1978, when Flxible was purchased by Grumman Allied Industries. Grumman discontinued production of "new look" buses and introduced its "advanced design bus" (ADB). The ADB had several innovations. Its driver windshield was angled to be glare-free. There were no external fastening devices, and floor heights were lowered to 30 inches for production models. The bus was also designed to provide easy access to the engine, transmission, and air conditioning equipment. The design made regular replacement of parts much simpler and faster than on previous buses.

In 1980, the New York City Transit Authority ordered over 1000 advanced design buses from Grumman–Flxible. Operation on New York City streets, however, caused a massive number of frame cracks to appear. Eventually, the entire fleet was pulled from service, when a steering shaft collapsed and dropped from the bus. The NYCTA was forced to lease and purchase many used buses from other systems to maintain basic service. After a long legal battle, the city could not prove that the cracks and other defects were the result of either a faulty design or a manufacturing defect. The NYCTA was forced to pay back $56,000,000 to the federal government (which had provided the money to buy the buses), and sold the buses back to Grumman–Flxible. The buses were refurbished and sold to a number of transit operators across the nation, including New Jersey. The NYCTA did not purchase Flxible buses thereafter, an action which weakened Grumman–Flxible's financial condition.

Fig. 12.17 Flxible Advanced Design Bus in New York

In 1983, Flxible was sold to General Automotive Corporation, and began manufacturing the Flxible Metro, the successor to the ADBs. It also launched a major program for the rebuilding and refurbishing of both Flxible and GM "New Look" buses. Despite these efforts, Flxible declared bankruptcy in 1995, and its assets were sold. A long-running legal battle over the handling of the bankruptcy and distribution of the assets continued into the early years of the 2000s.

12.3.6 General Motors Buses

General Motors entered the bus manufacturing business through cooperative arrangements with Yellow Coach. When it purchased Yellow Coach outright in 1943, GM became the largest American bus manufacturer for over 45 years.

In its history, GM produced a number of variations on three fundamental designs:

- The GM "old look" bus (1943–1959)
- The GM "new look" bus (1959–1977)
- The GM "advanced design" bus (1977–1986)

The "old look" bus was the continuation of earlier Yellow Coach designs. It was never referred to by the "old look" name until the introduction of the "new look" model in 1959. At that point, all previous designs were tagged as "old look" to differentiate them from newer models. Some of the smaller "old look" models were not actually built until 1969.

The buses were often referred to as "bread loaf" buses because of the general shape of the bus. The buses were built using a monocoque design, rather than the body-on-frame approach of most earlier buses. Virtually all of the "old look" buses were powered using a six-cylinder diesel engine, although some of the shorter versions were gasoline powered. Manual and automatic shift was available, but most buses were manufactured with automatic two-speed transmissions. From 1953 on, an air-ride suspension was standard, and air conditioning was offered as an option beginning in 1958. "Old look" GM buses were used throughout New York by most private and public operators.

Fig. 12.18 "Old look" GM Bus, circa 1956

The "new look" bus was introduced in 1959. It was also referred to as the "fishbowl" because of its six-piece rounded windshield. The "new look" bus used an airplane-like stressed-skin construction. In this design, an aluminum riveted skin supported the weight of the bus. Most of the "new look" buses were powered by two-cycle diesel engines. The entire engine, transmission, and radiator assembly was mounted on a cradle that could be quickly removed and repaired or replaced.

The GM "new look" bus was the single most successful model ever produced. There were over 41,000 "new look" buses in transit service between 1959 and 1977. In 1977, U.S. production of the "new look" bus ceased, and was replaced by the RTS bus. Canadian production, however, continued, as many Canadian transit operators had rejected the RTS design.

The GM "new look" bus was the most popular model in New York, and was extensively used by both private and public operators throughout its production period. The buses were found to be durable and efficient, and many remained in service a decade after the last bus had been manufactured.

Fig. 12.19 "New look" GM Buses, one operated by the NYCTA in Brooklyn, one operated by Queens–Steinway Transit on an express bus route

The Rapid Transit Series (RTS) bus is the last model manufactured by General Motors. The bus resulted from GM's participation in the USDOT Transbus project, through which subsidies were provided to develop an improved bus for urban transit use. The RTS had a more streamlined look, and featured a curved body side with curved window panels. The RTS bus was used throughout New York through the 1980s and into the 1990s. GM sold the RTS design and patent to the Transportation Manufacturing Corporation (TMC) of Roswell, New Mexico in May of 1987, and left the transit bus manufacturing business. TMS sold the design and patents to NovaBus in 1994. NovaBus left the U.S. market in 2002. Production was briefly revived by Millennium Transit Services, though the company went out of business in 2009.

Fig. 12.20 A new GM-manufactured RTS bus operated by Jamaica Bus Inc., circa 1985

12.4 The Bus Industry – Post-GM

With the departure of General Motors from the bus-building business in 1986, transit operators have become dependent on a number of foreign manufacturers, some of whom have set up small manufacturing facilities in the U.S., with very mixed success.

As early as 1976, MAN Nutzfhrzeuge AG, a major German bus manufacturer, formed a joint venture to manufacture articulated buses in the U.S. A plant was set up in Marshall, Texas, where bodies built in Germany were assembled. Sold under the name "AM General," ten non-articulated buses were purchased by Steinway Transit in New York. Their performance was deemed poor, and the buses were taken out of service.

Scania, a Swedish company, in conjunction with Saab, began assembling complete buses in Orange, Connecticut in 1984. The Saab–Scania factory closed after four years.

Volvo also set up a short-lived manufacturing facility for buses in Chesapeake, Virginia, producing buses under the name "Nova Bus." The factory lasted only two years, but continued its operations in Canada. In 2008, Nova Bus announced plans to open a new plant in the U.S.

Neoplan, a German company, established Neoplan USA in 1981, and opened a factory in Lamar, Colorado. Neoplans enjoyed some success in the U.S. transit market for some time, but the company folded in 2006.

New York has purchased buses from a variety of foreign manufacturers. The newest buses have been manufactured in Canada by Daimler Buses North America. Originally established as Orion International by the Government of Ontario in 1975, the company was privatized in 1993 and acquired by Daimler–Chrysler. When Daimler divested itself of Chrysler, the bus operation continued under the Daimler label, although the bus brand name remains Orion. The latest purchases from Orion have been diesel-electric hybrids.

Since 1995, new buses have been purchased from Nova Bus, New Flyer Industries (a Canadian manufacturer), and Daimler/Orion. A variety of equipment is in service, including standard diesel-powered vehicles, diesels using Compressed Natural Gas (CNG), and hybrids. NYCTA also operates a large number of articulated buses. For all of its express bus routes, NYCTA has purchased intercity-type coaches from Motor Coach Industries and Prevost.

The first NYC buses, as noted previously, were gas-electric hybrids. Certainly, the current hybrid technology is greatly advanced from the early days of the 20th century, but the fundamental concept was the same.

At this writing, New York is buying buses with three different technologies:

- *Compressed Natural Gas*: These buses generally provide reliable service and cleaner emissions. It has, however, proved costly to convert maintenance facilities to accommodate them.
- *Diesel-Electric Hybrid*: These vehicles offer advantages of requiring a smaller diesel engine, batteries that can recapture braking energy normally discharged as heat, and lower floors due to the elimination of bulky differentials. Higher capital and maintenance costs remain a problem, particularly the short service life of batteries and other electronic components.
- *Diesel*: The standard diesel engine has seen great improvements in both fuel efficiency and emissions.

New York continues to consider and analyze these alternatives. In terms of cost-benefit, the newer diesel engines may be the most cost-effective alternative over the next several years.

12.5 The Operating Companies

The corporate history of bus operators flows from the complicated, uncoordinated, and interlocking history of omnibus, horse-streetcar, and trolley operators. Like the rapid transit system, operations began as entirely private operations. As the bus business suffered the same financial difficulties as the subways, private operation gave way to public operation for most services. Along the path to public operation, several private companies continued their operations with direct subsidies from the city for a time.

The bus operators discussed in the sections that follow include all of the large operators with a substantial history. The list, however, is hardly exhaustive, as there have been numerous short-lived operators of only a few routes.

The boroughs have different histories, and thus the discussion is organized along borough-specific lines. Manhattan and the Bronx are treated as a unit, as there has been considerable overlap in the history of operators in the two boroughs.

(a) RTS bus manufactured by Nova Bus

(b) CNG-powered bus manufactured by New Flyer

(c) Articulated bus manufactured by New Flyer

(d) Orion VII electric-hybrid bus

(e) MCI express bus coach

(f) Prevost express bus coach

Fig. 12.21 Samples from the current fleet of NYC buses (2011)

12.5.1 Bus Operators in Manhattan and the Bronx

Fifth Avenue Coach Corporation: Fifth Avenue Coach was one of the earliest transit companies operating in New York. It began operations in October of 1885 as an omnibus operation along 5th Avenue. Since powerful residents consistently opposed the laying of rails along 5th Avenue, Fifth Avenue omnibuses had to compete with faster horse-streetcar lines on adjacent streets. To maintain operations, Fifth Avenue Coach charged a 10-cent fare, compared to the five cents charged by other omnibus and horse-streetcar operators. By 1901, Fifth Avenue had acquired additional franchises, and operated seven routes, most of which were in upper Manhattan.

After a failed attempt to operate battery buses, Fifth Avenue Coach became the first major operator of motor buses, having converted all of their routes to the motor bus by 1907.

In 1925, Fifth Avenue Coach was reorganized, becoming a subsidiary to the Omnibus Corporation, which was headed by a group of Chicago investors, including John Hertz. In 1954, Fifth Avenue Coach merged with the New York City Omnibus Corporation (Manhattan). In 1956, it purchased Surface Transit Inc. (the Bronx). In 1956, the merged operation was renamed Fifth Avenue Coach Lines, with New York City Omnibus operating as a subsidiary to it. In 1962, after a costly strike and a fight for corporate control with financier Harry Weinberg, the bus operations were placed under city control. The Manhattan and Bronx Surface Transit Operating Authority (MABSTOA) was formed for the purpose of the acquisition. MABSTOA continues to operate as a subsidiary of MTA Regional Bus operations.

At the time of the city takeover, Fifth Avenue Coach Lines was the largest operator of buses in NYC, and was still running at a profit.

Fifth Avenue Coach was famous for its operation of a fleet of double-decker buses along 5th Avenue until 1953, when poor acceleration and slow loading characteristics led to their replacement by single-level buses. Many of its buses in the late 1940s and 1950s sported a signature multicolored exterior, featuring a cream-colored roof and a two-tone green lower body.

(a) World-famous Fifth Avenue Double-Decker, circa 1940

(b) Traditional tricolor scheme of Fifth Avenue Coach on a GM "old look" bus

Fig. 12.22 Historic buses of the Fifth Avenue Coach Corporation

***New York City Omnibus Corporation*:** The New York City Omnibus Corporation was formed in 1923 as a holding company to acquire bus route franchises in New York. It took 14 years before the company began operations in 1936. Within a year, NYC Omnibus was the largest bus operator in NYC, with Fifth Avenue Coach in second place.

Despite being competing entities, NYC Omnibus and Fifth Avenue Coach shared owners, as both were subsidiaries of the Omnibus Corporation, and operations. Even the exterior of the buses were similar. In 1954, management of the two operations decided to buy the properties when the parent holding company offered them for sale. Each company operated separately and retained separate names on their buses. While the companies were originally merged under the name "New York City Omnibus Corp.," in 1956, the name was changed to "Fifth Avenue Coach Lines." The merged company was absorbed into MABSTOA in 1962.

Fig. 12.23 This "old look" NYC Omnibus Corporation bus is almost indistinguishable from Fifth Avenue Coach buses, as markings and colors are virtually the same

Avenue B and East Broadway Transit Corporation: The Avenue B and East Broadway Transit Corporation operated only two routes on the lower east side of Manhattan, but these were among the most heavily-used routes in the city. Its roots are in the horse-railways of the Dry Dock, East Broadway and Battery Railroad. Electrified in 1905, the Avenue B line was operated with storage-battery trolleys from 1911 to 1932, when the line was abandoned.

Bus operations were initiated in 1932. The franchises for the two routes were purchased by M. Bernard Greenberg in 1934, and the Avenue B and East Broadway Transit Corporation was born. The corporation outlasted its Manhattan rivals. Private operation ceased on March 29, 1980, when the franchises and operations were absorbed into MABSTOA.

Fig. 12.24 Ave B & East Broadway Transit Corp, Mack Bus 1955

Surface Transportation System: The Surface Transportation System can be traced back to 1853 and the Third Avenue Railway which eventually operated all of the streetcar lines in the Bronx. By the early 1920s, competition from the IRT significantly eroded the financial position of the trolleys. To keep the Bronx public transportation system under one operator, Third Avenue Railway created a subsidiary to operate buses. Surface Transportation was incorporated in 1924 for this purpose. By 1935, Surface Transportation obtained additional franchises that created a virtual monopoly for bus service in the Bronx.

The company ran into financial difficulties in the early 1940s. By 1948, the company entered receivership. In 1956, Fifth Avenue Coach purchased Surface, setting up a subsidiary, Surface Transit Inc., to operate its bus services. Private operation ended with the creation of MABSTOA in 1962.

Fig. 12.25 Surface Transit bus in the traditional red and cream markings; Mack Bus, circa 1951

***Manhattan and Bronx Surface Transit Operating Authority*:** MABSTOA was created in 1962 as a subsidiary to the NYCTA. It was created to take over operations from the financially defunct Fifth Avenue Coach Corp, which included the operations of the former NYC Omnibus Corp. and Surface Transit Inc.

The end of the Fifth Avenue Coach empire was quite dramatic. Harry Weinberg, a newly-elected board chairman precipitated a dispute with the city by threatening the elimination of all night and Sunday services if Fifth Avenue was not permitted to raise its fare from 15 to 20 cents. Within weeks, this led to labor disputes. On March 1, 1962, a strike was called, leaving most of Manhattan and all of the Bronx without bus service. The NYCTA, with the support of the mayor, pursued legal and legislative actions which resulted in a takeover of most of Fifth Avenue Coach's property and franchises by condemnation. By March 30th, NYCTA had reinstated service on all but two of Fifth Avenue's routes. These were restored to service on July 1, 1962.

MABSTOA was established as a separate subsidiary to allow Fifth Avenue's employees to avoid becoming part of the civil service system, and to allow the continuance of a separate pension fund for Fifth Avenue's employees.

By 1984, MABSTOA was no longer identified as a separate operating unit. Manhattan and Bronx bus services are currently operated under MTA Regional Bus Operations.

12.5.2 Bus Operators in Brooklyn

Brooklyn-Manhattan Transit Corporation: The Brooklyn-Manhattan Transit Corporation (BMT) was formed in 1923 when the former Brooklyn Rapid Transit Co (BRT) exited receivership after five years.

The BRT was initially established as a holding company with the objective of unifying elevateds, trolleys, and streetcars under a single operating agency. By 1902, BRT controlled all of the elevateds and all but two of the streetcar lines. Because of the restrictions requiring the five-cent fare and rapidly-rising costs associated with the new Dual Contracts subways, BRT was forced into receivership in 1918. While not the sole trigger, the Malbone Street wreck certainly hastened the trip to bankruptcy.

BMT created a subsidiary, the Brooklyn Bus Corporation, in 1924. Its objective was to obtain as many bus franchises as possible to maintain unified control over Brooklyn transit operations. Mayor John Hyland, at the tail end of his term, however, opposed giving BMT any substantial portion of the bus services in Brooklyn. His opposition to BMT and its predecessor (BRT) was deep-seeded, as he had been fired from his position as engineer after he nearly ran over a company executive on one of his runs.

The BMT operated for a year on temporary bus permits, and even offered free fares to demonstrate the need for their routes. In 1925, the NYC Board of Transportation decided to issue 23 permanent route franchises for the Borough of Brooklyn. BMT applied for the franchises.

In 1927, the Board granted all open bus franchises in all boroughs to the Equitable Coach Corporation. The action was specifically aimed at preventing street railways from obtaining control of bus operations. Equitable, however, refused to begin operations until they could be assured of profits from the proposed routes. The franchises were revoked by the Board of Transportation in 1929. After protracted negotiations and legal maneuvering, BMT was finally awarded permanent franchises to operate 20 bus routes in 1931.

***New York City Board of Transportation*:** As part of the process of unification, the BMT was acquired by the Board of Transportation on June 1, 1940. BMT was literally forced to sell, as it was bound by the five-cent fare, written into the Dual Contracts, and the ability of the City to acquire any part of the rapid transit system under the "recapture" provision of those contracts. In acquiring the rapid transit system, the Board of Transportation also obtained all of the bus operations of the BMT.

The 1940 purchase of the BMT was the city's first entry into the direct operation of transit services. Takeovers of some Queens and Staten Island bus lines soon followed. The period of the mid- to late-1940s involved a frenzied effort to update bus fleets and garage/maintenance facilities. Direct public operation of public transportation, however, would only last 13 years, with the creation of the New York City Transit Authority in 1953.

***New York City Transit Authority*:** In 1953, the legislature created the New York City Transit Authority to take over all transit operations and facilities in the city. The legislation also required that all trolley and bus lines be retained only until private buyers could be found. Since virtually no competitive bids were forthcoming, NYCTA retained virtually all of its bus operations. As time progressed, NYCTA would take over additional private bus operations, a trend unanticipated by the original legislation. Brooklyn bus operations are today part of MTA Regional Bus Operations.

12.5.3 Bus Operators in Queens

The development of Queens bus services differed somewhat from the other boroughs. Manhattan was dominated by a number of powerful private operators that sprung from the trolley routes. In the Bronx and Brooklyn, private and public interests worked to coordinate bus service under single operators.

Queens was originally served by many small operators. The city developed a plan to segregate the borough into four zones, with a single operator providing all services within each zone. In 1936, this "four-zone" plan was formally adopted. Triboro Coach was awarded all franchises in Zone A (Woodside and surrounding areas). Zone B (Flushing and surrounding areas) went to the North Shore Bus Company; Zone C (Ozone Park and surrounding areas) and Zone D (Jamaica and surrounding areas) was awarded to Bee Line. North Shore Bus Company took over Bee Line shortly thereafter, and was eventually the first private operator to be taken over by the NYCTA. After 1936, the general organization was preserved, with only a few exceptions, mostly in the Jamaica sector.

North Shore Bus Company: The North Shore Bus Company operated a number of bus routes in eastern Queens, many of which provided bus connections to rapid transit depots in Jamaica and Flushing.

The company was originally founded as the Rauchwerger Bus Company by Joseph Rauchwerger in 1923 to take over two streetcar lines operated by the New York and North Shore Traction Company. He founded North Shore Bus Company in 1926 when additional routes were added.

North Shore routes were generally long, and served areas with no other transit. As such, they were heavily utilized. During WWII, the company fell on hard times, as both new buses and parts were difficult to acquire. Its fleet of prewar buses fell into disrepair. After court intervention and arbitration, the NYCTA took over its routes in 1947. Its routes are currently operated by MTA Regional Bus Operations.

(a) North Shore operates one of only two REO buses that run in NYC

(b) One of the few White buses used in NYC transit operation

Fig. 12.26 Unique North Shore buses operating on Main Street, Flushing routes

Triboro Coach Corporation: Triboro Coach was originally founded in 1919 by Salvatore Fornatora as the Woodside-Astoria Transportation Company. Triboro was officially born on April 10, 1931.

Triboro was one of the most successful private bus operators, and provided most of the bus service in northwestern Queens for decades. In 1936, it acquired nine additional franchises from the city. The "hub" of its operation was the 74th Street/Roosevelt Avenue junction of the Corona/Flushing Subway and the Queens IND line. It operated out of a unique three-lane, multibay terminal located under an adjacent building, which is still used.

After WWII, Triboro was taken over by another private operator, Green Bus Lines, but continued as a separate operation under the "Triboro" name. Always looking to expand, Triboro extended the Q33 route into LaGuardia Airport when it opened. It was also the first private operator to enter the express bus field. The first "express" route in 1956 was a replacement for the abandoned Long Island Railroad Rockaway Branch. The service continued until the LIRR service was restored as an extension of the A-train subway route.

In 1961, Triboro actually took over a route being operated by the NYCTA, the B72, which was renumbered the Q72, as its entire route was within Queens. In the 1970s and 1980s, Triboro obtained franchises for six express bus routes.

Triboro Coach was one of the last companies to operate privately. While it had been receiving direct subsidies from the city for over a decade, it operated until February 2, 2006, when its operations were taken over by MTA Regional Bus Operations.

Fig. 12.27 Triboro Coach "old look" GM Bus

Jamaica Buses Inc.: In 1931, a city plan to widen Jamaica Avenue led to the demise of Jamaica Central Railways, which had operated several trolley lines using Jamaica Avenue. Jamaica Buses was formed in 1933 to replace the trolley lines with motor buses.

Jamaica Buses was never a large operation. It operated four local bus lines emanating from the trolley lines it replaced, and one express bus service initiated in the early 1970s. After WWII, it was purchased by Green Bus Lines, but continued to operate independently under the "Jamaica Buses" name. On January 30, 2006, MTA Regional Bus Operations took over all of its operations.

Fig. 12.28 A Jamaica Bus vehicle operating on an express route from Jamaica to Midtown Manhattan

Green Bus Lines: Green Bus Lines was incorporated on April 3, 1925, providing local bus service in several boroughs. In 1933, it acquired several Manhattan routes, but in a swap of franchises, turned these over to NYC Omnibus Corp in 1936 in exchange for franchises in south-central Queens.

In the period immediately after WWII, Green Bus Lines acquired Triboro Coach Corporation, Jamaica Buses Inc. and the Command Bus Company (a small private operator in Brooklyn). In each case, they continued to operate each under its own brand.

Like Triboro Coach, Green Lines was aggressive in providing new services. It established an express bus service in the early 1950s to replace Long Island Railroad service to and from the Brooklyn Manor station on the abandoned Rockaway Beach Branch. In the 1970s, it introduced four additional express bus lines. Green Line services were taken over by MTA Regional Bus Operations on January 9, 2006.

Fig. 12.29 A Green Lines RTS bus operating on Queens Blvd.

***Queens Transit/Steinway Transit*:** Both Queens Transit and Steinway Transit trace back to streetcar and trolley lines of the early 1900s. The Queens-Nassau Transit Corp. was formed to replace trolley lines of the New York and Queens Transit Corporation with buses in 1937.

The Steinway Railway Company was incorporated in 1892. It leased the Rikers Avenue and Sanford Point Railroad and acquired the Steinway and Hunters Point Railroad. The company was subsequently merged into the New York and Queens County Railroad. In 1932, a reorganization left the Steinway operations under the control of the New York and Queens Transit Corporation, that is, the same parent company that controlled Queens, Nassau Transit. In 1938, Steinway operations were bought out by the Queensboro Bridge Railway Company, which set up a subsidiary called the Steinway Omnibus Corporation. The Queensboro Bridge Railway Company was the last operating trolley in New York, controlling the single route that ran back and forth over the 59th Street Bridge until 1957.

In 1986, Queens Transit and Steinway Transit were merged into a single company, aptly named the Queens/Steinway Transit Corporation. In 1988, just two years later, the company was acquired by Lindon Bus Company. Before operations began, however, Lindon changed its name to the Queens Surface Corporation.

Unlike many of its competitors, Queens Transit began operations using a fleet of 41 ACF buses. After WWII, they were the only company to resist standardizing their fleet with GM or Mack Buses. They purchased 15 ACF–Brill buses in 1947, the only postwar ACF–Brills to be purchased in NYC. Steinway made a more conventional choice, and initially equipped themselves with Twin Coaches and Macks.

(a) A Queens-Nassau ACF–Brill bus, circa 1947

(b) A Queens-Nassau ACF H16 bus, circa 1945

Fig. 12.30 Unique buses used by Queens-Nassau Transit, predecessor to the Queens/Steinway system

The company operated local bus routes primarily in northwestern Queens. Like other Queens operators, it operated a number of express bus routes as well. When the company was taken over by MTA Regional Bus Operations on February 27, 2005, it was running 11 local bus routes, nine express routes to and from Manhattan, and one interborough local route over the Whitestone Bridge.

12.5.4 Bus Operators on Staten Island

Even today, Staten Island has the lowest population density of the New York boroughs. Until the opening of the Verrazano Bridge in 1964, the only connection between the other boroughs was via ferries, most of which operated out of the St. George area and connected to Manhattan.

Because of this, early omnibus and horse-streetcar operations, beginning in 1867, focused on bringing people from relatively remote areas and over long distances to the ferry docks in St. George. The current structure of bus routes in the borough still reflects this characteristic, although some other local services have been added, as have been a significant number of express buses operating over the bridge into Brooklyn and Manhattan.

Tompkins Bus Company: The company was originally formed in 1925 as the American Travel Corporation, with the intent of applying for bus franchises throughout the borough of Staten Island. The company was awarded 18 bus routes in 1927, and was permitted to charge a fare of 10 cents, double the citywide five-cent fare in the other boroughs. It is not clear whether the city allowed the 10-cent fare due to the much longer distances involved in Staten Island routes, or simply an oversight by the city. The company had great difficulty getting its routes certified for operation, with four routes taking until 1931 to begin operation. Tompkins operation was brief. By 1937, the company was in serious financial trouble, and its franchises were awarded to the Staten Island Coach Company.

Staten Island Coach Co./Isle Transportation Co.: Formed in 1902, the Richmond Light & Railroad Company operated streetcar services in Staten Island, although the actual operation was leased to the Midland Electric Railroad. When the streetcar system ceased operation in 1920, a plethora of private buses began operating over four routes. In 1927, the successor to Richmond Light & Railroad, Richmond Railways, set up the Staten Island Coach Company, and purchased 25 buses, expecting to be awarded franchises for 18 routes created by NYC. Unfortunately, the city awarded all 18 routes to Tompkins.

Richmond Railways continued to operate a number of trolley lines until 1933, when they reestablished the Staten Island Coach Company, and replaced the remaining trolley lines with motor buses. As noted previously, in 1937, they took over all operations of Tompkins. When this was done, the fare was dropped from ten to five cents.

Due to financial problems, Staten Island Coach did not renew their franchises in 1946. Isle Transportation took over the franchises and the operations, but proved to be unsuccessful. In 1947, the company was taken over by the Board of Transportation.

Until the opening of the Verrazano Bridge in 1964, most Staten Island bus routes made connections with the ferry. After the bridge opened, a significant number of express bus routes were initiated, as well as several local routes that used the bridge to connect with Brooklyn and Manhattan.

12.6 The Era of Public Operation

Virtually all of the bus services in New York City, and indeed throughout the nation, are now operated by local governments or government agencies. New York was indeed unique: large private bus operations continued in Queens until 2006, although direct public subsidies to these companies had been given for over a decade.

The economics of bus operation deteriorated steadily from the end of WWII. The costs of operation, including labor, equipment, maintenance, and fuel continued to rise, while fares were controlled by local agreements. The argument for keeping fares as low as possible has always been tied to affordability. Many bus systems had become the primary source of mobility for low-income residents. Faced with the practical inability to raise fares to meet expenses, private operations began to vanish.

With the deteriorating state of bus finances, it became almost impossible to keep up with increasing demands for service, and to keep an operable fleet of buses on the street.

Help for equipment costs began to flow from the federal government in 1961, when $25 million was authorized for mass transit demonstration projects as part of the Housing Act. The demonstration projects provided support for planning and loans for capital improvements. Demonstration projects were funded to test such ideas as express buses, alternatives to fixed-route transit services, and other ideas.

In 1964, the Department of Transportation became a cabinet-level agency. In the same year, the Urban Mass Transportation Act was passed. In 1968, the Urban Transportation Administration, housed in HUD (Housing and Urban

Development) was moved into the DOT. The 1964 Act providing funding for capital investments in transit covering up to two-thirds of the total cost.

Despite this, federal funding for transit depended on periodic legislation that had to be renewed. Changes in administration and the general political climate made planning for such funding difficult, if not impossible.

In the mid-1970s, the federal government set up the Transbus program, providing funds for manufacturers to develop a more efficient bus design. Three companies agreed to participate in the program, that is, GM, AM General, and Rohr Flxible. Each company produced a prototype, but the program was deemed unsuccessful, and the prototypes were scrapped. Nevertheless, both GM and Flxible produced new bus designs by 1977. The GM Rapid Transit System (RTS) bus and the Flxible Advanced Design Bus (ADS) were clearly descendants of the program.

Beginning in 1983, federal transit financing was folded into legislation and programs for highway finance (which had been in continuous existence since 1916). The Federal-Aid Highway Act of 1983 introduced the "interstate trade in" provision. This allowed, at state and local option, funding for uncompleted portions of the Interstate System to be "traded" for equivalent amounts of funding for public transit systems. Boston and New York were the first major users of this provision.

In 1991, the largest overhaul of federal-aid highway programs occurred with the Intermodal Surface Transportation Efficiency Act (ISTEA). In terms of transit, it provided significantly increased flexibility for state and local governments to divide their federal funds among highway and transit projects. The successor act, the Transportation Equity Act for the 21st Century (TEA-21), continued and expanded this trend in 1998.

At this writing, all federal funding for mass transportation is contained in the most expensive transportation act in history, that is, the Safe, Accountable, Flexible, and Efficient Transportation Equity Act – A Legacy for Users (SAFETY-LU), passed in 2005. While the initial act only extended through 2008, continuing resolutions have kept the program in place while a successor act is debated and written. As of late 2011, no successor act has been passed, although it is under congressional discussion as part of the broader efforts to attack generally-deteriorating infrastructure across the nation.

With the federal government providing substantial support for capital expenses involved in transit, including the purchase of buses, state and local operators could concentrate on operating issues. In New York, the city's takeover of private bus properties spanned decades. The city's first entry into direct operation of buses came in Brooklyn in 1940, with the "recapture" of the BMT. By 1947, it operated most of the buses in Staten Island. Manhattan and Bronx operations were acquired in 1962 and 1980. In Queens, subsidized private operation was maintained through 2005 and 2006.

12.7 Flirting with Double-Deckers

On September 15, 1976, the MTA conducted an experiment using new double-decker buses along 5th Avenue and Riverside Drive. The experiment was an attempt to recapture the nostalgia of Fifth Avenue Coach's double-decker fleet that operated along 5th Avenue until April of 1953. They purchased a fleet of eight new buses from Leyland, a British manufacturer (Model AN682K) and assigned them to the M4 and M5 routes operated by MABSTOA. The buses were air-conditioned and had a seating capacity of 69 passengers, 44 of them on the upper deck. If the experiment was deemed successful, additional buses would have been added to the fleet.

In many ways, the buses were highly successful. They had higher capacity, the public was very positive towards them, and they were a draw for tourists. Several problems, however, doomed the experiment. The buses were too tall for the doors of the 132nd Street bus depot, so they had to be reassigned to the 146th Street depot, and traffic lights along 5th and Madison Avenues had to be raised. The last runs were made in 1981, and the buses were sold to Gray Line of San Francisco, where they were used for tourist services.

The experiment with double-decker buses was financed through the Urban Mass Transportation Administration (UMTA) as a demonstration project.

12.8 Modern Problems for an Old System

As has been discussed, the current bus system in New York emanates from franchises originally awarded to omnibus and horse-streetcar operators, and other franchises awarded in a haphazard and uncoordinated fashion.

Particularly in Queens and Staten Island, buses cover long routes that in most cases became necessary when the expansion of subways was halted in the 1940s. The long-planned connection from Staten Island to Brooklyn's 4th Avenue Subway never happened. In Queens, plans to extend most subways and/or elevateds to the city line were also shelved. In these cases, buses are trying to service passenger volumes and distances that would be far more efficiently accommodated by rail.

In Manhattan, buses struggle to handle traffic on the East Side, particularly the Lower East Side, where the departure of elevateds left this part of the island without rapid transit service. Express buses attempt to provide service to riders who cannot easily access a rail line to get into and out of Manhattan.

There are many smaller cities in the nation where buses provide the only form of public transportation to modest ridership densities. New York is not one of these. Buses work most efficiently when they cover shorter distances and serve less intense ridership. In New York, this means as feeders to the rail system.

In Queens, this is what buses do, except it feeds to an incomplete rail system that covers only about one-third of the borough. In Brooklyn, many buses provide efficient feeders to a rail system that covers a great deal of the borough. Other routes serve direct connections within the borough that are either too short or inconvenient to make via rail. In Staten Island, express buses make up for the lack of a rail connection, while local buses provide feeders to the Staten Island Ferry. In the Bronx, feeder service is more efficient, but there are still areas not well-served by rail.

The NYC bus system is amongst the largest in the world. Over 4400 buses ply the streets every day carrying a daily load of 2.4 *million* people per day. A mammoth operation on its own, it is dwarfed by the New York rail system. Its role in keeping the city and its residents mobile, however, continues to be indispensible.

References

1. Martin, G.E.: New York City Transit Buses: 1945-1975 Photo Archive, Iconografix, Hudson WI (2005)
2. Luke, W.A., Metler, L.L.: City Transit Buses of the 20th Century, Iconografix, Hudson WI (2005)
3. Luke, W.A.: Yellow Coach Buses: 1923 Through 1943 Photo Archive. Iconografix, Hudson WI (2001)
4. Luke, W.A.: Buses of ACF Photo Archive. Iconografix, Hudson WI (2003)
5. McKane, J.: Flxible Transit Buses: 1953 Through 1995 Photo Archive. Iconografix, Hudson, WI (2001)
6. Exckart, H.: Mack Buses 1900 Through 1960 Photo Archive. Iconographix, Hudson WI (2000)
7. Luke, W.A.: Fageol & Twin Coach Buses: 1922 Through 1956 Photo Archive. Iconografix, Hudson WI (2002)

Figure Sources

12.1: *Courtesy of* Wikimedia Commons.
12.2: *Courtesy of* Wikimedia Commons.
12.3: *Courtesy of* William A. Luke, from Luke, W., and Metler, L., *City Transit Buses of the 20th Century*, Iconographics, Hudson, WI, 2005, pg 9.
12.4: *Courtesy of* William A. Luke, from Luke, W., and Metler, L., *City Transit Buses of the 20th Century*, Iconographics, Hudson, WI, 2005, pg 8.
12.5: *Courtesy of* William A. Luke, from Luke, W., and Metler, L., *City Transit Buses of the 20th Century*, Iconographics, Hudson, WI, 2005, pg 17.
12.6: *Courtesy of* William A. Luke, from Luke, W., *Fageol & Twin Coach Buses: 1922 Through 1956 Photo Archive*, Iconografix, Hudson WI, 2002, pg 23 and 30.
12.7: *Courtesy of* William A. Luke, from Luke, W., and Metler, L., *City Transit Buses of the 20th Century*, Iconographics, Hudson, WI, 2005, pg 18.
12.8: *Courtesy of* William A. Luke, from Luke, W, *Yellow Coach Buses 1923–1943 Photo Archive*, Iconografix, Hudson WI, 2001, pg 42.
12.9: *Courtesy of* William A. Luke, from Luke, W, *Yellow Coach Buses 1923–1943 Photo Archive*, Iconografix, Hudson WI, 2001, pg 60.
12.10: *Courtesy of* William A. Luke, from Luke, W, *Yellow Coach Buses 1923–1943 Photo Archive*, Iconografix, Hudson WI, 2001, pg 55.
12.11: *Courtesy of* William A. Luke, from Luke, W, *Yellow Coach Buses 1923–1943 Photo Archive*, Iconografix, Hudson WI, 2001, pg 62.

12.12: *Courtesy of* William A. Luke, from Luke, W., *Buses of ACF Photo Archive*, Iconografix, Hudson WI, pg 27 and 63.
12.13: *Courtesy of* Mack Museum, Detroit MI.
12.14: *Courtesy of* New York Transit Museum.
12.15: *Courtesy of* Motor Bus Society.
12.16: *Courtesy of* Motor Bus Society.
12.17: *Courtesy of* New York Transit Museum.
12.18: *Courtesy of* New York Transit Museum.
12.19: *Courtesy of* Motor Bus Society.
12.20: *Courtesy of* Motor Bus Society.
12.21: *Courtesy of* Wikimedia Commons, all photographs by Adam E. Moreira, used under the Creative Commons-Share Alike 3.0 unported license.
12.22: *Courtesy of* Motor Bus Society.
12.23: *Courtesy of* Motor Bus Society.
12.24: *Courtesy of* Motor Bus Society.
12.25: *Courtesy of* Motor Bus Society.
12.26: *Courtesy of* Motor Bus Society.
12.27: *Courtesy of* Motor Bus Society.
12.28: *Courtesy of* Motor Bus Society.
12.29: *Courtesy of* Motor Bus Society.
12.30: *Courtesy of* Motor Bus Society.

Chapter 13
Unification, War, and the New York City Transit Authority

By the time the Independent Subway's Eighth Avenue line was opened to the public in 1932, the economics of public transportation had radically changed. Anchored to a political policy of retaining the five-cent fare, public transportation, and rapid transit in particular, were doomed to be money-losing enterprises. It was clear that it was necessary for the city to provide massive funding to keep the system running, and that private entrepreneurs were no longer trying to get into the business. Those that were still in it, the IRT and the BMT, were looking for ways to get out. Between 1932 and 1953, the city would struggle to determine an effective means of dealing with the responsibility of providing and operating public transportation.

13.1 Introduction

In 1932, the IRT was just entering receivership, and the BMT was in deep financial trouble. Because the fare was regulated by political bodies, the companies had no means to increase their revenues in the face of rapidly-inflating costs. To survive the financial conditions of the late 1920s, the IRT and BMT introduced a number of new technological features, and some very unpopular policies to reduce their costs, primarily by reducing their workforce. Two measures were critical:

- Since the beginning of the elevateds and the subway, the train fare was collected manually by a ticket-taker as passengers entered the platform. Automated turnstiles that could be operated by dropping the appropriate coins into a slot were invented and placed throughout the system, eliminating many positions.
- Another great inefficiency involved the opening of doors on elevated and subway cars. Originally, a gatekeeper was needed between every two cars, as each gatekeeper could only directly operate the doors in the two cars adjacent to his location. When automatic door operation was developed, the crew of a six-car train went from a motorman and three gatekeepers to a motorman and one conductor.

These measures alone were not sufficient, however. Salaries of employees lucky enough to keep their jobs were reduced by as much as 10%, and they were forced to work longer hours. Station and train cleaning was severely curtailed, and the physical condition of the system deteriorated rapidly in very visible ways. Replacement of older, out-dated cars was deferred, and rolling stock was kept in service far longer than was prudent. Breakdowns and delays multiplied as a result.

R.P. Roess and G. Sansone: The Wheels That Drove New York, STTT 1, pp. 281–305.
springerlink.com

The city's new Independent Subway proved to a drain on the city's resources as well. It cost about nine cents per rider to operate the system, while the revenues were only five cents per rider. The city itself was in dire financial straights from the effects of the Great Depression. It was still required to pay off the bonds issued for construction of the Dual Contract lines, but was receiving no income from the IRT and BMT under the revenue-sharing agreement because there were no profits when "preferentials" were considered. The bond payments, therefore, had to be made directly from the city's treasury.

13.2 From Hylan to LaGuardia: A Change in the Political Climate

In 1925, Tammany Hall did not support Mayor Hylan for reelection to a third term. He was viewed by Hall leaders as too independent. With Tammany support, Hylan was defeated in the Democratic primary by Jimmy Walker, who easily won election as Mayor.

13.2.1 Jimmy Walker

Jimmy Walker was almost a polar opposite of Hylan. He was a flamboyant personality, handsome, and very much a "man about town." He had served four years in the State Assembly and four more in the State Senate prior to his election as mayor. Most importantly, however, he was a loyal Tammany Hall member.

As a young man, Walker dabbled as a songwriter, and penned several songs that became popular. He was well-known in Broadway circles, and, while married, he was frequently linked to various Broadway showgirls. Walker, insofar was possible, ignored prohibition, and "speakeasies" flourished during his administration. He quickly earned the nickname "Beau James." He became a fashion icon, with various publications following his attire in significant detail.

Fig. 13.1 Mayor Jimmy Walker in 1926

His first term was marked by investment in infrastructure, with schools, hospitals, and parks seeing much improvement. His involvement in transit was minimal, but he continued the tradition of maintaining the five-cent fare, and supported the completion of the Independent Subway. He was reelected in 1929 by a landslide over Fiorello LaGuardia and a Socialist Party candidate.

The Great Depression, however, ended the good times of the Roaring '20s and plunged the city into fiscal distress. With the financial difficulties of the depression came greater scrutiny of Walker's administration, supported by the Democratic governor, Franklin D. Roosevelt.

The initial investigation focused on the inferior courts of New York and Bronx Counties. It was alleged that Walker and his Tammany Hall cronies used the court system to enforce its will. Individuals, it was alleged, would be arrested for non-existent "crimes." Professional witnesses were employed to provide damning testimony. The victims would either pay bribes or be sent to jail. The immediate cause of the investigation was an allegation that a municipal judgeship had been sold for a $10,000 payment to Tammany Hall (1).

On August 26, 1930, Judge Samuel Seabury was appointed to referee the investigation (2). Seabury was a well-known jurist with a reputation for honesty and opposition to the influence of Tammany Hall. In December 1930, the investigation was expanded to include the State District Attorney, Thomas Crain, who was a Tammany regular (3).

The extension of the investigation to include Crain caused a serious and permanent rift between Roosevelt and Tammany Hall. Roosevelt, who was already angling for the Democratic nomination for President in the 1932 race, would be subject to fierce opposition from the Hall in this regard. It was, in effect, the beginning of a long feud between the Hall and the Roosevelt's that would continue into the 1960s and end in the Hall's ultimate demise.

In March of 1931, Walker took a trip to California, supposedly to address unspecified health issues. It was rumored that he was not going to return, and that he would shortly resign as Mayor (4). He was thought to be seeking various opportunities in the Hollywood movie business. The Mayor denied the rumors, and indicated that "... the stories of corruption are greatly exaggerated"(5).

When the City Affairs Committee levied additional corruption charges against Walker himself, Governor Roosevelt, under the State constitution, had to personally judge the case. He appointed Judge Seabury to assist as Special Counsel (6). The legislature, at the same time, established an investigating committee to consider the entire issue of corruption in NYC government, and named Samuel Seabury as its chief counsel as well (7). In reporting on the formation of this committee, the *Washington Post* observed:

> *"A city-wide investigation is a logical outcome of the limited inquiries that have been conducted recently. Corruption was discovered in the courts and the police department. The district attorney's office is under investigation. But all these attempts to strike out corruption are futile. No one individual or group is responsible for the deplorable conditions in New York. It is the system which Tammany Hall has built up that is at fault. Individuals holding out against the ruthless methods of the tiger have been crushed and pushed aside. A probe into the practices of the invisible government is the only hope of getting to the bottom of New York's corruption."* (Ref. 7, pg 6)

Tammany Hall was once again under scrutiny for a wide range of illegal activities. Moreover, the City Affairs Committee and other civic groups were providing for substantial public support for the investigation. Over 3000 people gathered at Carnegie Hall on March 30, 1931 to hear speeches from a number of civic leaders describing the alleged corruption of Tammany Hall (8).

On April 6, the Mayor indeed returned from his California sojourn to defend his office and his political allies from the charges levied against him. Amid additional rumors of a romance between the Mayor and showgirl Betty Compton, the Mayor's wife, who had been spending time in Miami Beach, returned to provide support (9). By April 18, the Mayor had prepared a response to the charges of the City Affairs Committee, and sent it to the Governor (10). Never one to hang around too long, Walker departed again on May 15th with a group of twenty-five mayors of American cities on a formal tour of France (11).

With the investigation unearthing more and more evidence against Walker and the Hall, the Mayor once again left New York in August 1931 for Europe to address still unidentified health issues (12). The famous political commentator and humorist, Will Rogers, wrote the following in a letter to the *New York Times* (13):

> *"France gave Jimmy Walker the Legion of Honor because he didn't bring nineteen other mayors with him."*
>
> *"Every time the Republicans start to pin a rosette of poison ivy on him in New York, he is sipping a stein of beer with Hindenburg, shooting craps with Premier MacDonald, or doing an adagio dance with Mussolini."*
>
> *"No man was ever so investigated on one hand and dined on the other. He keeps just three decorations ahead of the investigating committee."*
> (Ref. 13, pg 19)

On May 26, 1932, Walker finally appeared to testify before the Seabury Commission. As part of his testimony, he acknowledged that he had received a $246,692 profit without any investment from a brokerage account with Paul Block, who was a prominent publisher. It was alleged that this "profit" was payment for influencing the award of a bus franchise, a charge that he denied. The courtroom crowd cheered the Mayor like a celebrity. When he had finished his testimony, he was asked to sign a waiver of immunity, which he did with great flourish before the crowd (14). Within three days, the investigating commission began to assemble its case, which was to be presented to Governor Roosevelt, who had the constitutional duty to review it, and the ability to remove the Mayor from office (15). Even Walker's supporters expected such a dismissal, but were confident that Walker would be reelected.

What followed involved a number of tactics aimed at delaying Walker's trial before the Governor. At one point, Walker indicated that all of the evidence presented involved incidents from his first term in office, which he claimed was a "closed incident," and that there was no evidence of wrongdoing during his second term (16). The trial finally took place in August of 1932 (17, 18). Without waiting for Roosevelt to render a decision, Mayor Walker resigned his position on September 1, 1932, vowing to submit his record to the "the people who made me mayor, the people of New York" (19). He announced his intention to run for reelection in the Special Election to be held to complete his term. On September 12, 1932, Walker sailed once again for Europe for a 20-day trip, once again seeking improvement in his health (20). His wife accompanied him to the dock, but did not sail with him.

13.2.2 Joseph V. McKee

With the resignation of Jimmy Walker as Mayor, his place was taken by the President of the Board of Alderman, Joseph V. McKee, on September 2, 1932, in accordance with succession law at the time.

Fig. 13.2 Interim Mayor Joseph V. McKee

McKee's reign would last only four months. The governor and McKee hoped that the constitutional provisions on special elections were loose enough to allow McKee to serve out the remainder of Jimmy Walker's second term. The courts, however, ruled that a Special Election would have to be held on the next Election Day, in November of 1932 (21).

After a steady stream of statements and rumors that Jimmy Walker would run in the Special Election with the support of Tammany Hall, he eventually thought better of the idea and withdrew (22). Following his usual approach, Walker sailed for Europe on November 10, 1932. Betty Compton sailed on the same ship, accompanied by her mother (23). While in Europe, Walker divorced his wife, and married Compton. He remained in Europe until any threat of criminal prosecution dissipated.

During his short time as Mayor, McKee developed a reputation as a hard worker who attempted to attack the major problems of the city, most of which were heightened by the depression and the lavish spending of his immediate predecessors. His efforts while in office focused on reducing spending (24).

In terms of transit, he supported the five-cent fare, and was active in trying to preserve the city's recapture rights to the IRT and BMT. The IRT was already in receivership, but the city did not have the financial resources to exercise recapture immediately. The Eighth Avenue Subway opened during his tenure, but he did not take part in any formal activities marking the event. In fact, there were virtually no formal ceremonies associated with the opening. As the system had been running test trains for weeks, the doors to all stations were opened simultaneously at 12:01 on September 10, 1932 (24, 25).

His approach to reducing expenditures was to slash the salaries of city workers, beginning with his own, which was reduced from $40,000 to $25,000 per year. He reversed a $2,000,000 salary increase Walker obtained for higher members of his administration, and reduced the salary of all department heads to $12,000 per year (26). He also permanently garaged the Mayor's car, taking the subway to work, and reassigned its driver to another job (27), while at the same time reviewing the need for each of the 1800 cars and drivers provided to various city officials. He then ordered a 20% reduction in all city agency budgets (28) and additional cuts in the salaries of all city workers.

None of these measures made him popular with Tammany Hall. The Hall declined to support him in the Special Election, and nominated Surrogate John P. O'Brien for Mayor.

13.2.3 John P. O'Brien

The election of 1932 was a high point for both Tammany Hall and the Democrats. Franklin D. Roosevelt won the Presidency in a landslide over Herbert Hoover. In New York, Tammany Hall regained control of both the state and city with the election of Herbert Lehman as Governor and John P. O'Brien as Mayor (29). Less than two months after Jimmy Walker was forced to resign, Tammany Hall was back in charge in New York.

Fig. 13.3 Mayor John P. O'Brien

In December of 1932, before Mayor-Elect O'Brien took office, the Board of Alderman, still in the control of Tammany Hall Democrats, restored many of the pay cuts ordered by Mayor McKee (30).

On the subject of transit, O'Brien continued the tradition of support for the five-cent fare, and was in favor of recapture and unification. His administration continued discussions with the bankrupt IRT and soon-to-be bankrupt BMT without coming to any conclusion (31).

Budget-cutting, however, remained the number one issue for the city. After having overturned McKee's salary cuts, O'Brien and the Board of Alderman approved smaller reductions (32). He advocated having an "executive budget" that could not be changed or reduced by action of either the Board of Alderman or the Board of Estimate (33), but did not get it.

When his police commissioner resigned to accept a state post, O'Brien issued one of the more memorable quotes of any mayor in the city's history. When asked who his new commissioner would be, he supposedly replied "I don't know. They haven't told me yet." The actual quote was less spectacular, although the public response clearly interpreted its meaning to be the same as the more popularized version. When the question was asked, his actual response was "I have had no word on that" (34). O'Brien later contended that he was misquoted, and had actually said "I have no word for you," but any number of witnesses supported the actual quote (35).

One year after his election, O'Brien squared off in the regular mayoral election against Fusion Party candidate Fiorello LaGuardia and Joseph V. McKee, the former mayor of four months. The Fusion Party was a combination of Republicans and Progressive Independents, while McKee ran on a party made up of Roosevelt Democrats, who were opposed to Tammany Hall. LaGuardia, won the election with a plurality, but the Democrats continued to control a number of key city offices.

13.3 LaGuardia to the Rescue

Fiorello LaGuardia graduated from the New York University Law School, and opened a private law practice. He was active in lower Manhattan's Madison Republican Club. Although his philosophies were quite progressive, he joined the more conservative Republicans because he despised the corruption of Tammany Hall, and thought that his Italian-American heritage would hurt him in the Irish-dominated NY Democratic Party. He served several terms in Congress and earned a reputation as a fiery orator, opposing Prohibition and denouncing restrictions on immigration as racist. He was defeated in 1932 by a Democrat, riding the substantial coattails of the Franklin Delano Roosevelt landslide. Ironically, he and Roosevelt were close friends, and he would later benefit from having considerable influence with the new President.

LaGuardia won the election despite an extensive effort by Tammany Hall to defeat him. The Hall was not opposed to strong-arm tactics to persuade voters to support their candidate. As reported in the *Manchester Guardian* on November 9, 1933 (36):

> *"The election in New York passed off fairly quietly by American standards, since no one was killed. Tammany's desperate efforts to win, however, resulted in charges of wholesale vote-stealing. There were many cases of voters being intimidated and beaten or bodily ejected from the polling booths when they were believed to have the intention of voting for LaGuardia or McKee."* (Ref. 36, pg 4)

When he took office, the Depression was in full swing. At a time when New Yorker's needed increasing amounts of aid and assistance, the city's treasury was in serious trouble. The lavish capital spending of Mayors Hylan and Walker left the city with a veritable mountain of debt; the declining tax revenues of the Depression forced the city to borrow against future earnings to pay its bills.

Fig. 13.4 Mayor Fiorello LaGuardia circa 1934

LaGuardia was a man of almost infinite energy, who would immerse himself in the affairs of the city as no mayor before him had ever done. He despised organized crime and the negative image they created for Italian immigrants. Throughout his term, he fought "tooth and nail" to indict mobsters and remove them from influence in New York. He broke up Frank Costello's illegal slot machine empire, taking great glee in personally and publicly destroying several seized machines with a sledgehammer, and caused Lucky Luciano to be investigated and successfully prosecuted.

The war on slot machines had been started under Mayor McKee. They were associated with gambling, although they were not legally classified as "gambling equipment." Because of this, many local stores and businesses added to their revenue by having slot machines on their premises. LaGuardia began confiscating slot machines, particularly in locations where other crimes could be connected to their existence (37). On February 18, 1934, LaGuardia ordered well-publicized raids on three prominent locations, seizing 360 slot machines (38). The legality of the seizures was, however, challenged. On April 3, 1934, the U.S. Supreme Court agreed to hear arguments on the legality of local governments banning slot machines, and to seize them when found (39). When NYS laws were passed specifically outlawing slot machines (previously, they were only outlawed when actually used for gambling purpose), the Supreme Court ruled that the case was no longer relevant, and that the city had the legal power to seize slot machines under any circumstances (40). On October 12, 1934, 1155 slot machines which had been seized were smashed and dumped into Long Island Sound with great publicity, including the Mayor personally taking a sledgehammer to several of them.

LaGuardia was not overly interested in the subways. They were merely a problem that had to be solved. He supported the five-cent fare, and wanted the Independent Subway system completed as soon as possible. He also vigorously pursued recapture and unification of the rapid transit system under direct city ownership and operation.

In terms of the future, he believed that the automobile was the future, and it was under Mayor LaGuardia that Robert Moses rose to prominence and built the huge network of parkways and bridges that still form a major part of the modern New York highway system.

Completion of the Independent was by no means assured. Given the effects of the Depression on city finances, it was not at all clear that the there were sufficient funds to support additional construction bonds to complete it. Moreover, the city had no obvious way to finance recaptures on the IRT and BMT.

LaGuardia, however, had an ace in the hole. He had a close relationship with now President Franklin D. Roosevelt, as he shared the President's progressive philosophy. Because of this relationship, LaGuardia was able to obtain a number of grants under the Public Works Authority set up by Roosevelt to fund infrastructure projects and spur employment. While still Mayor-Elect, LaGuardia visited Roosevelt to lobby for a PWA grant to support the completion of the Independent system. As a result of the visit, the city received a $23,160,000 construction grant for use in building the remaining sections of the new subway (41).

He was also a strong proponent of unification. He believed that unification of all transit under direct city control was the only way to provide a system that could be financially solvent, while keeping the five-cent fare. While unification was under discussion through the terms of Hylan, Walker, McKee, and O'Brien, it would be LaGuardia who finally made it happen. Like everything else in New York politics, however, unification did not come easily.

13.4 The Road to Unification

LaGuardia believed that unification of the subways and elevateds under direct city operation would be a major step in the restoration of the city's financial situation. Unification, he believed, would:

- Save millions by eliminating the Dual Contract preferential provisions and allowing interest and sinking fund costs to be paid from subway revenues rather than out of tax revenues.
- Allow the city to refinance corporate securities at lower interest rates, significantly lowering the city's debt service.

- Lower operating and capital costs through the efficient management of a combined system, eliminating duplication and waste.

The problem was that unification required that the IRT and BMT be essentially purchased by the city. The Dual Contracts did include a "recapture" clause that could be used, but the price would have to be negotiated and/or litigated. Further, any reorganization of the companies would have to be approved by the state-controlled Transit Commission, which by 1932, had become (once again) dominated by Tammany Hall interests.

Both the IRT and BMT expected, and even supported unification. They wanted the best possible deal, but preferred a negotiated settlement to being taken over through the recapture clause of the Dual Contracts. In the latter case, the city could choose to take over parts of their systems, leaving them with a fragmented set of the most financially distressed lines or portions of lines.

While McKee and O'Brien had generally supported the concept of unification, and had started discussions with the IRT and BMT, both were wary of the enabling legislation for the Independent, which required a self-supporting fare after an initial operating period of three years. With the election of LaGuardia, and Tammany still in control of the Transit Commission, tepid support would turn to active opposition. Because of this, LaGuardia's initial battle was to obtain control over the process from the Transit Commission. This had to be done in the state legislature and signed by the new Governor Lehman, a Tammany regular.

LaGuardia proceeded by conducting his own discussions with the traction companies in January of 1934, initially led by city Chamberlain, A. A. Berle Jr. (42). The initial position of legal representatives for the bondholders of the IRT and BMT was that any "recapture" would have to be at par value of their holdings, an estimated $363,000,000 (43). On January 26, 1934, Samuel Seabury was named as Special Counsel for negotiations with the traction companies on unification and related issues (44). He would serve at no salary, and would team with Chamberlain Berle and a member of the city's Corporation Counsel, Paul Windels, to represent the city's interests in the unification effort.

Meanwhile, in February, Mayor LaGuardia ordered that the city decline to pay its portion of the salaries of Transit Commission members (45). The TC obtained a restraining order preventing the city from withholding its payment. In replying, the city claimed that the state law mandating city payment of salaries violated the "home rule" provision of the State constitution, which, as a result of a population greater than 1,000,000, allowed the city to control many functions that would normally fall to the State (46). This argument, however, failed, and the city was forced to continue its payments to the TC.

The city also pressed for state legislation to remove the Transit Commission, and give the city direct power over unification. By April of 1934, however, it was clear that this legislation would not be passed (47).

As 1935 approached, the end of the three-year operating period for the Independent loomed large. In essence, the enabling legislation allowed the city to use tax revenues to subsidize Independent operations for three years, after which fares would have to be raised to cover all expenses. LaGuardia, fully supporting the five-cent fare needed to find a legal remedy to the deadline. The city's legal arguments focused on the issue of when the three-year period began. They argued that the period did not start with the opening of the 8th Avenue subway on September 10, 1932, but when the last piece of the planned Independent system opened, an event that was still six years in the future. It was also argued that the limitation did not apply to the use of tax revenues to pay bond interest (48). When its legal challenges failed, state legislation was sought to expand the city's right to operate the Independent at a loss through 1937 (49).

On May 5, 1935, the State legislature passed the "Transit Unification Bill," which provided some of what Mayor LaGuardia wanted, but not all. It also contained difficult provisions. The city got the right to continue to subsidized operations to retain the five-cent fare, but also required a public referendum on any proposal to increase the fare. It established a 15-member Board of Transit Control, with eleven members appointed by the Mayor, with a three-fourths approval of the Board of Estimate (50). The new Board would actually operate the Independent system, and would operate any traction companies acquired by the city through recapture and unification. The Transit Commission still had to approve any negotiated deal for recapture of the IRT and/or BMT.

On November 1, 1935, a memorandum of understanding with the IRT and BMT was reached. The two companies would be purchased by the city for a gross total of $430,000,000, with a net cost of $416,861,000 when certain liquid assets of the properties were included. The new Board of Transit Control would issue bonds to raise the cash needed to complete the transactions (51). The agreement, however, had to be approved by the U.S. District Court, as both the IRT and BMT were in receivership, and by the Transit Commission. After studying it for three months, the Transit Commission rejected the plan on February 4, 1936 (52). In rejecting the agreement, the TC argued that it could not act on memoranda of agreement, only final contracts, which it had not seen. It also claimed that the deal was a bailout of the IRT and BMT, and that it would doom the five-cent fare.

The Tammany-controlled Transit Commission hoped to use the issue of unification to unseat LaGuardia in the 1937 elections, but he won in a landslide. The Fusion Party also captured control of the Board of Estimate in the election.

John Delaney, still chair of the city's Board of Transportation, led a new negotiating team to work with the IRT and BMT on a new agreement for unification. The delay in unification actually helped the city: Due to an economic downturn in 1937, the value of stock in the IRT and BMT had fallen precipitously. Purchasing their rapid transit lines had become a much more financially-attractive proposition.

On July 1, 1939, the Mayor signed a contract with BMT President William Mendon to sell the company's properties, including its street railways and bus lines, for $175,000,000 (53). On November 1, 1939, the IRT signed a deal for $151,000,000 (54). Moreover, the agreements gave direct control of the system to the city's Board of Transportation. The sole remaining obstacle was to find a way to either increase, or get around, the city's debt limit. This was accomplished through a state constitutional convention that was meeting at the time. A provision was inserted into the amended constitution excluding the amount borrowed to buy the IRT and BMT from the debt limit.

The actual acquisition process was complicated, as individual investors and bond-holders in the IRT and BMT had to deposit their securities, that is, hand them over for sale. After some complicated individual negotiations, the Transit Commission, now on board with unification after the 1937 elections, declared the IRT acquisition operative on November 22, 1939. Some BMT investors held out for better settlements, but the BMT acquisition was completed on May 31, 1940. The New York City Transit System had been created. Despite the fact that the system was to be operated in a unified fashion, IRT, BMT, and IND "divisions" kept the original identities alive.

13.5 The Decline of the Elevateds

Once unification was achieved, the Mayor had two other objectives with regard to transit: In Manhattan, he wanted to eliminate the elevateds and the trolleys. He viewed elevateds as unnecessary with the addition of the Independent system, and trolleys as unnecessary impediments to traffic.

The IND subway essentially duplicated service provided by the 9th Avenue and 6th Avenue Els in Manhattan, and the Fulton Street El in Brooklyn. Part of the unification plan was to eliminate costly duplications, and those three elevateds were the first major components to be eliminated. However, before they could be demolished, the city had to acquire them. The elevateds were not included in the IRT and BMT deals, as the city had no recapture rights over them. The IRT elevateds were held in a subsidiary organization, the Manhattan Railway Company.

The first El to come down was the 6th Avenue El, made superfluous by the new 6th Avenue Subway that would run directly beneath it. Moreover, there was reason to remove it *before* construction of the 6th Avenue Subway, as the construction cost would be reduced, and the process made substantially faster with the El gone. On April 27, 1938, the city agreed to purchase the 6th Avenue El for \$3,874,475 and forgiveness of \$8,625,525 in back taxes owed by the elevated (55). There was much discussion of the plan to eliminate Manhattan's elevated lines, and several lawsuits were filed against it. When the dust settled, however, the city's right to demolish the Els had been upheld, and the process began.

With great fanfare and media attention, on December 20, 1939, Mayor LaGuardia donned a construction helmet and proceeded to make the first cut through a beam using an acetylene torch provided to him by a construction worker (56). The end of the Manhattan elevateds had begun.

Over time, the environmental disadvantages of the elevateds would accelerate demands for their removal, and eventually, even the 3rd and 2nd Avenue Els would be torn down, despite not being replaced by an East Side Subway.

At the end of 1939, the city's rapid transit system was at its apex. The Sixth Avenue El had already been torn down, and the Sixth Avenue Subway would not open until 1940. Except for one extension of the Fulton Street Subway in

Brooklyn and the Sixth Avenue Subway, the Independent was complete, including the temporary World's Fair Line. All of the Brooklyn Elevateds, that is, Lexington Avenue, Myrtle Avenue, Fulton Street, and Fifth Avenue, were still operating, as were the Ninth Avenue, Third Avenue, and Second Avenue Els in Manhattan.

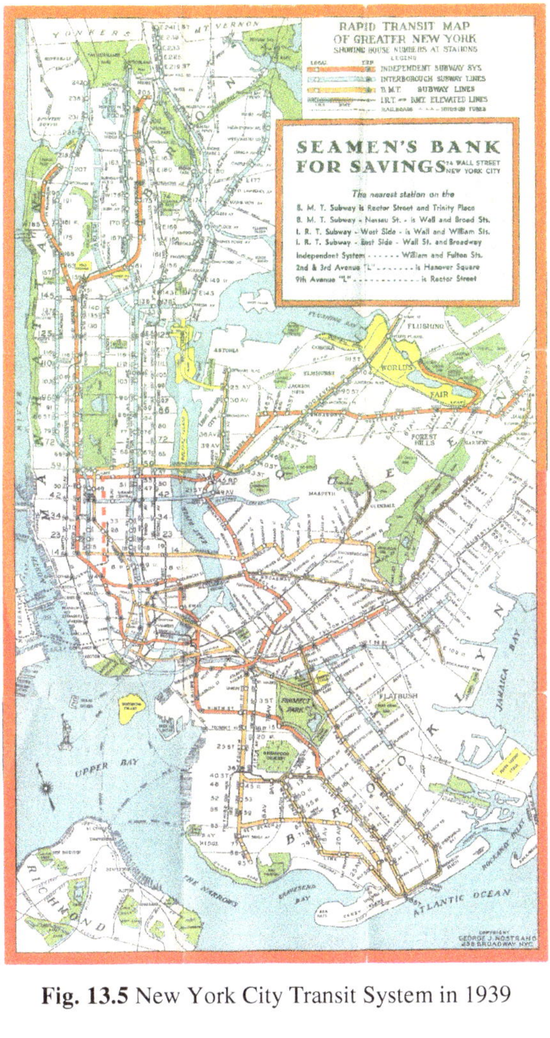

Fig. 13.5 New York City Transit System in 1939

The elevateds had pushed the development of New York to the north, allowing, indeed, the growth of upper Manhattan and the Bronx. They provided a vital lifeline to millions of New Yorkers who could now commute on a daily basis over distances that were simply unheard of in the early- and mid-1800s. Without them, New York as we know it would not have come to exist. Nevertheless, time marches on, and the faster subways replaced their function. The noise, blockage of light, steel and paint dust, and other environmental problems of the elevateds were eventually intolerable.

All of the Els in Manhattan, and some of the Els in Brooklyn, came down in bits and pieces over time. Some small sections of the elevateds had been eliminated earlier, when they were no longer needed or duplicated by other service:

- The old Park Avenue El segment of Brooklyn's Lexington Avenue El ceased operations before electrification on December 30, 1890.
- The 42nd Street Spur from the Third Avenue El to Grand Central was eliminated on December 6, 1923.
- The Sixth Avenue extension between 53rd Street and 58th Street was eliminated on June 16, 1924.
- The main section of the Sixth Avenue El from Morris Street to Ninth Avenue and 53rd Street shut down on December 4, 1938.
- The Fulton Street El in Brooklyn stopped service from Fulton Ferry and Park Row to Rockaway Avenue on June 1, 1940.
- Both the Ninth Avenue El (from South Ferry to 155th Street) and the Second Avenue El (from 57th Street to 129th Street) ceased operation on June 11, 1940.
- The remainder of the Second Avenue El, including the connection to Queens over the Queensborough Bridge, was eliminated on June 13, 1942.
- The entire Lexington Avenue El in Brooklyn was taken out of service on October 13, 1950.
- The Third Avenue El from South Ferry to Chatham Square shut down on November 14, 1951; the City Hall Branch to Chatham Square was discontinued on December 31, 1953.
- The main part of the Third Avenue El in Manhattan, from Chatham Square to 149th Street, shut down on May 12, 1955.
- The Ninth Avenue El branch from 155th Street to 167th Street and Jerome Avenue (which had served as a shuttle to the Polo Grounds while the New York Giants baseball team was still here), was taken out of service on August 31, 1958.
- The Bronx section of the Third Avenue El shut down on April 29, 1973.
- The Myrtle Avenue El in Brooklyn between Jay Street and Broadway shut down in 1969; the section from Jay Street over the Brooklyn Bridge to City Hall had ceased operation in 1944.

Not all of the elevateds outside of Manhattan were eliminated. There are many elevated sections of the rapid transit system in Brooklyn, Queens, and the Bronx that still operate. The "oldest" section of elevated still operating is the a segment of the Broadway El in Brooklyn, which was a part of the Brooklyn Elevated's original "main line." It has been structurally reinforced over the years to allow heavier cars to operate on it.

13.6 The War Years

World War II provided the transit system with a fleeting period of economic prosperity. Transit ridership had fallen during the depression years, as even the five-cent fare became a barrier for many. The war economy began to bring the nation out of depression, and civilian employment and salaries were rising. At the same time, the rationing of gasoline and rubber kept cars at home, and forced New Yorkers back to the subways. From a low point of 1.7 billion passengers in 1933, ridership rebounded to 1.9 billion in 1943, and peaked in 1947 with over 2 billion patrons. When LaGuardia left office in December 1945, declining to run for a fourth term, the New York City Transit System had a surplus of $25,000,000. The job of mayor, however, had taken its toll on LaGuardia, and within two years of leaving office, he died.

In a sense, New York City's transit system was following the same path as the nation's railroads. Both were in dire financial condition before the war, and both experienced a significant surge in patronage and revenues during the war. In both cases, however, the surge would not be sustained, and the financial crises of the prewar period would return.

When the war ended, the postwar economic boom began in earnest. The economy was rolling along, having recovered from the decade-long depression. Detroit began to churn out automobiles in record numbers, as Americans made the car their transportation of choice. Between 1945 and 1960, automobile registrations increased by almost 140% while ridership on the New York City Transit System slipped back to 1.3 billion passengers. By the end of the 1940s, the city was once again dipping into its tax revenues to fund mounting deficits, and by 1950, the fight to keep the five-cent fare would be over.

13.7 The Five-Cent Fare Succumbs

From the very beginnings of the rapid transit system, the five-cent fare had become an article of political faith. No mayor dared to advocate its increase, lest the stand end his political career. The issue, however, had been raised over the years, and Mayors Hylan, Walker, McKee, O'Brien, and LaGuardia had to answer frequent calls for a fare increase as a solution to the ever-mounting financial travails of the rapid transit system. Once it became clear that the five-cent fare would lead to deteriorating service and city subsidies to transit operators, pressure for its increase mounted.

13.7.1 Mayor William O'Dwyer

In 1945, Tammany Hall once again regained City Hall with the election of Mayor William O'Dwyer. An affable Irishman and Tammany Hall regular, O'Dwyer had lost the election to LaGuardia in 1941. He had been appointed as Magistrate by Mayor McKee in 1931, and was elected as Brooklyn District Attorney in 1939 (57).

As District Attorney, he became famous for successfully breaking up the infamous Murder Incorporated, and convicting its leader, Louis Lepke, for murder. The zeal with which he prosecuted the gang may have been influenced by an incident in 1918, which occurred while O'Dwyer was working as a policeman. Answering a domestic disturbance call, a young boy told him that his father was trying to kill his mother, and O'Dwyer was forced to kill the father who turned towards him with a drawn gun. He personally clothed and educated the son, and later, as Mayor, inducted him into the police force. After this incident, O'Dwyer developed a life-long aversion to guns and those who used them to commit crimes (57).

After losing the 1941 election to LaGuardia, he took a leave of absence to join the Army, and was commissioned as a Major in the Inspector General's office to lead an investigation into graft in military contracts. He left to run again in 1945, leaving with the rank of Brigadier General.

Fig. 13.6 Mayor William O'Dwyer

When O'Dwyer succeeded LaGuardia, he faced an unprecedented political situation with regard to the subways. Since the dawn of the elevateds, there was always a private company or companies to blame for whatever shortcomings in service, or fiscal difficulties arose. Throughout the Hylan and early LaGuardia eras, the IRT and the BMT had borne the brunt of criticism for the fiscal crisis, and for the failure to provide a remedy. With the city in full charge of the transit system, however, politicians now bore direct responsibility. The office of mayor had moved from being the source and leadership of transit criticism to the subject of it. While LaGuardia's general popularity provided some protection from public ire over the transit system, O'Dwyer would not be afforded this luxury.

The operating surplus of the unified system dropped steadily throughout the 1940s, and sustained its first deficit in 1947, a total of $18,000,000. If the required capital debt burden were considered, the actual deficit was much larger; in fact, deficits had been mounting since 1944.

13.7.2 Michael Quill and the TWU

Labor costs had risen rapidly after the war in the face of a strong labor union, now headed by a fiery Irishman who would dominate the transit labor scene until his death in 1967, namely, Michael Quill.

Mike Quill came to the U.S. from Ireland in 1926, arriving on St. Patrick's Day. He initially found work as a construction laborer, and worked on the construction of the 168th Street Station of the IND. In 1928, he became a ticket agent for the IRT, and worked at many stations where he came into contact with many transit workers.

After returning from a European trip in 1932, he learned that the company union had agreed to a 10% reduction in wages to help save the IRT, which was already in receivership. Enraged at this, he began to urge transit workers to form an independent union, which had to be done with the utmost secrecy. While he had some initial success, he did not make significant headway until he allied himself with the Communist-controlled Trade Union Unity League. On April 12, 1934, the Transport Workers Union (TWU) was formed. In 1935, Quill quit his job at the IRT to become the full-time president of the union.

By 1948, Quill was under a great deal of pressure to disassociate the union from its communist ties. He was also under pressure to win a substantial wage increase for transit workers, given that the depression and fiscal problems of the traction companies had badly eroded their position.

While Mayor O'Dwyer and Mike Quill represented adversarial interests, they were personally quite close. The two played off each other in a way that (a) allowed the City to raise the transit fare, (b) removed the communist influence from the TWU, and (c) left Mike Quill in firm control of the union.

13.7.3 O'Dwyer, Quill, and the Fare Increase

Mayor O'Dwyer's position on the five-cent fare was not particularly consistent. Robert Moses, who had been retained from the LaGuardia Administration and given additional powers by O'Dwyer, wanted an immediate fare increase to eliminate the use of any city funds to subsidize transit, a practice that limited the availability of funds to advance his highway projects. The Citizen's Transit Committee, led by Paul Windels, LaGuardia's Corporation Counsel, also wanted to see an increase. O'Dwyer initially put forth a plan to raise the fare to eight cents in February 1948, and needed the State to repeal the law requiring a public referendum on any fare increase (58). By March, he had reversed his position, and was opposed to a fare increase (59). He tempered his opposition by saying he would accept an increase if the State increased its contributions to the city, and if the state allowed the city to raise its property taxes. An article in the *New York Times* on March 16, 1948, pointed out the dilemma O'Dwyer and other supporters of the five-cent fare found themselves in (60):

> *"Now, for the first time, perhaps, those who support the five-cent fare come hell and high water will be brought face-to-face with the consequences of their position. They are the ones who will have to answer for it if there is no pay rise for city employees. They are the ones who will have to answer to 'the people' for a 'five-cent hospital and five-cent health system'...They are the ones who would be responsible to 'the people' if labor unrest in the subway system developed to a point where the people's own safety was in jeopardy."* (Ref. 60, pg 26)

On March 30, 1948, Governor Thomas Dewey signed into law a bill allowing the NYC Board of Transportation, with the Mayor's approval, to raise the transit fare without requiring a public referendum. After initially asking for such legislation, Mayor O'Dwyer now opposed it (61).

Meanwhile, Mike Quill was bombastically demanding a 30-cent per hour wage increase for transit employees (62). On April 14, Quill provided the Mayor with political cover to raise the fare. After years of opposing fare increases, Quill finally realized that without a fare increase, he could not get a decent wage increase for his workers. He called for a fare increase to pay for a wage increase of 30 cents an hour, and threatened to strike if such an increase was not provided (63).

On April 21, 1948, Mayor O'Dwyer officially announced a fare increase to become effective on July 1st. Its specifics were as follows (64):

> *"1. On the subway and elevated lines, 10 cents*
> *2. On the buses and trolley lines, seven cents*
> *3. For a combined surface and rapid transit ride, 12 cents*
> *4. For children in free elementary schools, no charge*
> *5. For children in free high schools, five cents*
> *6. Children under six years of age, accompanied by an adult, no charge*
> *7. Brooklyn, Queens, and Staten Island fare zones, eliminated*
> *8. Free transfers between BMT, IRT, and Independent systems at designated points*
> *9. All existing transfer privileges between surface lines – continued; and all present two-cent transfer charges – eliminated."*
> (Ref 64, pg 22)

The five-cent fare had survived over 50 years, from the earliest days of the elevateds until 1948. The stress of city ownership of transit, escalating costs, and the impacts of continued subsidies finally turned the tables.

Eight days after the fare increase was announced, the Mayor held a meeting with Quill and a number of his adversaries in the union. The Mayor asked Quill what he needed prevent a strike. He indicated that a 24-cent an hour increase would be sufficient, an amount even his union adversaries thought was outrageously high. O'Dwyer immediately agreed. Quill's position in the union was instantly strengthened, and his adversaries slowly faded from view.

Quill proceeded to purge the union of its Communist ties. On July 29, 1948, Local 100 of the TWU ordered all of it leaders to "withdraw from positions of leadership in the American Labor Party or quit their union jobs" (65). In September, two leaders of the national union, both of whom were American Labor Party leaders as well, called for a strike against private bus lines in the city. Quill denounced the call, and urged that the two leaders be dismissed at an upcoming union convention, saying "There will be no Communist-dictated bus strike at this time" (66). In December of 1948, a new union constitution was adopted, specifically barring "Communists and their supporters" from holding positions in the union (67).

The 10-cent fare did not save the system from financial woes for long, but it did break the tradition that had governed transit politics for over 50 years. Retention of the five-cent fare surely impeded the expansion of the subways, and improvements to service. New financial crises, however, were on the horizon. While there was only one fare increase in the 50 years between 1898 and 1948, there would be many over the subsequent 50 years.

13.8 Mayor O'Dwyer Departs

O'Dwyer won a landslide victory for a second term in 1949. By 1950, however, his health was failing, and a scandal concerning a protection racket and the NYC police was brewing. While citing his health as the reason, O'Dwyer resigned on September 1, 1950. Shortly thereafter, President Truman appointed him as Ambassador to Mexico. O'Dwyer returned briefly to testify, but immediately returned to Mexico. Even after he completed his service as Ambassador, O'Dwyer remained in Mexico, finally returning to the U.S. in 1960. Once again, a Tammany Hall politician left office under a cloud of suspicion.

13.9 The New York City Transit Authority Arrives

Unification of the transit system under direct city control took almost two decades to achieve, and involved six mayors. It only lasted for 13 years.

13.9.1 Support for a Public Authority

Paul Windels formed a group known as the Committee of Fifteen in 1940. A civic group formed with realtors and other businessmen, the group began promoting the view that New York's high real estate taxes put local businesses at a disadvantage. Wasteful spending had to be controlled, they argued, to avoid the loss of many private sector jobs, thereby destroying the tax base and endangering the recovery from the depression. Windels had been active in Fiorello's initial election, and served as his Corporation Counsel during his first term. He left, however, after LaGuardia's reelection in 1937, and became an outspoken critic of many of his policies, among them, LaGuardia's support for the five-cent fare. The Committee criticized other municipal actions as well, including wage increases for city workers, installing a three-platoon system for the fire department, expansion of welfare, and expenditures on parks, hospitals, and other services. The five-cent fare was its principal target, however.

Windels and the Committee believed that transit should be run as a private enterprise, and that no subsidies (i.e., direct city payments) for operating or capital costs was justifiable. From their formation in 1940, they argued strenuously for a 10-cent fare.

Their most vocal opponent was Stanley Isaacs, a City Council member, and former Manhattan Borough President. He argued that the capital costs of transit should be paid for by real estate taxes on the properties that had appreciated in value because of the system. He also argued that supporters of the 10-cent fare were simply trying to free up city money for highway expansion. A 10-cent fare, while making transit self-supporting, would also be providing a subsidy for the private automobile, at a time when only one of three New Yorkers owned one.

Supporters of a higher fare made little headway as long as transit subsidies were only related to capital costs and debt retirement. In 1947, the situation changed, when the system experienced its first operating deficit. When Mayor O'Dwyer finally raised the fare in 1948, public sentiment was running 52-1 against it. Even though Windels supported the fare increase, he used the widespread dissatisfaction of the public to argue for removing control of transit from the hands of the politicians. What he proposed was to place control of the transit system in the hands of an independent public authority.

There were successful models of such authorities in place. The Port Authority was already established, and was doing well in coordinating highway and other transportation facilities across two states. London Transport, a public authority, had been in place to run the underground since 1933, and was viewed as a great success at the time.

Windels' case was further supported when the 10-cent fare proved to be a failure in providing long-term fiscal stability to the transit system. By 1950, the operating deficit alone had already climbed to $1.2 million after only two years of surpluses under the 10-cent fare. In 1952, the system lost and alarming $24.8 million, but that would more than double in 1953. Rising labor and other costs and declining ridership could not be overcome. The argument for removing control of transit from direct political control gained support from a broad set of constituencies.

13.9.2 Governor Dewey and the New York City Transit Authority

On March 10, 1953, Governor Thomas E Dewey asked the legislature to establish an authority to run the New York City transit system. He argued that the authority would remove transit charges from the NYC capital budget, and would foster independent and business-like management and decision-making. Despite opposition from labor and liberal groups, the Republican-controlled legislature quickly approved the authority. The Governor signed the bill creating the New York City Transit Authority (NYCTA) on March 20, 1953, only ten days after asking the legislature for its approval (68).

The formation of the New York City Transit Authority was not without significant controversy. Pushed by Republican Governor Thomas Dewey and the Republican-controlled legislature, it was received by city officials with some coolness. The Board of Estimate, in initially approving of the creation of the Authority, sought to have the legislation modified to include significant new business taxes to subsidize transit operations. In his March 10, 1952 message to the legislature, the Governor harshly criticized the Board for attempting to "eviscerate the proposal," and assailed other major tax increases enacted by the city since 1945 (69).

The Board of Estimate was initially deadlocked. While the state could create the agency without the Board's approval, the Board had to approve the leasing of the system to the new Authority. It had the power to prevent implementation of the plan. City Council President Rudolph Halley and Manhattan Borough President Robert F. Wagner were opposed to the plan, as was City Controller Joseph Lazarus. The three controlled enough votes to prevent acceptance of the Authority. In truth, virtually none of the Board's members liked the plan, but some recognized that it would be necessary to approve it to save the city from financial disaster.

The Mayor who had to deal with this impasse was Vincent Impellitteri. He had been elected as President of the City Council in 1945, with the support of Mayor O'Dwyer and Tammany Hall. When O'Dwyer resigned, under the existing rules of succession, Vincent Impellitteri became Acting Mayor. Tammany Hall declined to support him when he ran in a Special Election to complete O'Dwyer's term, but he was elected on the "Experience Party" line. He thus became the first Mayor elected without the support of a major political party. His margin was small, however, and he only received 44% of the vote in a four-way race.

Fig. 13.7 Mayor Vincent Impellitteri

On March 26, 1953, the Mayor took to the airways to give a speech on the transit situation. In it, he argued that the New York City Transit Authority was a necessary, though not entirely desirable, measure. He attacked both Wagner and Halley, accusing them of providing no alternative to the city's fiscal problems. Failure to accept the Authority would, he declared, force significant reductions in fire and police services, the closing of hospitals, and other draconian measures. He attacked Wagner for suggesting that New York City become the 49th State, and

Halley for suggesting several measures that the city did not have the legal authority to implement. At the same time, he blasted Governor Dewey for essentially telling the city to accept the Transit Authority or be denied the legislation needed to raise local property taxes.

Comptroller Lazarus Joseph eventually agreed to accept the plan with the understanding that the new authority would immediately identify significant savings in operating costs, and implement cost-saving measures. He also insisted on some technical changes in the way depreciation costs were handled, reducing their impact on operating costs.

The Board of Estimate was hardly the only hold-out. The NYC Board of Transportation, which would now be stripped of its control of transit, was strongly opposed, and claimed that sufficient operating cost efficiencies could be made to improve the fiscal situation. Mike Quill, President of the Transit Workers' Union, threatened a strike if the Authority was implemented, but did not follow through with the threat.

In the end, with the reluctant support of the Comptroller, the city agreed to lease the transit system for a period of 10 years to the New York City Transit Authority on June 1, 1953. The NYCTA took over operation of the system on June 15, 1953 (70).

Subway historian Clifton Hood argues that the removal of the system from direct political control doomed it to the financial woes that would only get worse through the 1960s and 1970s. Without the power to tax and subsidize the system, the NYCTA was left with no option to operating deficits than to reduce costs through deferred maintenance and service curtailments and raise the fare to get more revenue (71).

The NYCTA would not work as Windels or Dewey had expected. The financial crises of the '60s and '70s would dwarf those of previous decades, and the system would suffer from them. The NYCTA could not insulate the transit system from the overwhelming fiscal crises that would hit the city. It is doubtful, however, that any management system could have.

References

1. Municipal Scandal in New York. The Manchester Guardian, 6 (August 22, 1930)
2. Seabury, Long Tammany's Foe, To Head Inquiry. The Christian Science Monitor, 2 (August 27, 1930)
3. Probe of Crain's Office Launched. The Atlanta Constitution, 2 (March 9, 1931)
4. Mayor Walker Runs Into Rumor Barrage Here. Chicago Daily Tribune, 5 (March 11, 1931)
5. Anti-Tammany Crusade. The Manchester Guardian, 9 (March 16, 1931)
6. To Rule on Mayor, Governor Insists. New York Times, 1 (March 25, 1931)
7. Probe of New York City. The Washington Post, 6 (March 26, 1931)
8. Mass Meeting Denounces Walker and His Regime; Ousting in 100 Days Seen. New York Times, 1 (March 31, 1931)

9. Link Names of Mayor Walker, Betty Compton. Chicago Daily Tribune, 5 (April 7, 1931)
10. Walker Will Rely Solely on Record; Reply Ready Today. New York Times, 1 (April 16, 1931)
11. American Mayors Sail For France. The Atlanta Constitution, 5 (May 16, 1931)
12. New York Mayor's Bank Account. The Irish Times, 10 (August 7, 1931)
13. Mr. Rogers Holds Our Mayor Has At Least One Distinction. New York Times, 19 (September 14, 1931)
14. Mayor Walker Bares $246,692 Stock Profit With No Investment. The Washington Post, 1 (May 26, 1932)
15. Review of the Testimony in Seabury's Investigation of Walker's Official Acts. New York Times, 10 (May 29, 1932)
16. Mayor Walker's Defense. Los Angeles Times, A4 (August 1, 1932)
17. Roosevelt to 'Try' Walker. Chicago Daily Tribune, 1 (August 7, 1932)
18. Walker Pleads with Governor in Own Defense. Chicago Daily Tribune, 1 (August 12, 1932)
19. Charges Hearing is Courtmartial; Prejudice Filled. The Atlanta Constitution, 1A (September 2, 1932)
20. Ex-Mayor of New York Coming to Europe. The Manchester Guardian, 13 (September 12, 1932)
21. Jimmy Walker May 'Come Back'. The Irish Times, 7 (October 7, 1932)
22. Mayor of New York. The Washington Post, 6 (October 10, 1932)
23. Jimmy Walker and Betty Compton Sail for Europe on Same Boat. The Atlanta Constitution, 6 (November 11, 1932)
24. M'Kee Promises to Slash Budget. New York Times, 1 (September 3, 1932)
25. City Subway Opens To Public Saturday Without Ceremony. New York Times, 1 (September 4, 1932)
26. McKee Swings Ax On Salaries of High Figure. Christian Science Monitor, 1 (September 7, 1932)
27. Salary Cut Program Widened By M'Kee. New York Times, 1 (September 8, 1932)
28. M'Kee Orders 80-100 Million Cuts In Budget. New York Times, 2 (September 10, 1932)
29. Tammany Is Elated But Not Surprised. New York Times, 17 (November 9, 1932)
30. Tammany Aldermen Defeat M'Kee Veto; Budget Cut Up Today. New York Times, 1 (December 23, 1932)
31. O'Brien Due to Act For Transit Unity. New York Times, 19 (January 2, 1933)
32. $19,112,068 Slash In City Pay Voted; Mayor Cut $10,085. New York Times, 1 (January 7, 1933)
33. Mayor O'Brien's Program. New York Times, 18 (January 4, 1933)
34. O'Brien Will Name Man From Force To Head The Police. New York Times, 1 (April 12, 1933)
35. Delay By O'Brien Leaves the Police Without A Head. New York Times, 1 (April 13, 1933)
36. New York's New Mayor. The Manchester Guardian, 4 (November 9, 1933)
37. Mayor Spurs War on Slot Machines. New York Times, 3 (February 18, 1934)
38. 350 Slot Machines Seized in 3 Raids. New York Times, 1 (February 25, 1934)
39. High Court to Hear Slot-Machine Case. New York Times, 7 (April 3, 1934)

40. LaGuardia Wins Slot Machine Victory; Highest Court Voids Old Ban on Seizures. New York Times, 1 (May 22, 1934)
41. $23,160,000 Grant to 8th Av Subway Won By LaGuardia. New York Times, 1 (December 13, 1934)
42. Berle Opens Talks on Unified Transit. New York Times, 2 (January 9, 1934)
43. $363,000,000 Set As Transit Price. New York Times, 17 (January 22, 1934)
44. Seabury is Named Transit Councel. New York Times, 2 (January 27, 1934)
45. City Votes Down Transit Board Pay. New York Times, 5 (February 3, 1934)
46. Backs Board Action On Transit Salaries. New York Times, 11 (February 8, 1934)
47. City Transit Fight Delays Unification. New York Times, 17 (April 30, 1934)
48. Seabury to Plead City Transit Case. New York Times, 19 (February 26, 1934)
49. City Bill to Save the 5-Cent Fare Rushed At Albany. New York Times, 1 (March 15, 1934)
50. The Unification Bill. New York Times, 18 (May 6, 1935)
51. $416,861,000 Net Price Set For Transit Lines in Pact; City and IRT In Accord. New York Times, 1 (November 2, 1935)
52. Unity Plan Barred By Transit Board. New York Times, 1 (February 5, 1936)
53. Strike Threatens on BMT System. New York Times, 4 (July 1, 1939)
54. Unification Aided By Action on IRT. New York Times, 33 (November 4, 1939)
55. City Agrees to Buy 6th Ave Elevated; Plans to Raze It. New York Times, 1 (April 27, 1938)
56. Elevated Razing Started By Mayor. New York Times, 21 (December 21, 1938)
57. Former Mayor O'Dwyer Dead; Prosecuted Murder Inc. Gang. New York Times, 1 (November 25, 1964)
58. Democratic Bolt Looms On Fare Increase. New York Times, 3 (February 6, 1948)
59. Mayor Holds Firm Against Fare Increase. New York Times, 29 (March 16, 1948)
60. Who Wants 5-Cent Fare Now? New York Times, 26 (March 16, 1948)
61. Fare Increase Bill Signed By Dewey As O'Dwyer Balks. New York Times, 1 (March 30, 1948)
62. Mayor Seen Facing TWU Pay Pressure. New York Times, 25 (March 22, 1948)
63. Quill Urges Rise In Fare, Threatens Tie-Up In Transit. New York Times, 1 (April 14, 1948)
64. Text of Mayor's Address Announcing Rise In Transit Fare. New York Times, 22 (April 21, 1948)
65. TWU Bars Leaders From Posts In ALP. New York Times, 12 (July 30, 1948)
66. TWU Executive Board Urges Removal of 2 Leftist Chiefs. New York Times, 1 (September 5, 1948)
67. TWU Bars Offices To Reds, Backers. New York Times, 29 (December 10, 1948)
68. Legislature Set to Approve Dewey Aid Program Today. New York Times, 1 (March 19, 1953)
69. State Turns Down City Plea To Ease Transit Act Now. New York Times, 1 (May 15, 1953)
70. Authority Leases Transit Lines; Fare Increase in Sight. New York Times, 1 (June 2, 1953)
71. Hood, Clifton, 722 Miles – The Building of the Subways and How They Transformed New York. The Johns Hopkins University Press, Baltimore (1993)

Figure Sources

13.1 *Courtesy of* Library of Congress, Prints & Photographs Division, NY World Telegram and Sun Collection, LC USZ62-121525.
13.2 *Courtesy of* New York Historical Society.
13.3 *Courtesy of* New York Historical Society.
13.4 *Courtesy of* Library of Congress, Prints & Photographs Division, NY World Telegram and Sun Collection, LC USZ62-132498.
13.5 *Courtesy of* New York Transit Museum.
13.6 *Courtesy of* New York Historical Society.
13.7 *Courtesy of* New York Historical Society.

Chapter 14
Prelude to Disaster: The NYCTA between 1953–1970

The creation of the New York City Transit Authority was intended to provide a layer of insulation between the politics and politicians of the city, and allow for more rational management of the transit system. The NYCTA has had mixed success on both counts. Mayors have continued to have major involvement in transit matters, particularly related to contract issues, although some mayors have been more involved than others. What creation of a public agency did not do, however, was insulate the transit system from the general economic travails of the city.

14.1 Introduction

When Governor Dewey signed the legislation creating the New York City Transit Authority on March 20, 1953, he kicked off several years of a difficult transition, during which a number of fundamental principles would be tested:

- While the legislation supposedly separated the transit system from the politics of the city, this would not be true. The complex relationship between the city and Authority would have to evolve over time.
- The relationship between the Authority and the Transit Workers Union would also be tested through the courts and in messy negotiations over many issues.
- The effect of the Authority on fares also evolved. The enabling legislation for the NYCTA mandated that operating costs be fully covered by fare and other revenues.

These and other issues had to be addressed in an atmosphere of almost constant crisis, as the transit system flirted with bankruptcy and deteriorating conditions.

14.2 Getting Started

The Authority was created over the objections of Mayor Impellitteri and the Board of Estimate. The Board of Estimate had to agree to the lease of transit properties to the new authority, and their opposition was strong. The enabling legislation, however, had a hammer: $50,000,000 of additional real estate taxing power was approved by the State, subject to making the NYCTA operational by June 1, 1953.

R.P. Roess and G. Sansone: The Wheels That Drove New York, STTT 1, pp. 307–332.
springerlink.com

On April 6, 1953, the city joined a taxpayer suit challenging the constitutionality of the enabling legislation. Once again, the suit alleged that the creation of the NYCTA by the State violated the home rule provisions of the State constitution (1). At the same time, the Mayor prepared a budget for next fiscal year assuming an operating NYCTA and an additional $50,000,000 in property tax revenues.

On April 9, Governor Dewey made a speech, condemning the city's handling of its finances. He alleged that the city balanced its FY1952 budget based upon $100,000,000 in borrowing and $48,000,000 in "accounting tricks." He pointed out that the city still had $140,000,000 in taxing power that it had not exercised; they didn't really need the additional $50,000,000 that had been requested and provided. The governor then suggested that the city might be better run by a professional City Manager, and threatened to initiate action to study this (2). Five days later, the Mayor responded with a speech of his own, complaining that the city, with over 50% of the State's population, had disproportionately low representation in state government, and subsidized the rest of the state. He complained that the world's greatest city is treated as a vassal of the state, frequently getting the short end of the stick (3).

With this battleground as a backdrop, the courts ruled that the NYCTA was indeed constitutional, and did not violate the home rule provisions of the State constitution (4). Under the terms of the NYCTA legislation, the authority was to be run by a five member board, two of which were appointed by the Governor, two of which were appointed by the Mayor, and the fifth chosen by the other four. The Governor named Henry K. Norton, trustee of the New York, Susquehanna, and Western Railroad, and Major Hugh J. Casey, a retired military engineer (5). The Mayor, acting under a deadline of April 20 to name his appointees, named William G. Fullen, member of the law firm of Baer, Bennett, and Fullen and a former member of the Transit Commission, and Ephraim F. Jeffe, president of the Kings County Lighting Company (6) at literally the last minute.

On April 28, the four appointed members met, and expressed optimism that they would soon agree on the fifth member (7). Less than three days later, they announced an impasse on this issue (8). The law mandated that in the case of such an impasse, the head of the Port of New York Authority would become the fifth member. Howard S. Cullman, the chairman of the Port Authority, declined to serve, however, citing poor health and too many competing commitments (9). In accordance with the law, the three New York members of the Port Authority would have to appoint the fifth member. On May 4, 1953, Eugene F. Moran, one of the New York members of the Port Authority, was appointed (10). He resigned from the Port Authority on accepting the position. He was 81-years old at the time.

The five members constituting the NYCTA were now in place. All of the positions were unpaid. They were about to step into period of controversy and new challenges as they undertook the staffing of the authority and the complex issues of labor relations and the fare.

The new board met for the first time on May 7, 1953, and immediately caused some controversy. The five elected Major Casey as Chairman on a vote a vote of three to two, split along city-state lines (11). A number of other key issues were discussed. While most members agreed that a fare increase would be necessary to meet the mandate to cover operating costs, Mr. Jeffe wanted to look into cost savings and efficiencies first. A controversy also arose as to whether or not "operating costs" included depreciation of the physical plant. The City Controller, Lazarus Joseph, believed it did. This was a critical issue, as the fare would have to be higher if it had to support both operating and depreciation costs.

Recognizing that a fare hike would cost the system riders, the NYCTA tasked its engineering consultants with investigating how various levels of fare increase would impact ridership and overall financial stability (12). This was critical, as the 10 month deficit for the transit system now run by the NYCTA was already $32,500,000, with two months left in the fiscal year (13).

On June 1, 1953, the Board of Estimate formally approved leasing of all city transit properties to the NYCTA. In the final analysis, the Board had no options should it refuse to do so. The city would have forfeited the $50,000,000 in additional real estate taxing power, and had no real plan for dealing with such a loss (14).

As a result, of the Board of Estimate action, they NYCTA became fully operational on June 15, 1953. With a full NYCTA board in place and an operating lease for all city-owned transit, a new era was about to begin.

14.3 The 15-Cent Fare

Because the NYCTA had to cover all operating expenses from the fare, it was generally assumed that a fare increase was in the offing. Despite this, there, was much controversy.

Prominent merchants opposed any fare increase. The Retail Dry Goods Association filed a report with the Mayor and Board of Estimate claiming that the increase from five to a 10-cent fare had resulted in continuing ridership losses, and even greater losses in retail sales throughout the city (15).

While four members of the NYCTA were on board with a fare increase, Ephraim Jeffe was not. He preferred handling the deficit through service efficiencies, with a fare increase as a last resort.

Democrats in the State legislature attempted to pass bills banning any increase in the fare until the next Mayoral election. The Republican legislature defeated all attempts to postpone or prevent a fare increase (16).

A practical issue was also critical. Along with a fare increase, the NYCTA, for the first time, introduced the use of tokens. Recognizing that all turnstiles had to be reset each time a fare increase went into effect, they introduced tokens that could be sold at whatever the fare was. However, to introduce this, they had to acquire a huge number of tokens in a short period of time. Finding manufacturers was a difficult task that threatened a quick introduction of a higher fare. Eventually, three manufacturers were found who could deliver 16,500,000 tokens by July 30th, the statutory deadline for implementing a new fare during that fiscal year (17).

On July 14, 1953, the NYCTA board voted to increase the fare to 15 cents on July 25 by a vote of four to one, with Mr. Jeffe opposing. It also officially announced the change to a token system for collecting the fare (18). The action was met with universal outrage from the usual political quarters. The Mayor called the decision a "hasty action without the benefit of mature consideration." City Council President Rudolph Halley, a candidate for Mayor on the Liberal Party line, called the fare "the bitter harvest resulting from eight years of incompetence in city government." City Controller Lazarus Joseph said that the increase was "another example of state domination over city affairs" (18). In its first public issue, it was clear that the NYCTA was not very successful at insulating the transit system from city politics.

As a result of the fare vote, Ephraim Jeffe resigned from his position on the NYCTA board. After trying to convince him to stay, Mayor Impellitteri finally accepted the resignation on July 21, 1953 (19). The Mayor appointed Harris J. Klein, an attorney and vocal opponent of the 15-cent fare, to take his place (20).

Implementing the token system went relatively smoothly. During the initial weeks of the 15-cent fare, riders could only purchase two tokens at the same time. This was gradually relaxed to allow purchase of tokens in 10-token packs.

The five-cent fare had held firm for over 50 years. The 10-cent fare lasted only 15 years. As time progressed, the time between fare increases would get shorter and shorter.

14.4 Service Cuts – The First Battle with the TWU

Even with the fare increased to 15 cents, it was clear that cost reductions would be necessary to fully close the expected first-year operating deficit. On its first day in power, the NYCTA cut seven administrative positions and downgraded others to save $74,315 (21).

The TWU was wary of any service cuts, as they would involve a reduction in union positions. It was also approaching the end of its current contract, which expired on December 31, 1953, and needed to stake out a strong negotiating position with the new authority. Michael Quill fired the first volley on June 28, 1953, stating that any attempt "to hand economies to the operating force" would be met with a job action. He further stated that he expected the NYCTA to announce service reductions at its July 15th meeting, reductions "which we don't intend to let them make" (22).

The NYCTA, acting on the recommendations of its engineering consultants and on previous reports issued by the Board of Transportation, announced a series of service reductions, including eliminating night service on bus lines that paralleled subway or elevated rail lines, elimination of some bus lines serving ferry operations that no longer existed, and increasing headways on rail lines by up to five minutes during off-peak hours, particularly after midnight. It was also considering elimination of the Third Avenue El in Manhattan and other measures.

Quill and the TWU fought each and every service change with almost missionary zeal. On September 5, 1953, Quill protested the plan to add an 11th car to trains on the Flushing Line without adding another conductor (23). Calling the request "featherbedding," the NYCTA ignored him and went forward.

On September 18, 1953, Quill threatened a "rule-book slowdown" unless the NYCTA began to bargain with him over both the service cuts and a new contract with the TWU (24). The rule-book slowdown was an old tactic of the union in which its workers would meticulously follow all operating regulations, invariably resulting in a significant delay in service while not staging a strike, which was illegal under the Condin–Wadlin Act. The NYCTA issued a stern reply, that is, it would not give up any of its managerial powers to the union nor discuss the exercise of such powers with the union. In effect, the NYCTA stated that it had the unilateral right to change schedules in any way it saw fit (25).

With a stalemate looming, the TWU asked Mayor Impellitteri to appoint an arbitration panel to settle the dispute. The Mayor responded by saying that he would not consider such a panel if the union took a job action that "in any respect inconveniences the riding public." The issue of whether the Mayor was empowered or even required to appoint an arbitration panel was also in dispute. The union claimed that under its existing "Memorandum of Understanding" between the TWU and the former Board of Transportation (which had been formally accepted by the new authority) required the Mayor to act on the request. Neither the Mayor nor the NYCTA were sure that this was the case (26).

On September 25, 1953, the NYCTA implemented its new service schedules, ignoring the union. Quill once again called on the Mayor to appoint a committee, with the NYCTA taking the position that schedules did not fall under the provisions of the memorandum of understanding, and was, therefore, not subject to the Mayor's intervention (27). On September 29, the Mayor appointed an arbitration committee (28). The NYCTA refused to recognize the committee, causing Mike Quill and the TWU to seek an injunction against the new schedules in State Supreme Court (29). The union obtained a "show cause" order, forcing the NYCTA to show why it should not be enjoined from putting new schedules into service (30).

If the ongoing controversy were not enough to deal with, on October 3rd, Eugene Moran resigned from the NYCTA for personal reasons. He was replaced by Douglas M. Moffat, an attorney. He was appointed on November 5, 1953 (31).

On October 5th, Justice Walter Hart urged the parties to find an acceptable settlement, and offered to make specific recommendations for such a solution. Lawyers for both the authority and the union agreed that they would take the Judge's suggestions back to the full authority and to officers of the union for their consideration (32). Both sides agreed to let Judge Hart make recommendations as an "impartial adviser." The TWU agreed to make the recommendations binding on the union; the NYCTA agreed to consider the recommendations, but believed they could not treat them as binding (33).

Judge Hart presented his recommendations to both the TWU and NYCTA on November 17th. The session also served as the first personal meeting between Mike Quill, TWU chief, and Hugh Casey, the NYCTA chairman. The meeting was an important event, as the NYCTA had been expressing the view that it could not, under the law, bargain with the union (34). On November 19, 1953, the union and the authority agreed to end the dispute by accepting Judge Hart's recommendations. The resulting cuts would reduce the number of employees on the IRT and

BMT by 349, and save the NYCTA an estimated $1,434,400 per year in operating costs (35). While the cuts were somewhat less than what the NYCTA had originally planned, the biggest financial impact was the delay in implementation by almost half a year, reducing the savings to the NYCTA to $700,000 in its first fiscal year.

14.5 The First Labor Contract

Mike Quill used the occasion of the service dispute settlement to begin his quest for a new contract. He announced that he would be submitting a proposal for a $50,000,000 package of wage and benefit improvements for the union's 44,000 NYCTA workers. Having recently won an arbitration award for an increased wage rate for bus drivers of five private operations, he indicated that their settlement would be "a floor, not a ceiling" in the forthcoming negotiations. Quill also stated that the union would not help the private companies get a fare increase to pay for the settlement (private bus lines were still on a 10-cent fare). If they did not pay the increases on their scheduled dates, he said, "we will find ways and means of making them do it." One of the nine operators, the Third Avenue System, was already bankrupt. Quill quipped, "If Third Avenue can't get up the cash, let them turn their franchises and property over to the Transit Authority. Where they get their cash is their problem; we will insist on getting our cash."

14.5.1 A New Player: Mayor Robert Wagner

With the battle lines clearly drawn, a new player was about to enter the stage. Mayor Impellitteri was opposed in the 1953 primary by Manhattan Borough President Robert Wagner Jr., who was backed by a fragmented Tammany Hall, now led by Carmine DeSapio, the first non-Irish leader in memory. Wagner easily defeated Impellitteri in the primary, and won an impressive victory over the Liberal and Republican Party candidates in the general election.

Fig. 14.1 Mayor Robert Wagner Jr.

The election of Robert Wagner as Mayor was part of Tammany Hall's last stand. It had survived the corruption of Boss Tweed, Jimmy Walker, and the allegations against William O'Dwyer. It survived the 12-year reign of Fiorello LaGuardia and the presidency of Franklin D. Roosevelt, both of whom reduced the Hall's influence to a virtual zero at the time. Their ultimate downfall was their last success. One year after electing Robert Wagner as Mayor, the Hall followed up by electing Averill Harriman as Governor. In doing so, they also defeated Franklin Roosevelt Jr., who was running for State Attorney General. Their success enraged Eleanor Roosevelt, who teamed with two prominent liberals, former Governor Herbert Lehman and Thomas Finletter, to form the New York Committee for Democratic Voters. Their primary mission was to fight the influence of Tammany Hall. The group eventually succeeded by discovering a new set of scandals, forcing Carmine DeSapio to resign. Tammany Hall lost its visibility and influence, and within a few years, ceased to exist as a formal organization.

14.5.2 Sorting Out the Negotiation Process; Achieving a Deal

With the dispute over service cuts just settled, the fight over a new contract was enjoined in earnest. The existing memorandum of agreement was set to expire at midnight on December 31st.

On November 23rd, the TWU asked for a meeting with the NYCTA at the earliest date to avoid a last minute struggle for a new contract. It claimed to have sent the NYCTA a letter containing this request. The NYCTA indicated that it had not yet received the letter, but would begin discussions with representatives of the NYCTA. The five board members refused to meet personally with Quill, but would send representatives (36). A critical aspect of the negotiations would involve the NYCTA board's view that while the agency could negotiate concerning wages and other issues, it could not, by law, sign a contract for a specific period of time.

On November 27th, the NYCTA board members changed their position and agreed to a face-to-face meeting with Quill and other union officials, but only for the purposes of establishing a process for the negotiations (37). With talks set to begin, Quill issued another set of threats on the radio program "Searchlight." He repeated his demands, and indicated that he didn't care where the Transit Authority got the money. He said that they should "let Dewey worry about it, he gave us the Transit Authority" (38). He also warned that he expected to deal directly with Hugh Casey, the TA chairman. Despite all the bluster, negotiations were in progress by the beginning of December.

On December 9th, Mike Quill described the talks as "deadlocked." He warned that the city would be "thrown into a tailspin" if no agreement was in place by December 31st. He chided the NYCTA for conducting what he called "treadmill negotiations." Meanwhile, the TA chair released a statement indicating that they would take no action that would jeopardize the 15-cent fare (39). On December 15th, Quill stalked out of the negotiations, claiming that the NYCTA had said "no" to every one of their demands, and had refused to guarantee retroactivity if the talks went beyond December 31st. The TA refuted these claims, and indicated that they were waiting for details from the union on some of their requested work rule changes (40).

With talks rapidly getting nowhere, and a strike deadline approaching, Mayor-Elect Wagner proposed to appoint a fact-finding panel to help mediate the dispute. The union immediately agreed; the TA agreed several days later, but with a key reservation (41):

> *"We must state, however, that the authority cannot, under the statutory law of its creation, agree to be bound by any recommendation of such a committee, although any recommendation will be given prompt and respectful consideration.*
>
> *It should be remembered that the New York City Transit Authority is not a management group involved in a contest with labor unions for a division of profits between shareholders and employees.*
>
> *It is a public body, serving without compensation, to direct the operations of the transit system in the most efficient manner possible for the benefit of the citizens of New York and to see that its employees enjoy proper working conditions and adequate compensation for their labors. The authority has no other interests."* (Ref 41, pg 32)

With the strike threat removed, the three-member panel was named by Mayor Wagner shortly after taking office (42).

On July 11, 1954, TA board member Harris Klein gave a speech to the Brooklyn Civic Council. In it, he accused Michael Quill of being the single biggest cause of inefficiency in the TA, and indirectly, the reason for the fare increase. He charged that Quill "has flagrantly encouraged and fostered waste, featherbedding, and inefficiency." As a result, Klein indicated, the TA was in financial chaos and wages were not as high as they could be. Klein cited specific examples of inspection and maintenance tasks that were conducted by private bus operators took as much as five times longer in the NYCTA (43).

In the midst of this inflamed environment, the fact-finding panel opened its inquiries on January 13, 1954, thirteen days after the expiration of the previous memorandum of agreement. The panel indicated to all parties that they might not complete their work within the 30 days allotted (44).

Things got worse once the panel began to work. TA representatives testified that granting the union demands would result in a 20-cent fare. Quill, obviously angered by Harris Klein's statements to the Brooklyn Civic Council, charged that Klein offered to support his wage demands in exchange for Quill's support for Klein's appointment as a Justice of Special Sessions. Quill further indicated, "I am now prepared in the presence of Mr. Klein and the New York City Commissioner of Investigation to make this statement under oath" (45). Kings County District Attorney Edward Silver convened a grand jury to determine if an attempted bribery or criminal libel had been committed (46). On February 11, 1954, the grand jury indicted Mike Quill on three counts of criminal libel for maliciously defaming Klein (47).

The mediation panel finally issued its findings on May 17, 1954. Decrying a "disgruntled labor force and a hapless management," the panel called for a radical change in attitude of each side to each other, warning that a failure to do so would result in further deterioration of the situation (48). Its specific recommendation included a wage increase of 14 cents per hour, split into two parts, with 6.5 cents retroactive to January 1, 1954 and 7.5 cents to begin on March 15, 1955. Other recommendations included:

> *"Signing of a binding contract between the authority and the unions of its employees until June 30, 1956."*
>
> *"Determining by a vote or check card which union or unions were entitled to exclusive representation for groups of workers, subject to legal limitations."*
>
> *"Spelling out in the contract by 'appropriate and vigorous' language the authority's unquestioned right to exercise normally accepted management functions."*
>
> *"Providing in the contract for pledges by unions of complete acceptance of the agency's basic management rights as well as commitments to cooperate with the authority to keep the transit system on a sound operating basis."*
>
> *"Providing for impartial review of all grievances and factual disputes by appointment of an arbitrator with final powers for the duration of the contract."* (Ref. 48, pg 13)

Two days later, the TWU accepted the recommendations of the fact-finding panel (49). On May 24, 1954, the NYCTA rejected the recommendations (50). Instead, it ordered wage increases of 5 to 12 cents per hour, retroactive to January 1st, at a cost that was slightly higher than the 6.5-cent increase recommended by the fact-finders. The authority again brought up its contention that they could not legally bind themselves to anything beyond the present, in accordance with their enabling legislation.

After the union issued a strike threat for 12:01 AM on June 14th, the Mayor called the sides to separate sessions to discuss the situation (51). On June 4, 1954, the TA softened its stand, allowing the union to call off the strike. In doing so, the authority agreed to:

> *"Discuss wages and working conditions with the union."*
>
> *"Accept the fact-finding report as the main basis for new negotiations."*
>
> *"Join with the union in seeking a court ruling on whether it could make labor contracts; if it was determined to be legal, the authority would enter into such agreements."* (Ref. 51, pg 1)

In a June 12th editorial, the *New York Times* summed up the substance of the chaotic search for an agreement (52):

> *"The Transit Authority had a great deal to learn about labor relations, and it learned the hard way, with Michael J. Quill as an unpleasant teacher. Actually, as a study of the Fact-Finding Report and of the authority's responsive resolutions reveals, the two sides were not too far apart on substance. It was the absence of tact, of good manners on both sides, that was provocative. The courtesies of labor relations today demand consultation, the exchange of confidences, respect for opposing points of view. There are some inexorable limits to what the authority can give away, but the bargaining table is the place to talk them over. The era of the unilateral edict has passed."* (Ref. 52, pg 14)

The two sides reached agreement on July 8, 1954. The settlement called for retroactive pay increases on 6.5 to 11 cents per hour, with an additional, unspecified, wage increase to occur on March 15, 1955. Other aspects of the fact-finders recommendations were incorporated into the agreement. With the courts confirming the legality of the authority entering into fixed agreements, the two sides signed the agreement several days later (53).

Mike Quill was not finished testing the system, however. On August 12, 1954, he filed 18 grievances under the new contract (54). On September 16, the arbitrator ruled against the union in all but one minor issue regarding meal times (55). On November 18, 1954, Quill called upon Governor-Elect Harriman to support legislation abolishing the Transit Authority (56). A day later, Harriman indicated that he would not support such a move (57).

14.6 A Difficult Horizon

As the TA ended its first year of operation, storm clouds were beginning to gather. The 15-cent fare had not provided the benefits anticipated. In what would become a continuing pattern, ridership had dropped after the new fare was imposed, and service cuts had still been required to meet the TA's mandate to be "self sufficient." While it ended its first year of operation with a small surplus, ridership drops threatened to produce a deficit in the second year (58).

At the same time, it faced a strong labor union that was becoming very used to getting its way. Isolated from both politics (in form maybe, but never in substance) and private enterprise, the TA would have to make significant decisions on wages and benefits. Board members, however, were relatively insulated from the consequences of their decisions, that is, they were not subject to election; they had six-year appointed terms, and could not be dismissed midterm; they had no personal financial interest in the transit system. The future would prove that fare increases could not cover every financial problem. The law of supply and demand could not be ignored, and fare increases inevitably led to reductions in ridership.

The union and the TA had forged a working relationship that was not comfortable, but at least the legal limitations had been determined. The first year of the authority also produced a number of capital improvements to the deteriorating system, including 200 new subway cars, repainting and rebuilding of additional subway cars, improved track and signal maintenance, and refurbishing of many subway stations.

The first labor negotiations clearly damaged the perception that the authority insulated the system from city politics. The Mayor had still been a leading and important presence, and the union was still able to pull some political strings to support its positions.

The first years of authority operation also produced a few interesting twists. In 1955, for example, the authority briefly flirted with the idea of using atomic power to generate its electricity as a cost-saving measure. The idea was eventually scrapped, primarily because the business of atomic generation was in its infancy, and there were more questions about it than there were answers (59).

Through the late 1950s and early 1960s, problems continued to arise, including steady loss of ridership, escalating costs, increases in crime in the subways, and general deterioration in equipment. Labor peace was the rule, as Quill and Mayor Wagner enjoyed a cordial working relationship.

14.7 The Transit Contract Two-Step

Throughout Robert Wagner's three terms as Mayor, not one transit strike was called. Mike Quill actually helped the city in ending two "wildcat" strikes by motormen, who were trying to organize their own separate union, in June of 1956 and December of 1957. But having no strikes did not mean that they weren't frequently threatened.

Like clockwork, every two years, there would be a threat from Quill to call a transit strike. At the last minute, Mayor Wagner would step in and the strike would be called off, and a contract signed based upon his recommendations. An article in the New York Times shortly after Mike Quill's death described the relationship between the union and Mayor as follows (60):

> "*It was said that City Hall was always willing to come up with a money package for Mike Quill, but it had to act under the threat of a strike to justify the expense to the voters. Mr. Quill obliged with the threats. Moreover, he relished the tough-guy role, like a seasoned actor, and carried it off with Irish panache.*" (Ref. 60, pg 30)

The same article also pointed out that the public Mike Quill and the private Mike Quill were often two very different men:

> "*Off stage, in private, when he dropped his pose as the transit world's Terrible Tempered Mr. Bang, he was sweetly reasonable, cool, efficient, businesslike, and devastatingly funny.*" (Ref 60, pg 30)

Through three mayors, namely, O'Dwyer, Impellitteri, and Wagner, Mike Quill played the game with uncommon skill, winning many benefits for the members of his union. When he started the union, typical workers had 6-day, 12-hour shifts.

He always considered that his greatest achievement was obtaining a 40-hour work week for transit workers. Workers also achieved significant improvements in compensation, medical coverage, work rules, vacation time, and other benefits under his leadership.

Mike Quill's finest, or worst hour, depending upon which side of the fence you are on, came with his first and only confrontation with a fourth mayor, John Lindsay, who would not, and did not play the game, with devastating results to the city.

14.8 Private Enterprise Anyone?

In April of 1959, the city was presented with an interesting and unexpected opportunity. Headed by New York and Washington D.C. financier O. Roy Chalk, a group of financiers proposed to buy the New York City Transit System and operate it as a private enterprise. There were several stipulations that had to be met by the city, including an immediate fare increase to 20 cents, and a long term agreement that the fare would be set to provide the corporation with a 6.5% return on its investment. The city would have to continue to cover the existing capital debt service, but the company would assume responsibility for future capital spending, except for the Second Avenue Subway, which the city would build and lease to the corporation (61, 62).

Chalk already owned D.C. Transit, which he had taken over from public operation in Washington D.C. He had significant initial success in managing the system to produce a profit, while not causing any significant decline in service. Already wealthy, he was also famous for owning the world's largest emerald, which he had purchased abroad for his wife.

The politicians in general reacted with horror to the proposal, fearing an immense political backlash if the fare should rise precipitously. The main supporter of the plan (besides its proposers) was the Chairman of the Transit Authority, who believed it would simplify the system's operation and management, and relieve the city of a substantial drain on its resources.

Both the Board of Estimate and the Mayor were opposed to the plan, particularly because the sale would lead to fare increases. Mayor Wagner clearly stated that he would only support the sale if the 15-cent fare were preserved (63, 64). When a revised offer of $615,000,000 was formally received on April 24 (66), it was rejected after a few days of intense discussions amongst the Mayor, the TA, the Board of Estimate, and Chalk's representatives (67).

14.9 The Gathering Storm

Mayor Wagner tried to be an advocate for public transportation. One of his early campaign themes was that "New York needs more than a token mayor." He took advantage of every opportunity to show himself riding the subways with prominent friends.

Mayor Wagner, however, was not the principal player in New York City transportation matters. The man in charge was Robert Moses. Through an interlocking set of authority appointments won from Mayors LaGuardia and O'Dwyer, Moses was at the peak of his power in the 1950s, and for Robert Moses, transportation meant only one thing, that is, the automobile.

When he started, Moses was content to build highways through outlying regions. After the war, he set about bulldozing highways through developed areas of the city. One highway, namely, the Cross Bronx Expressway, was particularly damaging, destroying several vibrant communities over sustained opposition. The highway devastated the portion of the Bronx it cut through, leaving abandoned buildings, new slums, and rampant crime in its wake.

It is said that Robert Moses' postwar highway program displaced 250,000 New Yorkers from their homes, and countless others from their businesses. Moses spent over $32 billion on highways and bridges after the war, draining all sources of federal aid, and a substantial portion of the city's capital budget.

Among the prices to be paid for highway expansion was the rapid deterioration of the transit system. By 1960, not a single mile of new trackage had been laid since 1940 (the Rockaway extension doesn't count, as it replaced a LIRR line). In his landmark book on Moses, *The Power Broker*, Robert Caso notes that "people using the subways and railroads were riding on tracks laid between 1904 and 1933, the last year before Robert Moses came to power."

The cracks literally began to show. Track fires became more common, as did derailments. Stations and trains got dirtier. Most of all, crime was on the rise. As more New York neighborhoods were gutted by highway construction, the middle class fled to the suburbs, and the chasm between rich and poor became more apparent, and more inflammatory.

Through all of the years of rapid transit ups and downs, and all of the many controversies and scandals the system encountered, there had been one constant: people felt safe on the trains and in the stations. Isac Freilander's famous etching, "3 AM 1939," depicted a group of subway riders at 3:00 AM in 1939. There are men and women and people of different races, most sleeping, unafraid to be on the city's transit system at 3:00 AM in the morning.

As the 1960s began, and crime on the subways increased, this critical bastion disappeared. Ridership continued to decline, not only due to the fare and the impact of the automobile, but because people no longer felt safe riding the New York city transit system. Mayor Wagner responded by placing a policeman on every train, expanding the ranks of the Transit Police by a factor of three, to a force of almost 3000 officers. However, this wasn't enough to stem the tide.

Meanwhile, in the repair shops, parts were becoming scarce. Often, repairs would be made using used or reconditioned parts. Working conditions in the shops also deteriorated as dirt, leaks, and other problems occurred. The system was about to enter a twenty-five year period, roughly between 1960 and 1985, in which steady, accelerating decline would take over. These were to be the darkest days in the system's history.

14.10 The Rockaway Beach Extension

While the 1950s were a period of gathering travail for the transit system, one significant expansion did occur. On September 5, 1952, the Board of Transportation (the NYCTA was not yet in control) purchased the Rockaway Beach and Rockaway branches of the Long Island Railroad. The railroad had been trying to sell this line to the city for years, and when a devastating fire destroyed a major section of the wooden trestle over Jamaica Bay, the railroad simply abandoned the line.

After purchase, the city had to rebuild the wooden trestle (with a steel structure), refit elevated sections for use by IND trains, and build connections to the IND Fulton Street line. Service began on June 28, 1956. When opened, the Rockaway Branch was operated as a separate division of the NYCTA, and a "double fare" was collected from riders.

Initial service was provided as part of the "E" line. At the time, it was the longest continuous rapid transit route in the world at over 36 miles.

14.11 Mayor John V. Lindsay and the Transit Strike of 1966

John Lindsay was born on November 24, 1921 to a prosperous, but not wealthy, family. His mother, Florence Vliet, could trace her origins to a Dutch family at the time of the Revolutionary War.

After serving as a gunnery officer on a destroyer during the invasion of Italy in WWII, he studied law at Yale and graduated in 1948. He became active in Republican politics. He was a founding member of "Youth for Eisenhower" in 1952, and attended the convention that year, where he attracted the attention of Brownwell, who became Eisenhower's Attorney General. In 1955, he appointed Lindsay as his executive assistant.

In 1958, with Brownwell's encouragement and support, Lindsay ran for Congress in Manhattan's 17th District, which was known as the "Silk Stocking District" because it included the city's major entertainment and upscale residential areas. Running for the seat, which had been vacated by a Republican, Lindsay had to win a primary against the Republican Party nominee, and a Democratic-Liberal candidate in the election. He won both, but by relatively small margins.

John Lindsay established a very liberal record in Congress, often supporting Democratic positions on civil rights, immigration, housing, school construction, foreign aid, and other issues. By 1965, even though he was popular and strong in his district, he was becoming isolated in Congress, as the Republican Party moved strongly to the right. Because of his strong interest in urban affairs, he decided to run for Mayor of New York City.

At the time, there had not been a Republican mayor in New York since Fiorello LaGuardia. The Republican nomination was easily won, as there were few who were even interested in running under the party banner. He ran against a Democrat, Abe Beame, who was then the Comptroller of New York City, and a Conservative, William F. Buckley, a noted columnist and author. He won in a very tight race by emphasizing his liberal views and "maverick" standing in the Republican Party. He had been one of the few Republicans to vote for the Civil Rights Act of 1964, vocally opposed the war in Viet Nam, and refused to support Barry Goldwater in the presidential race of 1964.

Fig. 14.2 Mayor John V. Lindsay in 1966

As John Lindsay prepared to assume the office of Mayor, he had created a dynamic and vibrant public image of himself. Tall, young (only 43 on his election), handsome, and energetic, he was often referred to as the "Republican John F. Kennedy." At the same time, he projected the image of an urbane aristocrat, more at home with the great philosophical issues of the day than with the daily personal interactions needed to effectively run the city.

His eight-year term as Mayor is still controversial today, as historians try to assess and balance his lofty goals and objectives with the reality of the calamity that struck the city during and immediately after his term.

14.11.1 The Transit Strike of 1966

One of Lindsay's most enduring problems was the basic clash of cultures between the aristocratic reform-minded Mayor and the old guard of municipal labor leaders. As Lindsay's inauguration moved closer, Mayor Wagner and the TWU were engaged in their usual "pas de deux" over the upcoming transit contract. Wagner, who was leaving office, however, had no incentive to settle; some historians believe that he deliberately let it fall into Lindsay's lap to embarrass him as he took office.

On December 1, 1965, Mike Quill sent a telegram to Mayor-Elect Lindsay to join in the stalled contract talks. He also rejected a set of 10 names Lindsay had submitted as potential mediators (68). Lindsay's reply sent Quill into a rage. The Mayor-Elect indicated that he would not enter the talks personally until he had evidence that both sides were bargaining in good faith (69). Quill's retort was testy, to say the least. He claimed that Lindsay had "shot to hell" any hope of an agreement before December 31, 1965. He also claimed:

> *"It's all in the lap of Mr. Lindsay. We won't move an inch. His telegram is the height of stupidity."* (Ref. 69, pg 1)

When Lindsay calmly responded that he stood by his telegram, Quill declared:

> *"He can stand on the telegram, sit on it, or do anything else with it that is handy."* (Ref. 69, pg 1)

On December 3rd, Quill moved up his strike threat from 12:01 January 1, 1966, the time the existing contract expired, to December 15, 1965, claiming that the Mayor-Elect had broken contact (70). Mayor Wagner rejoined the fray, and Quill called off the December 15th deadline when the Mayor and Mayor-Elect indicated that they had convinced labor negotiator Theodore Kheel to join a three-man mediation panel (71). On December 23rd, the three-man mediation panel reported that good progress was being made (72). On December 16, the union reiterated its January 1st strike deadline, claiming that no money offer had yet been made by the TA (73).

On December 28, 1965, Lindsay made his first personal contact in the matter, holding separate meetings with Quill, the mediation panel, and TA representatives (74). He reportedly told the union that there must be no strike and that he expected them to negotiate a fair, reasonable, and equitable settlement. His stance merely further enraged Quill, who believed that the Mayor-Elect was giving lectures, not negotiating.

The TA obtained a court order against the strike under the Condon–Wadlin Act on December 29th. Quill responded by publicly ripping up the court order, declaring that such injunctions were (75):

> *"…Gatling guns on paper. They'll have to put 36,000 transit workers in the clink."* (Ref. 75, pg 1)

On December 31st, Quinn stormed out of negotiations, demanding the personal intervention of Mayor-Elect Lindsay. Lindsay replied that he would indeed step into the discussions "today," but did not indicate what role he would attempt to play (76). The last-minute participation of the Mayor-Elect did not help. He placed a $25,000,000 package on the table, which was less than half what the union was demanding. Quill stormed out of the talks once again, calling the offer a "peanut package" (77). On January 1, 1966, as John Lindsay took the oath of office as Mayor, Quill began is first and only system-wide strike, shutting the transit system down at 5:01 AM.

On January 2, 1966, everyone was scurrying to figure out next steps. The mediation panel called for new negotiations. Quill warned that the strike could last a month or more, and referred to the new Mayor as a "pipsqueak" and a "juvenile." Lindsay warned commuters that only about 25% of the Manhattan workforce would be able to get to work. Massive traffic jams were expected, and did occur (78).

On January 4th, Mike Quill and five other officers of the TWU were ordered to jail (79). On surrendering for arrest, Quill issued one of his more famous quips, saying that "the judge can drop dead in his black robes, and we would not call off the strike" (80). Two hours after his incarceration, Quill suffered a heart attack, and was taken to the hospital.

On January 6th, Mayor Lindsay offered amnesty from the penalties of the Condon–Wadlin Act if the union would go back to work. Union leaders, after conferring with Quill in the hospital, declined (81). He also offered to spring union leaders from jail, believing that appeasement might soften their stance (82).

Lindsay appealed to President Johnson for assistance. Johnson used the request as evidence the city could not handle its own affairs. He sent Secretary of Labor W. Willard Wirtz to confer with the two sides and make recommendations on how to speed the negotiations. His finding was that the two sides should just keep bargaining. He returned to Washington, having accomplished nothing but a few days diversion (83).

On January 13, 1966, the union and the NYCTA agreed to accept the recommendations of the mediation panel, and the strike was ended (84). The settlement was a costly one:

- During the strike, the city lost $1.5 billion in business revenues.
- The cost of the settlement was estimated to be $52,000,000, more than double what Lindsay's initial offer was, and almost twice that of any previous settlement with the TWU.
- TWU workers were now able to retire after 20 of service, with a retirement income of half their final years of active service, including overtime.

The retirement provision was particularly difficult. Because work schedules of transit employees were selected in order of seniority, workers approaching retirement could load themselves up with overtime for a few years and wind up earning more in retirement than their base pay while working. The provisions also encouraged large numbers of employees to retire. Within five years, over 50% of the transit work force had retired, leaving the system with very few "senior" workers and their extensive experience.

The settlement would also provide a model for other city employees when their negotiations came up. Thus, the police, fire department, sanitation department and others won similar settlements, and suffered similar losses of senior employees.

The cost of the contract forced the city to press for a rise in the city income tax and higher water taxes for residents. A tax on commuters, who worked in the city but lived outside it, was also enacted. The State approved all of these measures. By 1970, the average NYC resident was paying $384 per year in city taxes, more than 50% higher than the next highest-taxed city, Chicago.

When he was released from the hospital on January 16, Mike Quill was quite complimentary to Mayor Lindsay, claiming that his previous criticisms had been offered "in the heat of battle and under war conditions." He heavily criticized President Johnson for playing "cheap politics" with suffering New Yorkers. He actually called for a congressional investigation of the strike to determine who was to blame for its occurrence (84). Less than two weeks later, Mike Quill died of a massive heart attack, and became a virtual martyr and hero to his followers.

Mike Quill had taken an immediate and intense dislike to John Lindsay, but much of his bombastic behavior was an act. In truth, Quill had another reason for calling his one and only system-wide strike in 1966. He was fighting a number of uprisings within the union itself. The membership of the TWU in 1966 was more than half African-American and Hispanic. Quill was desperately trying to hold power for the traditionally Irish leadership. As the newer members were calling for more militancy in the union, Quill was also playing to what he knew his changing audience wanted.

14.11.2 Other Travails and Crises

Mayor Lindsay's first term was marked by numerous incendiary and damaging labor disruptions.

True to his liberal tradition and beliefs, in 1968, Mayor Lindsay responded to complaints from minority parents concerning the New York City schools. With the support of the Ford Foundation and the NYS Legislature, he put into effect a pilot program to test school decentralization in several minority districts, including Ocean Hill-Brownsville, where the protests had begun. The district administrator immediately transferred 13 teachers and six administrators, all White, without pedagogical reasons. The United Federation of Teachers denounced the involuntary transfers as illegal, and called a strike that shut down 85% of the city's 900 schools for 55 days. The strike became extremely bitter, with charges of antisemitism from the union, and charges of discrimination from minority communities. The strike did not end until the courts determined that the "transfers without cause" had violated valid union contracts, and dismissed the offending district administrator. The state legislature later structured a compromise system, decentralizing the city's schools into 32 districts, giving locally-elected boards power to run their elementary and junior high schools. The law also adopted strong protections for teachers and their jobs.

In the same year, police staged a slow-down, firemen threatened to strike, and the sanitation workers went on strike for two weeks, leaving 10-foot tall "mountains" of trash to grow on New York City sidewalks. Many blamed the Mayor for not making any counter-offer to the union's prestrike proposal.

On February 9, 1969, New York was struck by a blizzard that left it buried in 15 inches of snow. While major thoroughfares were quickly cleaned, side streets in Queens and the other "outer boroughs" were buried for almost a week. When Lindsay toured the Queens in an open truck, he was pelted by snow balls hurled by angry residents and store owners. The national coverage of the event was used to paint an image of a mayor unconcerned with the problems of the middle class.

14.11.3 Reelection and More Travails

As the 1969 elections approached, Lindsay was in significant political trouble. He lost the Republican nomination in a primary to a conservative Republican from Staten Island, that is, John Marchi. The Democrats nominated their most conservative candidate as well, Comptroller Mario Procaccino. The two split the conservative vote, allowing Lindsay to be reelected on the Liberal Party line with a plurality of only 42% of the votes cast.

His second term didn't start much better than his first. In the summer of 1970, 8000 city workers belonging to the AFSCME District Council 17 walked off the job. These included draw bridge operators and sewage plant workers. For two days, many drawbridges were stuck in the "up" position, most with mechanisms that had been vandalized. At the same time, hundreds of thousands of gallons of raw sewage was discharged into city waterways. When the strike ended, there were no prosecutions against workers who were guilty of sabotage.

1970 was also the year of the famous "Hard Hat Riot" in downtown Manhattan. Youthful protestors of the Viet Nam War clashed with construction workers from the World Trade Center worksite. Students were waving Viet Cong flags and defiling American Flags in front of the statue of George Washington on Wall Street, in protest over the Kent State killings by national guardsmen. Construction workers stormed the statue, setting up American Flags, and pursued the fleeing protestors through the streets. The resulting chaos reached Pace University and City Hall. Surprisingly, the construction workers were joined by workers from the New York Stock Exchange in their pursuit. The Mayor had lowered all city flags to half-mast in observance of the Kent State deaths. The rioting construction workers threatened to overwhelm City Hall unless they were raised to full height. Lindsay capitulated, raising the flag over City Hall to full height. As the flag was raised, construction workers demanded that 15 officers remove their helmets in respect, although only about half reportedly responded. The Mayor was later forced to apologize for the lack of response from the NYCPD, which had made virtually no attempt to stop the riot.

His second term also featured a broad investigation of the Police Department that uncovered endemic corruption involving "a substantial majority" of officers. The investigation forced a major organizational overhaul of the department. A major issue also arose when he tried to build low income housing in the mostly Jewish area of Forest Hills. Mario Cuomo mediated the dispute, and a modified plan for public housing in the area was finally approved.

14.11.4 Legacy

Mayor John Lindsay's two terms as NYC mayor were among the most controversial in the city's long history. The image of a city largely out of control endured, although the last years of his second term were comparatively quiet.

He avoided additional labor problems by routinely agreeing to generous settlements in the last years of his term. Transit workers got an 18% wage increase and an extra week of vacation. District Council 37 got retirement at half pay after 20 years, along with other municipal unions. Teachers were awarded raises between 22 and 37%. The money had to come from someplace.

Lindsay was forced to borrow huge amounts and resorted to a variety of budget gimmicks to make ends meet. Salaries were paid on July 1st instead of June 30th to get into a new fiscal year; borrowing was propped up by inflated estimates of future revenues; money for noncapital items was paid from capital accounts, and so forth.

When he left office in 1973, the city was in virtual shambles. Crime had become a major problem. From 1961 through 1965, the homicide rate in New York City was 7.6 per 100,000 people. From 1971 through 1975, it had grown to 21.7 homicides per 100,000 people. Racial tensions were literally boiling. The white middle-class associated the increase in crime with minorities, and continued their flight to the suburbs.

At the same time, New York was the only major city in the nation to escape major race-riots during this turbulent period. The Mayor tried to address the needs of the city's poor and minorities, and while he was not always successful, he kept tempers under control. He vigorously promoted the city, and it was during his term that New York became known as "Fun City." He had thought that New York would be the perfect place to implement a broad range of liberal policies, only to find that the city was far more conservative than he anticipated.

John Lindsay switched his party affiliation in 1971, becoming a Democrat. He made a short-lived effort to run for President in 1972, and lost a primary to run for the Senate in 1980.

By 1970, the city's overall finances were in a perpetual crisis. For the transit system, it meant deferred maintenance, service cuts, and a dirtier, less reliable system. At one time, if a New Yorker feared they were threatened, they might have run to the nearest transit station for safe haven; as the 1970s dawned, they would have to run the other way, with no safe haven outside the system in sight.

On his death on December 20, 2000, the *New York Times* said this about the legacy of the city's most controversial mayor (85):

> *"Mr. Lindsay's death Tuesday, at age 79, brings back many indelible images of that time, including garbage piled high on the streets, angry picket lines at city schools, and a transit strike that began the second he took office on New Year's Day of 1966. There was the way he led cyclists and pedestrians back into the city's crime-ridden parks for concerts, festivals, and 'happenings,' offset by the debacle of snow clogging the streets of Queens long after a blizzard, a frozen symbol of a city government in collapse. But surely the most memorable image is of a mayor in a trench coat, heading bravely into the city's tormented black communities to assure them that someone did care after the murder of Rev. Martin Luther King in 1968. That singular gesture helped keep the city together in an era when other cities across the country had gone up in flames."*

"Mr. Lindsay's 1965 mayoral campaign slogan had been: 'He is fresh, and all the others are tired.' Eight years later, he was as tired of the city as it was tired of him. It is easy in hindsight to see the mistakes that Mr. Lindsay made. They include failed liberal experiments like school decentralization, 'maximum feasible participation' of the poor in antipoverty programs, and giant housing projects for minorities in middle-class neighborhoods, as well as unwieldy government superagencies and a mistaken belief that the city could tax itself out of financial troubles..."

"But his legacy is enormous. He inspired a whole generation of idealistic young people, many of them minority members and women, to join the cause of public service. Look around the city, and you can also see a separate legacy, brought by a new generation of gifted architects and planners from his day who created vest-pocket parks, Model Cities programs, and innovative building projects. But the greatest legacy of all is the memory of a time when the city was sorely tested and even experts wondered whether it would ever see great days again. John Lindsay may have overreached, but he embodied the city's combative spirit, the honor of public service, and ultimately the city's ability to renew itself." (Ref. 85, pg A38)

14.12 Oh Yeah! There Was a Fare Hike Too

The first casualty of the 1966 transit strike settlement was the 15-cent fare. It had lasted for 13 years, two years less than its predecessor, the 10-cent fare. At 12:01 on July 5, 1966, the fare was raised to 20 cents. It would not survive the decade.

14.13 The Metropolitan Transportation Authority (MTA)

In 1959, the Kingston Trio, a popular folk group, cracked the Top 20 with their amusing song about a poor soul named Charlie who takes a ride on the MTA, only to have the fare increased, and be trapped riding the train forever. Of course, they were referring to the MTA of Boston, now known as the MBTA, but it could have been an apt prophesy on the coming of New York's very own MTA. The song's most famous lines are in its haunting (and hilarious) prelude:

"These are the times that try men's souls. In the course of our nation's history, the citizens of Boston have rallied bravely whenever the rights of man have been threatened. Today, a new crisis has arisen. A burdensome tax has been levied on the population in the form a subway fare increase. Citizens, hear me out! This could happen to you..."

Poor Charley, unable to come up with another nickel to get off the train, is trapped. His wife meets him every day "at the Scully Square Station" and hands him a sandwich as the train passes through. Perhaps, she should have given him a nickel…but, that would require another song.

In 1965, the state legislature created the Metropolitan Commuter Transportation Authority. Its role was to regulate and arrange subsidies for the states two largest commuter railroads, that is, the Long Island Railroad and what is now known as the Metro-North Railroad. In 1968, with the problems of the New York City Transit Authority continuing to mount, the state passed additional legislation changing the authority's name to the Metropolitan Transportation Authority (MTA), and placing the NYCTA and the Triborough Bridge and Tunnel Authority under its purview. Despite the new umbrella organization, the NYCTA would continue to have significant discretion in directing its own operations (86).

As originally established, the governing board of the MTA consisted of nine members, all appointed by the Governor, two from nominees submitted by the Mayor. The most shocking aspect of the new agency was the inclusion of the Triborough Bridge and Tunnel Authority. The agency was the last bastion of power for the then 79-year old Robert Moses, and its tolls produced a significant surplus for it. In a case of sublime irony, the plan was to utilize the profits from the TBTA to subsidize the rapid transit system. After spending most of his career insuring that public transportation got as little money as possible from any arm of government, the success of Robert Moses' last empire would be used to prop up the New York City subways and surface lines.

Robert Moses' power base began to erode in the mid-1960s, and accelerated after the 1964-1965 World's Fair, which he had headed. The Fair lost money, and was seen as a failure. At the same time, Mayor Lindsay had made no bones about his intention to strip Moses of his powers, and Republican Governor Nelson Rockefeller was happy to assist. With the formation of the MTA, Moses lost his last directorship, although he was given a job as a consultant to the TBTA, where he continued to be served by a driver and a car for many years. In yet another irony, for all of the roads he built, Robert Moses never had a driver's license, and never drove himself on any of his facilities.

As the decade of the '60s drew to a close, the MTA was mired in yet another cliffhanging dispute with the Transport Workers Union. Mayor Lindsay declined to participate, essentially leaving the ball in the hands of the MTA. A last-minute $120,000,000 settlement forced consideration of yet another fare hike. Governor Rockefeller and Mayor Lindsay, despite both being among the few liberal Republicans left at the time, did not like each other, and feuded constantly. The fare issue was no exception. Lindsay wanted to limit the fare increase to 25 cents, and proposed bridge and tunnel toll increases, and a handful of tax increases to cover the rest. Rockefeller was unwilling to fight with his own party to get the hikes, and eventually backed the MTA's recommendation of a 30-cent fare. The 30-cent fare went into effect on January 1, 1970, and was a harbinger of many more such increases to come with increasing rapidity. The five-cent fare lasted 45 years; the 10-cent fare 15 years, the 15-cent fares lasted 13 years; the 20-cent fare died in only 3.5 years. The 30-cent fare would not last that long.

Rockefeller and Lindsay both harbored national political ambitions which neither would ever fulfill, a major factor in their many feuds and arguments. The only thing they ever agreed on was a massive proposal for subway expansion hatched in 1968. It required a massive bond issue to be passed by the voters. It failed by a large margin, with many voters indicating that anything Rockefeller and Lindsay agreed on had to be wrong! The plan, of course, was never implemented. Rockefeller did go on to serve as Vice-President under Gerald Ford, appointed by the Senate after President Nixon resigned as a result of the Watergate scandal in 1974.

The MTA has expanded its scope since 1970, and now oversees the Metropolitan Suburban Bus Authority (Nassau and Suffolk Counties) and the Staten Island Rapid Transit Operating Authority (SIRTOA). Its current board has 17 members, all nominated by the Governor and approved by the New York State Senate. The Mayor of New York recommends four of the nominees, Nassau and Suffolk County Executives nominate one each, and the County Executives of four northern counties, namely, Dutchess, Orange, Rockland, and Putnam, nominate one each, although they collectively have only one vote. Six nonvoting rotating members are held by representatives of organized labor and a Citizen's Advisory Committee. The Authority maintains control of its subsidiary organizations largely through placing its own members on their governing boards as well.

14.14 The Wretched '70s

There is general agreement that the 1970s and early 1980s represent the worst period in the history of the city's transit system. Financial crisis after crisis continued to take their toll on the city itself, and on its public transportation system in particular. The system would rebound, but it had to hit rock bottom first.

References

1. City to Join in Suit Assailing Validity of New Transit Act. New York Times, 1 (April 6, 1953)
2. Text of Dewey's Speech Condemning City's Handling of Its Finances. New York Times, 21 (April 9, 1953)
3. Text of the Mayor's Reply to Governor Dewey on New York City's Finances. New York Times (April 14, 1953)
4. The Transit Authority. New York Times, 26 (April 21, 1953)
5. Dewey Fills 2 Jobs On Transit Agency; City Weighs Choice. New York Times, 1 (April 19, 1943)
6. Transit Authority is Upheld by Court; 2 Members Added. New York Times, 1 (April 21, 1953)
7. Transit Authority Meets in Harmony. New York Times, 1 (April 28, 1953)
8. 4 On Transit Board At Odds On Fifth. New York Times, 17 (May 1, 1953)
9. Cullen Declines City Transit Post, 3 Will Get Offers. New York Times, 1 (May 2, 1953)
10. Moran of Port Unit Named 5th Member of Transit Board. New York Times, 1 (May 4, 1953)
11. Transit Body Picks Casey as Chairman by a Divided Vote. New York Times, 1 (May 7, 1953)
12. Transit Authority Orders Fare Study. New York Times, 17 (May 15, 1953)
13. City's Transit Authority In A Race Against Time. New York Times, E9 (May 17, 1953)
14. Authority Leases City Transit Lines; Fare Rise in Sight. New York Times, 1 (June 2, 1953)

15. Merchants Warn On Fare Increase. New York Times, 43 (June 4, 1953)
16. Transit Bills Pass State Legislature By Party-Line Vote. New York Times, 1 (June 26, 1953)
17. Transit Authority Assured of Tokens. New York Times, 1 (June 9, 1953)
18. Fare Goes To 15c On July 25; Tokens For Subway and El; $9,000,000 Surplus Seen. New York Times, 1 (July 14, 1953)
19. Jeffe Resignation Finally Accepted. New York Times, 1 (July 21, 1953)
20. Foe of 15c Fare Is Appointed To Transit Agency by Mayor. New York Times, 1 (August 5, 1953)
21. New Transit Body Cuts Costs $74,315. New York Times, 43 (June 16, 1953)
22. Transit Authority is Warned By Quill. New York Times, 44 (June 28, 1953)
23. Quill Protests Plan to Lengthen Trains. New York Times, 17 (September 5, 1953)
24. Quill Threatens to Slow Subways. New York Times, 1 (September 18, 1953)
25. Transit Board Rebuffs Quill; Will Cut Trains as It Sees Fit. New York Times, 1 (September 22, 1953)
26. Arbitration Asked In Transit Dispute. New York Times, 50 (September 24, 1953)
27. New Transit Runs Set, Ignoring Quill. New York Times, 23 (September 25, 1953)
28. Mayor Intervenes In Transit Dispute. New York Times, 23 (September 29, 1953)
29. Authority Rejects New Board to Arbitrate Transit Dispute. New York Times, 1 (October 2, 1953)
30. Transit Body Hauled To Court By T.W.U. New York Times, 1 (October 3, 1953)
31. Transit Authority Gets New Member. New York Times, 1 (November 6, 1953)
32. Jurist Puts Off Transit Suit, Urges Out-of-Court Accord. New York Times, 1 (October 6, 1953)
33. Dispute on Transit Is Left To Jurist For Adjudication. New York Times, 1 (October 10, 1953)
34. Transit Disputants Get Hart Proposals. New York Times, 11 (November 18, 1953)
35. TWU and Authority Settle Dispute Over Schedule Cuts. New York Times, 1 (November 20, 1953)
36. Transit Talks Asked. New York Times, 26 (November 24, 1953)
37. Authority To Meet Quill On New Pact. New York Times (November 28, 1953)
38. New Strike Threat Is Made By Quill. New York Times, B1 (November 30, 1953)
39. Transit Deadlock Charged By Quill. New York Times, 58 (December 10, 1953)
40. Quill Warns of Tie-Up Jan. 1 As He Quits Transit Meeting. New York Times, 1 (December 15, 1953)
41. Transit Authority Accepts Fact Unit to Avert Strike. New York Times, 1 (December 25, 1953)
42. 3 Named To Sift Transit Dispute. New York Times, 56 (January 4, 1954)
43. Klein Calls Quill Wasteful 'Bully'. New York Times, 1 (January 12, 1954)
44. Transit Fact Unit Opens Its Sessions. New York Times, 33 (January 13, 1954)
45. 20-Cent Fare Seen In Union Demands. New York Times, 1 (January 20, 1954)
46. Jury in Klein Case Hears 4 Newsmen. New York Times, 28 (January 23, 1954)
47. Quill Is Indicted On Three Charges of Libeling Klein. New York Times, 1 (February 12, 1954)
48. Fact-Finders Ask Transit Pay Rise. New York Times, 13 (May 18, 1954)
49. Transit Pay Plan Accepted By Quill. New York Times, 1 (May 20, 1954)
50. Transit Body Bars a Union Contract, But Sets Pay Raise. New York Times, 1 (May 25, 1954)
51. Transit Parleys Called By Mayor. New York Times, 1 (May 29, 1954)

52. Threat of Strike on Transit Eased. New York Times, 1 (June 4, 1954)
53. No Transit Strike. New York Times, 14 (June 12, 1954)
54. Quill Wins 2-Year Contract With Transit Authority. New York Times, 1 (July 9, 1954)
55. Quill Union Files Set Of Grievances. New York Times, 25 (August 12, 1954)
56. Transit Workers Fail In Protests. New York Times, 31 (September 16, 1954)
57. Quill Spurs Fight On Transit Board. New York Times, 1 (November 29, 1954)
58. Harriman Deflates Quill's Transit Idea. New York Times, 1 (November 30, 1954)
59. Transit Agency Finds It Did Well In First Year, Then Deficits Began. New York Times, 1 (January 10, 1955)
60. No Atoms For Subways. New York Times, 16 (March 5, 1955)
61. Tough-Talking Mike Quill Was Gentle Off Stage. New York Times, 30 (January 29, 1966)
62. Chalk, 51, Is Owner of Capital Transit and Air Cargo Line. New York Times, 45 (April 12, 1959)
63. Investor Group Seeking To Buy All City Transit. New York Times, 1 (April 12, 1959)
64. Mayor Declares Any Transit Sale Must Limit Fare. New York Times, 1 (April 14, 1959)
65. Transit Proposal Revised By Chalk. New York Times, 1 (April 17, 1959)
66. Chalk Bids $615 Million For Part of N.Y. Transit. New York Times, B1 (April 25, 1959)
67. New York City Rejects Chalk's Proposal To Buy Its Transit System. Wall Street Journal, 3 (April 30, 1959)
68. New T.W.U. Plea Made To Lindsay. New York Times, 93 (December 1, 1965)
69. Quill Calls Halt To Transit Talks. New York Times, 1 (December 2, 1965)
70. Quill Threatens De. 15 Walkout On Transit Lines. New York Times, 1 (December 3, 1965)
71. Quill Calls Off Dec. 15 Walkout; 3 Will Mediate. New York Times, 1 (December 10, 1965)
72. Talks With Quill On Right Track, 3 Mediators Say. New York Times, 1 (December 23, 1966)
73. Transit Walkout Voted For Jan. 1 If Parleys Fail. New York Times, 1, December 27
74. Lindsay Demands Fair Settlement In Transit Talks. New York Times, 1 (December 28, 1965)
75. Union Head Rips Up Court Order In N.Y. Chicago Tribune, 1 (December 31, 1965)
76. Transit Union In N.Y. Quits Talks. Los Angeles Times, 2 (December 31, 1965)
77. Quill Quits Talks. New York Times, 1 (January 1, 1966)
78. New Talks Today. New York Times, 1 (January 2, 1966)
79. Orders Jailing of Quill. Chicago Tribune, 1 (January 4, 1966)
80. The Judge Can Drop Dead. New York Times, 30 (January 5, 1966)
81. Lindsay Strike Plea Rejected. Los Angeles Times, 1 (January 6, 1966)
82. Mayor Lindsay Gets Weak Knees. Chicago Tribune, 8 (January 8, 1966)
83. Wirtz's Finding: Keep Bargaining. New York Times, 10 (January 8, 1966)
84. Union Sees Gains. New York Times, 1 (January 13, 1966)
85. Quill Praises Lindsay's Role In Ending Strike. Los Angeles Times, 20 (January 26, 1966)
86. The Lindsay Legacy. New York Times, A38 (December 21, 2000)
87. M.T.A. Takes Over Transit Network. New York Times, 36 (May 2, 1968)

Figure Sources

14.1: *Courtesy of* Wikimedia Commons, World Telegram Collection, Walter Alberlin, photographer.

14.2: *Courtesy of* Library of Congress, World Telegram & Sun Collection, LC USZ62-133401.

Chapter 15
Things Have to Get Worse before They Get Better: NYC Transit – 1970 to the Present

The 1970s were the transit system's worst years. The impacts of underfunding and deferred maintenance made for a difficult time, as the city dealt with its own financial disasters. Yet, the system survived the many crises encountered, and began to improve over the last decade of the 20th century.

15.1 Introduction

In a fitting opening to the 1970s, the transit fare was raised from 20 cents to 30 cents on January 4, 1970. The fare increase resulted from a costly labor settlement and declining ridership that had left the system in dire financial condition.

As the 1970s dawned in New York, the city was in the throws of its own financial crisis. To establish labor peace, Mayor Lindsay continued to approve lavish settlements that fueled budget increases without a rising tax base to support them. The city was forced to continue to cut services, and the ranks of the police and fire departments continued to diminish. The streets were dirtier, and crime was up and continuing to rise.

The transit system was part of the mess. The results of long-term deferred maintenance policies were evident throughout the system. Cars with inoperable doors and minor mechanical problems were left in service. Nothing was repaired until it was irretrievably broken. The mean distance between failures (MDBF) was approximately 24,000 miles, an unacceptably low number that would get many times worse by the end of the decade. Instead of fixing track problems, if the problem wasn't critical, the affected zone would be "red flagged," instead, forcing trains to travel at less than 10 mph for safety. Like the rest of the city, stations and trains were dirtier, and the assault of graffiti had begun.

John Lindsay left office in 1973 as an unpopular mayor, but still in possession of the flair that had made him popular earlier in his career. He left office with national ambitions that would not be fulfilled. He was succeeded by Mayor Abraham Beame, who had been the city's comptroller. He defeated Republican John Marchi in a relatively tame election campaign. Unfortunately, Mayor Beame was about to be hit by a rhetorical steamroller.

R.P. Roess and G. Sansone: The Wheels That Drove New York, STTT 1, pp. 333–375.
springerlink.com

15.2 Mayor Abraham Beame and the Financial Crisis of 1975

No two people or mayors could have been as different as John Lindsay and his successor, Abraham Beame. Lindsay was tall, handsome, aristocratic, charismatic, and a reform liberal. Abe Beame was as unassuming as Lindsay was flamboyant. He was short, only 5 ft 2 inches tall, and a product of the Brooklyn Democratic machine. He was generally liberal, but was not from the reform wing of the party. His manor was low-key and businesslike, and he was not comfortable dealing with media. He was also the city's first Jewish mayor.

Fig. 15.1 Mayor Abraham Beame

He served for only four years, but they were among the most tumultuous in the city's history. In announcing his death in February 2001, the *New York Times* described the difficulties of his time (1):

> *"The mayor, 67 when he took office, had to cope with terrorist bombs exploding in office buildings and department stores; a citywide blackout that led to arson and 3000 arrests; and a psychopathic killer called Son of Sam, who shot six people dead, wounded seven others, and terrorized the city."* (Ref. 1, pg 1)

In the midst of the general turmoil of the times, he faced a financial crisis of unprecedented gravity. As 1975 dawned, the city was on the brink of bankruptcy.

In a December 2001 article reviewing the period, the New York Times posed the question "How close was New York City to bankruptcy in 1975?" They provided the answer (2):

> *"So close that the city's lawyers were in State Supreme Court filing a bankruptcy petition. So close that police cars were mobilized to serve the papers on banks. So close that aides to Mayor Abraham D. Beame had written a statement announcing the default along with an emergency effort to save the city's dwindling cash for vital services like police and fire protection."* (Ref. 2, Pg 1)

In an age where federal bankruptcy laws had to be changed to ward off a runaway increase in personal and business bankruptcies, it is legitimate to question why a New York City default would have been so horrendous. A default by the city would have left it unable to borrow money to finance capital improvements, or to sell bonds without paying exorbitant interest rates. Further, New York was not the only city in financial difficulty, and a default by New York would have led to defaults in other major cities across the country. Every unit of government borrows money on a regular basis for two general purposes: (1) to fund long-term capital projects, and (2) to pay bills that occur before tax money is received in the same fiscal year (called "tax anticipation" borrowing). In 1975, the city's total debt was $14 billion. A default would have cost investors an estimated $6 billion of losses, which would have had a significant impact on local, state, and federal taxes collected. Lastly, the city's spending would have been under the control of a bankruptcy court.

Mayor Beame reacted to the crisis by making massive budget cuts. The cuts included significant reductions in manpower for the police, fire, and sanitation departments and other services he considered to be vital to the city's safety and security. In mid-August, the city was forced to notify the Municipal Assistance Corporation that it could not be expected to raise more than half the $900 million needed to fund its September obligations. Governor Hugh Carey, an old ally of Beame's, released a bristling attack on the Mayor, claiming that he had not done nearly enough to win investor confidence (3). Shocked by the vehement criticism of someone he considered to be an old friend, Beame angrily replied:

> *"Instead of criticizing the city, the governor should follow our example. This year, our budget showed near-zero growth. The state's budget went up more than 9 percent."* (Ref. 3, pg 5)

By August 22nd, the governor had softened his stance a bit, and indicated that he was willing to call a special legislative session to consider a massive loan to NYC to overcome its immediate cash-flow needs (4). A last-minute agreement with several municipal labor unions to invest pension funds in NYC bonds avoided a default in early September (5).

On September 9, 1975, the state legislature passed a "bail out" plan that was signed by Governor Hugh Carey later the same day. The measure provided $2.3 billion to ward off financial collapse for at least a month, during which the city was required to balance its budget and get its financial house in order. The bail out plan had some unique and difficult requirements (6):

- Control of the city's budget was transferred to a new state-dominated board, the Emergency Financial Control Board. Members of the board were the Governor, the State Comptroller, the Mayor, and three members appointed by the Governor with the approval of the State Senate.
- Through the Municipal Assistance Corporation, the State would provide direct aid of only $750,000,000 in two installments. The city had to raise the rest of the $2.3 billion through having various municipal union pension funds invest in New York City bonds.

The state measure incorporated a wage freeze which the city had imposed a month earlier. It mandated the freeze for at least one year, and gave the Emergency Financial Control Board the power to extend the freeze for the duration of the fiscal crisis.

The law also provided strict procedures to be followed in the event that the city defaulted despite the other measures.

The measure left the city without control of its money, and dependent upon unions whose wages were frozen to approve investment of their pension funds in New York City bonds. Unhappy with cuts in personnel and services, the teacher's union had gone on strike as the bail-out was announced (6). To make matters worse, within a month, Moody's Investor Service, Inc. downgraded both New York City and New York State bonds to "Ba," an unprecedented move that would increase the interest both would have to pay to attract investors (7).

Most of the city's unions went along with the plan, and made substantial investments in NYC bonds. The United Federation of Teachers, however, refused to invest. Lacking the teacher-union share of bond investment, the city was once again set to go bankrupt on October 17, 1975 (8). The next day, the union relented after an angry meeting with Governor Carey, and purchased $150,000,000 in municipal bonds just two hours before the banks closed (9).

On the federal front, Beame had requested a loan from the government, which was turned down by William Simon, the Secretary of the Treasury. He later mounted a lobbying effort to convince President Ford to provide loan guarantees. Ford initially turned him down, prompting the famous Daily News headline: "Ford to City: Drop Dead."

After suggesting that the city in fact go into bankruptcy, Ford eventually relented, and the federal government provided the requested loan guarantees (10).

The cuts made necessary by the crisis were often draconian in nature. They included (11):

- A reduction in the city's labor force of 30,000 people, with 70,000 more cuts to be implemented over the three-year period from 1975–1978.
- Strict wage freezes for all city employees.
- Reductions in the capital budget of $600 million, including the suspension of all housing programs.
- Increases in commuter and NYC transit fares. The transit fare, which had been raised to 35 cents in 1972, was increased to 50 cents on September 1, 1975.

Despite such cuts, the Governor and the Mayor often clashed, with the Emergency Financial Control Board often rejecting city projections and labeling cost-cutting measures as insufficient. On March 10, 1976, Governor Carey appointed the state's Social Services Commissioner, Stephen Berger, as chief administrator of the Emergency Financial Control Board, with the objective of providing stronger oversight of the city's fiscal affairs and putting more pressure on the city and the Mayor to make even deeper cuts.

In 1977, Beame was opposed by five prominent Democrats in the primary election for mayor. With all of the problems he had faced, Beame also had to deal with the blackout of 1977, which had occurred only months before, and the widespread looting and lawlessness that came with it. He finished third in the primary, which was won by U.S. Representative Edward I Koch of the 18th District, which was part of John Lindsay's old "silk stocking" district. In fact, Ed Koch had been initially elected to the House from Lindsay's 17th District. A later redistricting moved him into the 18th District, which retained parts of the former 17th District. Future Governor Mario Cuomo ran second.

No mayor in previous history had had to deal with such a difficult set of circumstances, circumstances that had developed over decades of faulty fiscal management on the part of several administrations. In an editorial upon the death of Mayor Beame, the *New York Times* summed up his contributions (12):

> *"Mr. Beame was an honest and unpretentious public servant at a time when history demanded more than he, and perhaps most mayors, could bring to the job. He cut the budget, won concessions from the unions and pushed through a program of extraordinary austerity only to find that the markets, the state and the federal government felt more was needed. A quarter-century later, New York City is better for the lessons learned painfully while Mr. Beame was mayor. The fiscal crisis could not have been solved without sacrifice shared by everyone – the taxpayers, the municipal work force, the banks and politicians like Mr. Beame. The mayor faced these realities with grim acceptance and grace. In the process, he contributed greatly contributed to what turned out to be a new understanding of the limits of government, the dangers of fiscal irresponsibility, and New York City's capacity for self-renewal."* (Ref. 12, pg A22)

15.3 Transit Plans Go Awry

In 1968, Mayor Lindsay and Governor Rockefeller had proudly presented a massive plan for subway improvements, including the oft-planned but never built Second Avenue Subway. It was commonly referred to as the "new routes" plan. Amazingly, neither Lindsay nor Rockefeller recognized the serious financial times that were rapidly approaching. The massive $2.9 billion plan, to be implemented in two phases, included (13, 14, 15):

Phase 1: 10 years, $1.6 billion:

- Queens "superexpress," running along the LIRR mainline right-of-way for the Queens Blvd. Subway;
- Extensions of the Queens Blvd. subway to serve northeastern and southeastern Queens;
- A new Second Avenue Subway in Manhattan from 34th Street to the Bronx, with connections at 63rd Street to Queens and the westside;
- A new 63rd Street crosstown line;

- A new express line in the Bronx connecting the new Second Avenue Subway with the Dyre Avenue and Upper Pelham Bay lines;
- Extension of the Nostrand Avenue Subway to Avenue U;
- Extension of the New Lots Avenue line;
- Purchase of 500 high-speed air conditioned subway cars for the new lines, and upgrading of shops;
- Rehabilitation of the Staten Island Rapid Transit line.

Phase 2: Indefinite Time, $1.3 billion:

- Completion of the Second Avenue Subway from 34th Street to lower Manhattan;
- Extensions of the new northeastern Queens subway and of rapid transit east of Jamaica and removal of the Jamaica Elevated in downtown Jamaica (Archer Avenue Line);
- Replacement of the Third Avenue Elevated in the Bronx with a new line adjacent to the Penn Central right-of-way along Park Avenue;
- Extension of the Pelham Bay and Grand Concourse lines in the Bronx;
- Purchase of 500 additional subway cars, and additional improvements to shops;
- Extension of the Flatbush Avenue LIRR line into lower Manhattan.

The "new routes" program was met with general approval by most political and civic organizations, including the influential Regional Plan Association (16). Some debate ensued concerning the appropriateness of individual routes. In particular, the issue of whether the new Second Avenue Subway should be a two-track or a four-track line was discussed, as was the specific routing of the downtown portion of the subway. There was also some debate over whether the Second Avenue Subway should be converted into a First Avenue Subway.

Remarkably, with all of the fiscal problems the city would face, some of the projects actually got started:

- Ground was broken for the Second Avenue Subway at E 103rd Street on October 27, 1972 (17). Work was also started on a section between 110th and 120th Streets and on a 700-ft section near Chinatown's Grand Street Station at the same time, although there were no formal groundbreaking events at these locations.
- A formal ground-breaking for the 63rd Street Tunnel occurred on November 24, 1969, with both Governor Rockefeller and Mayor Lindsay in attendance (18).

The *New York Times* observed that the joint appearance of Rockefeller and Lindsay, who did not get along well was unusual:

> *"Wearing hard hats, Governor Rockefeller and Mayor Lindsay stood in an unaccustomed pose – side-by-side..."* (Ref 18, pg 49)

The tunnel was actually "holed through" on October 10, 1972. Work on the 63rd Street Tunnel connection under Central Park began in July 1971, and was completed on January 20, 1973. Work on another part of the 63rd Street Tunnel connection between 5th Avenue and Park Avenue was started on August 30, 1974. Work on the Archer Avenue Subway began on August 15, 1972.

On March 21, 1971, Governor Rockefeller announced a $2.5 billion bond issue that would fund both the first phase of the new routes plan as well as upstate highway projects (19). On November 6, 1971, with the full support of Governor Rockefeller, Mayor Lindsay, Republican and Democratic parties, the Regional Plan and other civic groups, the bond issue was defeated in a landslide (20). Some pundits speculated that the voters distrusted anything that Rockefeller and Lindsay actually agreed on. Rockefeller claimed its defeat was because of the public's caution at taking on new debt and new taxes. Lindsay blamed the defeat on the inclusion of "concrete" funds for more highways. Whatever the cause, however, the defeat spelled doom for the new routes program. As if to emphasize the situation, a $3.5 billion bond issue was offered three years later, and went down to defeat by an even larger margin.

The financial crisis of 1975, together with the defeated bond issues, killed the Second Avenue Subway project, which was abandoned leaving several stations and subway tunnels in place. Work stopped in April of 1975 after the expenditure of $63 million. Remarkably, work on both the 63rd Street Tunnel and Archer Avenue both survived the crisis, although progress was certainly slowed. Trains would not run through the 63rd Street Tunnel until 1988, and the Archer Avenue project, once planned to be a long route serving southeastern Queens, would be limited to two miles.

15.4 The Transit Malaise of the 1970s

15.4.1 Loss of Route Mileage

The 1970s were the only decade in the transit system's history to lose route mileage. The 1960s had escaped that fate by the opening of the Rockaway and Far Rockaway lines, which had been taken over from the Long Island Railroad. The following route miles were taken out of service during the decade:

- Third Avenue Elevated in the Bronx ,April 29, 1973 (21)
- Culver Shuttle, May 11, 1975 (22)
- Bowling Green–South Ferry Shuttle, February 13, 1977 (23)
- Jamaica Elevated between Queens Blvd and 168th Street, September 10, 1977 (24)

15.4.2 Effects of Deferred Maintenance Escalate (25)

The condition of the subways continued to deteriorate. Deferred maintenance ruled the day, with disastrous results. The decade started with the MDBF at approximately 24,000 miles, an unacceptable number in itself. In July of 1980, the MDBF dropped to 5937 miles, a disaster by any definition, and the lowest value ever recorded. The length of off-hour trains was substantially cut. Further cuts in the number of scheduled trains, even on the most crowded routes, were implemented. Crime and vandalism increased dramatically. In 1977 alone, there were eleven incidents of vandals throwing debris at trains running in the open cuts of Brooklyn, two resulting in serious injuries to motormen.

15.4.3 Subway Crime

In 1978, there were nine murders on the New York City Subways. The Lexington Avenue Express was generally referred to as the "muggers' express" (26). By September of 1979, felonies were occurring on the transit system at a rate of 250 per week, far higher than any other transit system in the world (25).

The crime situation was so bad, it inspired Curtis Sliwa to form the "Magnificent 13," which later became known as the "Guardian Angels." Sliwa and his group began to ride subways and elevateds, making citizen's arrests and generally protecting riders from marauding felons. Despite initial objections to this vigilante group, they eventually worked out a stable relationship with the police department, and expanded to patrol most trains and stations in the most dangerous parts of the system. In the public's perception, they were more effective than the TA Police (27, 28).

15.4.4 The Graffiti Wars

No one is sure when the first bit of graffiti showed up in the New York City transit system. It began on the walls of subway stations in 1966–1967. By early 1968, news reports and articles concerning the problem began to appear, some of which treated the subject as a form of "urban art" (29). In a May 1, 1971 *New York Times* article, D. Keith Mano noted the change in station "graffiti" from the age-old one- and two-line obscenities to more elaborate works of colorful "art" (30). By 1972, the graffiti problem was a well-recognized epidemic that had affected both stations and rolling stock (31, 32).

It was a gathering snowball, and a symbol of a city and a transit system that was out of control. Starting as small acts of vandalism, graffiti grew and grew, as vandals painted intricate murals that covered the entire sides of subway cars. By the end of 1972, virtually every one of the NYCTA's 7000 railcars was covered in graffiti (32). In that year alone, 1562 youths had been arrested and charged with vandalism in connection with transit system graffiti (33).

Graffiti "artists" grew bolder as time went on, and produced car-covering "artwork" that could only be done overnight, when most transit cars were in shops and yards. They had "monikers" and signed each piece of work. Gradually, it became a contest to see who could produce the largest and most outrageous works (34). The city, with all of its fiscal constraints, could not keep up with the damage, and was forced to run trains covered in graffiti. Lax policing of major train yards contributed heavily, as many of the paintings clearly took many hours to create.

Fig. 15.2 Rolling stock in service was frequently covered with graffiti, both inside and outside in the 1970s and early 1980s

In 1971, the city spent $800,000 removing graffiti from its subway cars. In 1972, the budget jumped to $1.3 million. In March 1973, Mayor Lindsay estimated that it would cost $24,000,000 a year to make a visible dent in graffiti (35). As the TA and the city came up with new ways to fight graffiti, the graffiti "artists" found new ways to beat the system. By the mid-1970s, the city was spending ever-increasing amounts of money to fight graffiti, but the graffiti got worse. There seemed to be no way to eradicate the blight, and virtually every train on every line was covered in it.

It didn't help that graffiti was becoming a counter-culture art form, receiving significant attention as serious art in some segments of the media (36). Andy Warhol loved graffiti, and gave numerous interviews extolling its virtues as art. Norman Mailer wrote a book about it, *The Faith of Graffiti*, that also helped give it credence (37). The MTA tried everything it could think of to stem the tide, even introducing dog-patrols in the yards overnight (38). Nothing, however, seemed to make a dent in the problem. In the ultimate gesture of defeat, in August of 1975, the senior executive officer of the NYCTA, John de Roos, announced that the authority was abandoning its efforts to attack the problem (39).

The only thing that occasionally covered the graffiti was dirt. Car cleaning was another casualty of the fiscal crisis. Instead of a regular washing schedule, cars were washed on an as-needed basis, although it was hard to tell what was considered to be "needed."

15.5 The R-44 and R-46 Mess

In the 1970s, even the good went bad. In anticipation of some of the new routes announced in 1968, the TA made two large orders for new cars to service the new lines. In both cases, the cars were filled with new technology, and could reach speeds of 75 mph, which would have been possible on some of the new routes that were planned, but never built. The cars had complex electronics, and were built using light-weight materials in many components. Noise insulation was used extensively to produce a quieter ride, and car suspension was "softened" to provide a smoother ride.

The first order for new cars, given the model designation of R-44, went to St. Louis Car, Inc. The TA ordered 300 cars to be in service by the end of 1973. The R-44s experienced a wide variety of problems in service, so that at any given time, over one-third of the fleet was in maintenance, and the failure rate was three times the NYCTA average. Because the problems were many and varied, it was difficult to come up with a plan to remedy the situation. When the time came to order more cars, dubbed the R-46, although there was little change from the R-44s, St. Louis Car would not even bid on the order. The R-44s were the last rail cars made by the company, which subsequently went out of business. (25)

Despite their poor performance, the NYCTA claimed that initial problems were the result of "introductory gremlins" (40). A consultant report based upon performance data supplied by the NYCTA found a number of problems with the new cars. In particular, Dr. Martin Huss of the Polytechnic Institute of New York contended that a problem in electrical circuitry could lead to fires, several of which had occurred while the cars were in yards (41). The NYCTA indicated that Huss had misinterpreted much of the data, and continued to support the introduction of new technology in the R-44s.

In general, the R-46 cars were significantly better than the R-44s, except for one glaring systemic problem, namely, cracking of trucks. In April of 1972, Pullman-Standard submitted a low bid of $273,000 per car (42). Competing bids were submitted by General Electric, Rohr Industries, and Westinghouse Electric Corporation. The TA placed an order for 752 cars with Pullman-Standard in September of 1972 (43). Because of the pressure to renew the TA's aging fleet, the usual practice of ordering a few trains for significant testing before a production run was not followed. The R-46 was the first subway car Pullman had built since 1938, and it would be the last passenger rail car they ever built.

In March 1977, the first structural crack in an R-46 truck was found (44). While some public reports indicated that the crack caused a motor to drop onto the tracks, internal reports state that under certain circumstances, the crack could have led to dropping a motor, but no actual drop occurred.

In January of 1978, 264 more trucks were found to have structural cracks. By February, the number had grown to 889 structural cracks on 547 trucks. At times, the entire fleet had to be pulled from service to effect repairs. Structural cracks in truck units were the most significant, but not the only defect in the R-46s. In July of 1979, hand brakes on the R-46 were declared defective, and the cars were pulled from Manhattan Bridge and several other runs where the use of hand brakes might become necessary (25).

Negotiations with Pullman-Standard, and the subcontractor who built the trucks, Rockwell Inc., went on for months. Because repairs made by welding did not perform well, the TA rejected all proposals to fix the trucks, and sued Pullman-Standard and Rockwell for the costs to replace all trucks (45). In July of 1979, the companies countersued the NYCTA. They blamed the cracks on the poor conditions of tracks and roadbeds in the New York City subway (46). Further complicating the situation was the allegation that at least one TA employee had accepted bribes from Pullman-Standard to accept delivery of substandard equipment (46).

In September of 1979, the MTA decided to move forward with the replacement of all trucks on the entire fleet of R-46 cars, at a cost of approximately $50,000,000 (47). By April of 1980, the TA cut the use of R-46 cars by 47%, given the toll of heavier use on various conditions (48).

On December 24, 1980, the city won its suit against Pullman-Standard and Rockwell, and was awarded $72,000,000 in damages (49).

With the R-44 and R-46 orders, two American rail car builders were forced out of business, one completely, the other out of the passenger car-building business. Pullman-Standard's subcontractor, Rockwell, went bankrupt. The situation left only one American manufacturer of rail cars, that is, Budd, in operation. Subsequent new car orders would shun "new technology" and stick to the basics. Only in the late 1990s would the TA once again venture into the area of "new technology" on its trains.

15.6 A Bond Issue Passes

A small note of hope ended the 1970s. In November of 1979, voters finally approved a much-trimmed transportation bond issue of $500,000,000, with most of the money going to the New York City transit system (50). It passed by a slim margin, and would provide money for fixing up the subways and remaining elevateds, as well as to improve highways in upstate and downstate regions. Of course, by 1979, both Mayor Lindsay and Governor Rockefeller were long gone, and the issue was more-or-less considered on its merits, not as part of a presumed political maneuver.

15.7 Mayor Edward Koch and the Transit Strike of 1980

The 1980s were a turbulent period for New York City transit. As the decade dawned, the TA's problems were still growing worse. By January of 1981, one-third of the entire fleet of subway cars was out of service at any given time. The MDBF had slipped to 6639 miles. Over 500 runs a day had been cancelled, and there had been 30 derailments in 1980. Not a single repair was done until a failure, and not one of the 2637 IRT cars in service had ever been overhauled, despite the fact that many of them were running well past their expected retirement dates (51).

In 1977, Edward I. Koch was elected the 105th Mayor of New York. His election formed a unique line of succession from Lindsay to Beame to Koch. Ed Koch was elected after serving as U.S. representative from the same 17th "silk stocking" district as John Lindsay. He was a Democratic liberal, although he would cross party lines a number of times during his career. He had served as the Democratic district leader in Greenwich Village between 1963 and 1965. During that time, he took the unprecedented step of crossing party lines to support John Lindsay's candidacy for Mayor. From 1969 until his election as Mayor in 1977, he served in the House of Representatives. He would serve three terms as mayor, and in 1981, he ran as both a Democrat and a Republican.

Fig. 15.3 Mayor Edward I Koch, wearing a sailor's hat at the commissioning of missile cruiser U.S.S. Lake Champlain in 1988

His political battles with Mario Cuomo became legendary. During the 1977 primary campaign, Koch had to deal with issues surrounding his sexuality. A lifelong bachelor, he was attacked as a homosexual. While the Cuomo campaign denied any involvement, Koch was never convinced of this. In 1982, Koch challenged Cuomo in the primary for Governor, this time losing. Many believe that a Playboy interview in which Koch disparaged the lifestyle of upstate New York led to his defeat. From 1982 on, Koch (as Mayor) and Cuomo (as Governor) would have to deal with each other, but the relationship was always frosty.

Koch's great love of New York provided some interesting moments. In 1984, he opposed the creation of a second area code for New York City because it would "divide New Yorkers." When the football Giants won the Super Bowl in 1987, he refused to allow a victory parade in the city, quipping "If they want a parade, let them parade in front of the oil drums in Moonachie." Moonachie is a small New Jersey town bordering the Meadowlands home of Giant's Stadium.

As Mayor, he is generally credited with restoring fiscal stability to the city. He placed the city on a balanced budget using generally accepted accounting principles. He was also noted for installing a merit selection system for criminal and family court judges, and for establishing extensive housing programs. He was a champion of gay rights and issued several executive orders banning discrimination against gays. When the AIDS epidemic hit, he ordered that gay bathhouses be closed. When several went to court claiming discrimination, because heterosexual "sex clubs" (like Plato's Retreat) were allowed to stay open, Koch ordered them closed as well.

Ed Koch was similar in some ways to John Lindsay in style. He was very comfortable with the media, and was not above using them to promote his positions. While he lacked the aristocratic demeanor of Lindsay, he was (and is) charismatic in his own right. Like Beame, however, he believed in getting down to business. Unlike both, he had a real sense of humor, and frequently used it to his advantage.

By 1980, the city had a balanced budget and was no longer under Emergency Financial Control Board supervision. Unfortunately, no good deed goes unpunished. With a TWU contract expiring on April 1, 1980, contentious negotiations began in February with the MTA. The TWU, seeking to recoup what it considered to be its losses over the period of financial crisis, presented demands that included a 21-month contract, a 30% wage increase and a new holiday honoring former TWU President Mike Quill. They argued inflation since the last contract had raised the cost of living by 53% (52, 53).

With a strike deadline of April 1st rapidly approaching, the MTA obtained a restraining order from court on the basis of the Taylor Law, which made strikes by public employees illegal (54). Negotiations with the MTA became confrontational, and a counter-offer was not made until March 31st, on the eve of the strike deadline. The offer of a 34-month contract with three and one-half percent annual increases, which was increased to 5% in the final hours of negotiation, was deemed insufficient, and the TWU went on strike on April 1, 1980 (55).

In an April 7, 1980 article, the *New York Times* pointed out that one of the problems in resolving the strike was a deeply divided union, primarily along racial lines. The ethnic make-up the TWU union was rapidly shifting, and, in 1980, was 50.8% Caucasian, 40.3% African-American, and 6.5% Hispanic (56). Within the union, there was great controversy over disciplinary hearings, which often pitted white supervisors against minority workers. Minority workers accused the union of not defending their cases with the same determination and dedication that was given when the accused was white. In this atmosphere, it was difficult to get a clear view of what kind of offer the union would accept.

After taking a hands-off approach during the negotiations leading up to the strike, Mayor Koch played an active and visible role in the aftermath. The city was actually pretty well-prepared for the strike. As the strike coincided with the Easter/Passover holidays, schools were closed, reducing the normal load of commuters. Most Manhattan businesses had been reserving hotel rooms for weeks in anticipation of the strike, with an estimated 500,000 workers temporarily living in

Manhattan. It is estimated that up to 200,000 people commuted to work via bicycle. Driving was discouraged, with strict "car pooling" rules limiting Manhattan access to vehicles with three or more passengers during the usual rush hours.

Droves of commuters came to work by walking over the Brooklyn Bridge. They were met every day by the Mayor himself, who would walk over the bridge himself, talking to his constituents as he did. It was during these morning walks that he first started using the phrase that became his tag line, asking "How'm I doin'?" He also urged commuters to let the rush hour pass before heading home in the evening, taking some time to have a Martini after work in Manhattan.

The strike was also the first under the state's Taylor Law. The Public Employees Fair Employment Act (referred to as the Taylor Law) was enacted in 1967 in response to the 1966 transit strike. Under the law, public employees were (and are) prohibited from striking. Penalties include fines to each individual striker of two days pay for each day of the strike, jail time for union leaders, heavy fines against the union, and loss of union dues "check-off" deductions for a period of time. The Mayor and the MTA joined in petitioning the courts to implement all of these provisions (51).

After 11 days, the strike was called off and the dispute was submitted to mediation. The resulting settlement was a two-year contract with a 9% wage increase in Year 1, and an 8% wage increase in Year 2, along with a one-time cost-of-living adjustment, which brought the package to about 20% (57).

While agreeing to go back to work, union dissidents were strongly opposed to the settlement, and mounted an effort to defeat its ratification. In a mail ballot, however, the settlement was approved by a two to one margin (58). The Mayor and the city harshly criticized the settlement as far too expensive.

The courts did impose the full penalties called for in the Taylor Law (59). Workers complained that they lost over half of the first year's increase in fines. The union was also fined, and lost union dues check-offs for several years. Given the size of the settlement, however, it is fair to say that the union "won" the dispute, although the MTA and the city made its point by not negotiating away any of the Taylor Law penalties. The settlement also forced an increase in the fare to 60 cents on June 29, 1980. The 60-cent fare did not even survive the TWU contract, and it was raised to 75 cents on July 3, 1981 (51).

New York City transit began to improve in the early 1980s, although the visible effects were not apparent until the middle of the decade. Mayor Koch did not have much to do with the system, except for providing a stable balanced budget for the city which provided an encouraging background.

Ed Koch served three terms as Mayor. During his third term, several members of his administration were found to be corrupt, and several served short jail sentences. The most heavily publicized was a scandal in the Parking Violations Bureau (PVB). Several key administrators were found to be skimming money from contracts with private parking operators of New York City parking garages. The scandal led all the way to the Borough President of Queens, a Koch protégé, Donald Manes. Manes was considered to be the odds-on favorite to be the next mayor, and was clearly being groomed for a run by the Democratic Party. Manes may be the only person known to commit suicide twice. As the scandal broke, Manes was found in what was thought to be an abandoned vehicle in Flushing Meadow Park with his wrists slashed. He was taken to the hospital, where he recovered. A few months later, while on the phone with a relative, he stabbed himself in the stomach with a large knife, and died at the scene. The manner of the second suicide attempt caused many to suspect that he had been a victim of foul play, perhaps from unknown participants in the PVB scandal.

With the scandal hanging over him, Ed Koch lost the 1989 primary to David Dinkins, who narrowly defeated Rudolf Giuliani to become the city's first Black mayor.

Ed Koch remains an active commentator, and is a frequent guest on talk shows. He has written 14 books, and was a judge of the People's Court between 1997 and 1999. He has frequently endorsed Republican candidates, including Rudy Giuliani and Michael Bloomberg for Mayor, Al D'Amato for Senate, George Pataki for Governor, and George Bush (the younger) for President. More recently, he has endorsed Democratic candidates, including Eliot Spitzer for Governor and Hillary Clinton, for Senate and President.

15.8 The Grumman Flxible Bus Fiasco

In 1980, before the R-46 lawsuit had been settled, the TA purchased a fleet of 837 buses from Grumman/Flxible. The Flxible Bus Company had been an independent manufacturer for many years. In the late 1970s, it was bought out by Grumman. At the time, only General Motors and Grumman/Flxible were manufacturing buses in the U.S. Under federal laws, since federal mass transit aid money was being used to support the purchase, the TA was forced to accept the low bid of Grumman for the bus orders.

The buses were controversial even before they were delivered. Both the NYCTA and the American Public Transit Association indicated that the Grumman/Flxible bus was significantly less fuel efficient than the older buses it was replacing, and that their air conditioning systems were neither efficient nor reliable. Because the buses had sealed windows, malfunctioning A/C systems in the summer would make them virtually inoperable. Transit officials in Houston, Washington D.C., and San Francisco indicated that their experience with the A/C left 50% of their fleets inoperable during the summer months (60).

The first buses arrived on June 13, 1980, and were pressed into immediate service (61). By August, 57 of 217 buses were out of service due to a variety of problems related to air conditioning, brakes, steering, electrical or fuel systems, body panels, or transmissions (62). The most serious problem, however, was first discovered on November 22, 1980. During routine inspections, 13 of 131 buses examined were found to have cracks in the understructure. Grumman/Flxible initially accepted responsibility, indicating that a design flaw might the cause (63).

New York was hardly the only city to experience these problems: Flxible buses delivered to the Southern California Rapid Transit District, Houston, Atlanta, and Santa Monica were also found to have cracks in the undercarriage, often shortly after being placed into service (64).

On December 13, 1980, a Grumman/Flxible bus with a full load of passengers suffered a collapsed chassis as a result of a structural crack. The MTA immediately pulled all Flxible buses out of service, creating a shortage of buses for daily service (65). Los Angeles took the same action, pulling all of its Flxibles from service (66).

Grumman/Flxible claimed that the cracks in NYC bus frames had occurred "randomly,' and were the direct results of poorly-maintained city roads and overloading of the buses (67). At the same time, however, Grumman offered to repair cracks in all 2600 buses it had sold in New York, Los Angeles, Atlanta, Houston, Santa Monica, and Hartford (68). By the end of December 1980, Chicago, which had experienced a variety of problems with Flxibles, started to discover structural cracks as well (69).

With the Grumman/Flxible fleet out of service, the city had to purchase or lease older buses from other transit systems to operate a full schedule. The first wave included 110 surplus buses from Washington D.C. (70).

On January 5, 1981, the MTA and Grumman reached an agreement. MTA would continue to make payments on the new buses (it had suspended payment when the buses were pulled from service), and Grumman agreed to make repairs to all affected buses (71). On January 24, yet another structural problem was noted. Metal fatigue was noted in a portion of the undercarriage called the "trunnion." As a result, repairs to the first 205 Flxibles were halted, while new inspections for the new problem took place (72). By the middle of July, however, Grumman was returning repaired buses to New York City (73). In August, yet another problem was found; rubber insulated engine mounts were deteriorating (74). The entire fleet of Flxibles was once again pulled from service.

In January of 1984, five Grumman/Flxibles were discovered to have the problems with the steering column. Fearing that up to 182 additional buses might be affected by the problem, the TA once again shut down the entire fleet (75). In February, the last straw was a Flxible bus that burst into flames on a midtown street after being removed from service for another problem. The TA decided to remove all Flxible buses from service permanently (76). The move cost the city one-fifth of its entire fleet. In March, the MTA ordered 400 new buses from General Motors to replace about half of the mothballed Flxible fleet (77).

The MTA sued Grumman, but could not sustain its charge that the problems involving the Flxibiles were a result of design errors or fraud. In a final insult, the MTA was obliged to repay the Federal government $56,000,000, 80% of the amount of federal aid used to purchase the buses (78). The MTA eventually resold the buses to General Automotive Corporation, which had purchased Flxible a few years earlier, for $4,500,000 (79). They were refurbished, and sold to other transit agencies, including New Jersey Transit. The refurbished buses operated without significant problems.

Fig. 15.4 Grumman Flxible Bus of the type that experienced numerous operating deficiencies in New York and elsewhere

15.9 Richard Ravitch and the MTA: The System Starts Its Recovery (51)

In the early 1980s, a number of steps were taken that began to improve the system. Led by its chairman, Richard Ravitch, the MTA announced an $8.1 billion capital improvement program for the transit system. By the end of 1982, the MTA had signed contracts for $3 billion for a variety of improvements, more than had been spent during the entire decade of the 1970s. Some of the contracts let were for:

- Reconstruction of the Coney Island and 207th Street maintenance facilities.
- Track, signal and switch replacements and improvements.
- Station renovations.

These projects would take time to implement. Richard Ravitch retired in 1984, before most of his work had been fully implemented.

Not everything would be successful. In 1982, a program to eradicate graffiti was initiated. A number of subway cars were painted in a special white paint that supposedly could be cleaned of graffiti easily. The program was only partially successful. Vandals still had access to cars in yards, and the "special" paint wasn't quite as easily cleaned as had been claimed.

Rising crime was still a problem. In 1982, reported crime on the subways increased by 60% over the previous year. At a speaking engagement, Ravitch said that he would not allow his children to ride the subways at night, and that he personally did not feel comfortable riding during off-hours. Ravitch responded to the increase in crime by placing a transit policeman on every train between the hours of 8:00 PM and 4:00 AM.

15.10 Bernhard Goetz – The Subway Vigilante

On December 22, 1984, shortly after Ravitch had retired, the most famous criminal incident on the New York City subways occurred. Four young Black men approached Bernie Goetz on a New York City train, with approximately 20 passengers in the car. They approached Goetz in what he believed was a menacing manner, one with a screwdriver, demanding $5.00. Goetz pulled a gun and shot each of the men once, and then went over to one (Darrell Cabey) and shot him again. The uproar over the incident caused deep racial divisions in the city.

In the early days after the shooting, Goetz was not immediately identified as the shooter, nor had he been identified. The depth of reaction to the incident is difficult to imagine in today's setting. Against a backdrop of a transit system riddled with crime for over a decade, the range of reactions is more understandable. The Congress of Racial Equality offered to pay for the legal defense of the shooter. Its Director, Roy Innis, said (80):

> *"I can't imagine any reason why any citizen on his way on the subway should be surrounded by... young men who, no matter how quietly they speak, make clear the threat of impending violence by their presence if the man dares to reject their request for $5."* (Ref. 80, pg 5)

Public officials, including Governor Cuomo and Mayor Koch, decried vigilantism, and spoke of the dangers of the incident setting a precedent for future incidents. Friends and family of the injured youths were appalled at the public reaction.

On January 1, Bernie Goetz, who had fled to Concord, New Hampshire after the incident, surrendered to the Concord police (81). He claimed that he had been mugged three years prior to the incident, but that he had been denied a gun permit when he sought one. After a grand jury returned only one indictment for possession of an illegal weapon, a second grand jury added four counts of attempted murder (82).

At the time of the incident, all four of the injured youths had been arrested or convicted of at least one violent crime, and each one was facing a hearing or a trial on additional criminal activities (83). In June of 1985, one of the youths was arrested for raping a young woman in Bronx housing development (84).

Goetz was a hero to some, and a villain to others. Over the course of the next several years, the case would be reviewed from virtually all angles; a trial in the press took place long before the trials in the courts. Eventually, the Goetz trial went forward. In June of 1987, he was found guilty of a single count of illegal weapons possession, and acquitted of all attempted murder charges (85). During the trial, columnist Jimmy Breslin testified that Darrell Cabey, one of the four attackers, had told him that the four intended to rob Goetz because he looked like "easy pickin's." The post trial reaction was predictably mixed. As he left after the verdict, Goetz was surrounded by a group of Guardian Angels, who were protecting him from a group of demonstrators from the Revolutionary Communist Party. Goetz was sentenced to six months in prison on the possession charge (86).

Darrell Cabey, who was paralyzed by the second shot, successfully sued Goetz, who immediately filed for bankruptcy. The incident left an indelible mark on most New Yorkers, and refocused attention on the problem of crime in the subways.

The *Wall Street Journal*, in its editorial of June 18, 1987, commented on the Mayor and Governor's views decrying "vigilantism" (87):

"*Yet one might wonder why the mayor used the word "vigilantism." The jury did not think Mr. Goetz was acting as a vigilante. The four young victims of his 1984 subway shooting sought him out, in all probability to rob him. He was not seeking them. His only crime, in the jury's view, was illegal possession of a weapon, for which he faces a sentence.*"

"*We suspect that the mayor and other New York public officials who are unsympathetic to the jury's verdict have vigilantism on their minds for a good reason. They know, as we all do, that if the people of a city begin to take law enforcement into their own hands, as has happened to some degree in New York, it is not a reflection on the people but on the government. Mr. Goetz was so widely applauded for defending himself because the city had so thoroughly failed to defend its citizens.*" (Ref. 87, pg 30)

15.11 New Subway Cars (51)

After the problems of the R-44 and R-46 cars, the MTA made steady progress on replacing older rolling stock. In March of 1982, an order for 325 new IRT cars, dubbed the R-62, was placed with Kawasaki. It was the first order for subway cars ever placed with a foreign manufacturer. Under new state laws, the MTA was not required to accept a low bidder, although there was some question as to why the contract did not go to Budd, the last remaining U.S. maker of transit cars. The design of the R-62's was devoid of the high-tech components of the R-44's and R-46's, although their stainless steel exteriors would set the pattern for the future. The R-62s entered service on November 29, 1983.

Fig. 15.5 R-62 cars manufactured by Kawasaki, 1983

The MTA was very conservative with the R-62s, placing an order for test trains before the production order, and including performance guarantees in the contract.

When the MTA wanted to order another 825 cars, Kawasaki, whose manufacturing facilities had been taxed by the R-62 order, declined to bid. The order went to Bombardier (of Canada), primarily because they agreed to use the Kawasaki design. The cars were dubbed the R-62A, and experienced a number of electrical problems on initial service. Eventually, the electrical problems were fixed, and the cars performed acceptably.

In October of 1982, Westinghouse-Amrail, a consortium, received a contract to produce 225 cars for the BMT/IND divisions. Dubbed the R-68, these cars did not arrive until 1986.

As Richard Ravitch retired in 1984, clear signs that the system had turned the corner were in place. A program of car replacement was well underway. Major maintenance yards were themselves being rebuilt. Station, track, and signal improvements were taking place. By 1984, the number of derailments had been decreased to 15.

Some highly visible problems remained: crime was still on the rise, and graffiti was unabated. These problems would be addressed later in the decade.

15.12 The Robert Kiley/David Gunn Era: The System Rebounds (51)

Robert Kiley succeeded Richard Ravitch as chairman of the MTA in 1984. His first task was to find a new head for the New York City Transit Authority. David Gunn, head of the Philadelphia Transit System, took a tour of the Coney Island Yards, where he found conditions to be "atrocious." When asked by the press about the presidency of the NYCTA, he replied "It would be a suicide mission." Robert Kiley himself was quoted as saying "The subway system is having a nervous breakdown." The press and the public were greatly surprised when David Gunn accepted Robert Kiley's offer and began his "suicide mission" on January 11, 1984 (88, 89).

David Gunn became a hands-on manager, involving himself in every aspect of the TA's operation. He instituted a number of very controversial measures, but the crisis conditions allowed him to prevail.

By far the most controversial action came when Gunn removed union and Civil Service protection from top supervisors. At the time, virtually every level of supervision in the TA had its own union, and most managers held Civil Service status. The system, Gunn claimed, effectively removed the prerogatives of the TA President to act. In *The City Beneath Us*, Vivian Heller comments on the results of this radical change in the power structure (90):

> *"Some lost their jobs, others moved up; some were forced out of the system, others were given breaks that made their careers. Some claimed they were victims of the Kiley-Gunn regime; others believed that the chaos had finally been replaced with a rationally organized system of management."* (Ref. 90, pg 74)

And:

> "*It was Gunn, with Kiley backing him up, who saw to it that workers were given what they needed to do their jobs – parts, tools, and materials – and above all, decent working conditions.*" (Ref. 90, pg 74)

It of course helped that as he took office, the shop and yard reconstructions of Richard Ravitch were beginning to come on line. Gunn established a strict Scheduled Maintenance System (SMS) under which cars and their components were replaced and/or repaired *before* they broke down. Cars were to be cleaned, inspected, and put together again in accordance with a strict schedule.

David Gunn also recognized that while much needed to be done to improve the system, it would take time before the public would notice the improvements. He decided that a highly visible program would also be needed. He instituted an all-out war on graffiti. Recognizing that the chief thrill to the graffiti artist was to see his/her work circulating throughout the city, he instituted the "Clean Car Program" in May of 1984. Under the program, no car in the program would be allowed to leave the yard if it had graffiti. If a car in service was found to have graffiti, it would be pulled from service within 24 hours. The program was implemented on a line-by-line basis. The policy took time, but by May of 1989, the subways were clean, and graffiti was gone. The policy, now system wide, remains in effect today.

The war against graffiti was a public demonstration that even the worst conditions could be handled with a well thought-out management plan. It brought a sense of visible control back to the system.

In addition to regularizing car maintenance and cleaning schedules, Gunn instituted a massive program of rebuilding and reconditioning subway cars, many in the TA shops. In 1984, he noted that 4000 of the 6400 subway cars in the system needed "a lot of work." He blamed years of deferred maintenance as the culprit, not the age of the fleet. He instituted a program in which 200 cars per year would be completely rebuilt, a number which increased over time. The program started with the aging IRT fleet of R-36 cars. All of the rebuilt cars would be painted inside and out with a new scheme, and thus the rebuilt cars became known as the "Redbirds" because of their dark red exteriors. In addition, cars that were too old to rebuild would be spruced up as long as they were still in service.

Fig. 15.6 Rebuilt R-36 cars, dubbed "Redbirds" due to their dark red exteriors

Gunn's overhaul of the maintenance system was extremely popular with the workers, who rediscovered a sense of pride in keeping the system in top shape. The improvement in their working conditions was dramatic. Shops were clean, they had tools, and they had parts. The rebuilding program was a reward and an inducement. For decades, major reconstruction of subway cars was farmed out to private contractors. Gunn brought it inside, challenging TA workers to do the job. With major rebuilding work now in-house, the union was able to maintain a larger workforce. Gunn's faith in the workers paid off, and the rebuilding program was a great success. It was cheaper than having outside contractors do the job, and TA workers accelerated the number of cars that were reconditioned each year.

The success of Gunn's maintenance programs was evident. By 1986, derailments decreased to three, one-tenth of what the number at the start of the decade, and were a rare occurrence thereafter. By 1988, 94% of the fleet was graffiti-free, and by the end of 1989, graffiti had virtually disappeared from the system. By 1988, 86% of the subway cars in NYC were air conditioned.

15.13 The Manhattan and Williamsburg Bridge Problems (51)

While rail service over the Brooklyn Bridge had been discontinued with the death of the elevateds, regular service over both the Manhattan and Williamsburg Bridges formed (and still forms) key parts of the transit system. Thus, when structural problems occurred on each, significant service disruptions occurred.

The Manhattan Bridge problems were actually caused by the transit system. The bridge carries four tracks, with two located on each side of the bridge. Their location on the "outside" of the bridge caused severe problems. The great weight of trains running on outside tracks caused extreme torsion loads on the bridge. By 1982, these loads were resulting in 8-ft structural displacements.

A routine inspection in May of 1982 revealed a major crack in a beam. D-service over the bridge was suspended, with a rerouting that added 20 minutes to the run, while the beam was repaired. In the Summer of 1985, the northside tracks were taken out of service for emergency repairs. They reopened in Fall 1985, at which time the southside tracks were closed down. The southside repairs took until 2001, and caused extensive rerouting of subway services that normally used the Manhattan Bridge. Further periodic disruptions of service on the Manhattan Bridge continued until 2005, when a more extensive rebuilding project was completed.

The Williamsburg Bridge was another story. Subway tracks run on the inside of the bridge, as they had on the Brooklyn Bridge. Thus, damage to the structure was not the result of train traffic. Deferred maintenance, this time on the bridge, had taken its toll. Pigeon droppings, rust, and salt had built up on the cables, causing severe deterioration. In April 1988, on a structural rating scale of one to seven, the bridge's cables were rated a 1.6, barely allowing its continued operation. On April 11th, the bridge was closed for emergency repairs. J and M service could run only as far as Marcy Avenue in Brooklyn. A special ferry service was instituted to take passengers across the river to Manhattan. The bridge reopened on June 12, 1988, although reconstruction of the bridge went on for many years.

15.14 Other Service Disruptions (51)

On April 13, 1985, the Jamaica El service was cut back further, terminating at Queens Boulevard instead of 121st Street. This was to prepare for the connection of the El to the Archer Avenue Subway. The disruption, however, would last for three years before the new service became available.

On May 13, 1985, express service on the No. 7 Flushing Line was suspended for 4.5 years while a major track replacement project was underway. The No. 7 is one of the busiest lines in the system, and the change added significant time to its users. After completion of the $60 million project, service was restored on August 21, 1989. The original plan called for removal of the express stop at 61st Street due to excessive delays caused by across-the-platform transfers from locals to express trains. After many local protests, the express stop was retained.

While both of these projects caused long-term problems for many commuters, both were harbingers of a system on its way to recovery. The discontinued portion of the Jamaica El would eventually be replaced by a new subway, and the No. 7 track work, long deferred with the rest of system maintenance, would result in faster and smoother rides for commuters.

15.15 The Strange Story of the 63rd Street Tunnel

The 63rd Street Tunnel was one of two projects under the 1968 "New Routes" program to survive the fiscal crisis of 1975. It was to be part of a much larger system expansion, with connections to a new Second Avenue Subway in Manhattan. A massive project, the underwater tunnel contained four tracks on two levels, that is, an upper level for the Long Island Railroad, to be used to access a new East Side LIRR terminal, and a lower level for the subway.

The tunnel itself was an interesting construction project. Sections of the tunnel were prefabricated at Port Deposit, Maryland. They were barged to the site and sunk into dredged channels that had been prepared for the tunnel. After being put in place, divers connected the sections, which were then back-filled to provide separation between the tunnel and the river itself. Started in 1969, the tunnel was "holed through," i.e., all sections were connected, on October 2, 1972. The tunnel, however, was not ready for service. Connections under Central Park and local streets had to be built to provide access to the 6th Avenue Subway. On the Queens side, the tunnel was to be connected to a new subway serving northern Queens, which was never built.

By the mid-1980s, the tunnel and Manhattan side connections had been complete for over a decade. Work had stopped, and periodic inspections began to reveal leaks and other problems due to neglect (91). For a time, the MTA considered abandoning the tunnel, but when structural evaluations came back indicating that the basic structure was sound, plans were developed to complete the tunnel and establish service.

To begin service, a Queens-side station had to be provided. The 63rd Street Tunnel already included plans for a station on Roosevelt Island. A station was built at Queensbridge/21st St. It would not be connected to any Queens subway, and would be a terminal station. If passengers wished, they could get off and walk several blocks to access IND or BMT/IRT services. When completed, the station became part of a unique, but strange situation in the vicinity of the Queens side of the Queensborough (59th Street) Bridge.

The general area contains three stations that have no interconnections: Queensborough Plaza, an elevated station serving the Flushing Line (IRT) and Astoria Line (BMT); Queens Plaza on the IND; and Queensbridge/21st Street on the new 63rd Street Line. Because these lines were all built separately, no attempt was made to provide for a coordinated, interconnected terminal, which would have included many lines.

The Queensbridge/21st Street station was put into service on October 29, 1989 – and the first train to use the 63rd Street Tunnel began its run. Between the completion of the tunnel itself and its first use, 17 years had elapsed. Because of its isolation, the Queensbridge/21st Street station was lightly used, and the 63rd Street Tunnel service became known as the "train to nowhere."

Shortly after its opening, the MTA recognized the need to provide a connection to existing Queens subways from the Queensbridge/21st Street station. A connecting tunnel to the existing IND subway was built, and was opened for service in 2001. The F-train currently uses the 63rd Street Tunnel as part of its route.

The LIRR level of the 63rd Street Tunnel has yet to be used. A project providing the LIRR with an East Side terminal in Grand Central, however, has been approved and construction is underway. The tunnel would be used as a connector.

(a) 21st Street/Queensbridge Station (b) Roosevelt Island Station

Fig. 15.7 Stations on the 63rd Street Tunnel Line

15.16 The Archer Avenue Subway

The 63rd Street Tunnel was not the first new subway since the 1940s. That distinction goes to the Archer Avenue Subway, which opened on December 11, 1988. Originally planned as a major subway to serve southern Queens, the Archer

Avenue Subway ends at Parsons Blvd., within a block of the Jamaica Long Island Railroad station. The subway is connected to the remaining portions of the Jamaica Avenue El near 121st Street. It is also connected to the IND subway at Queens Blvd. near the Van Wyck Expressway.

At its completion, the Archer Avenue Subway included only two route-miles and three stations. It cost five times the original estimates to build. Nevertheless, it was the first new subway in New York City since the completion of the IND system in the 1940s. The J and E trains now terminate at the Jamaica Center at Parsons Blvd. and Archer Avenue.

(a) Upper-level platform at Jamaica Center terminus of the Archer Avenue Subway

(b) Van Wyck Boulevard Station on the Archer Avenue Subway

Fig. 15.8 Stations on the Archer Avenue Subway

15.17 The Westway Trade-In and Other Funding

The deterioration of the subway system in the 1970s and early 1980s was not the city's only pressing transportation problem. In 1973, a four-mile section of the West Side Highway, an old elevated parkway, began to collapse of its own accord. Deferred maintenance would claim another victim. The section of the highway between Battery Park and 42nd Street had to be closed, and the structure eventually dismantled.

With the demise of the West Side Highway, the westside of Manhattan was left without a limited access facility serving it. Eventually, a grandiose plan that was known simply as "Westway" was developed. Westway was not, in the strictest sense, a highway project. It was a massive urban renewal effort that happened to be conveniently wrapped around a highway to qualify for federal funding as a highway project.

The plan provided for 4.2 miles of highway, half of which was to be a tunnel running under 169 acres of land fill in the Hudson River. The plan included full development of the land fill, including housing, recreational facilities, new docks, and commercial development. The project was immediately the subject of great controversy, all of which focused on the federally-required Environmental Impact Statement for the project. Under federal law, the EIS had to identify all environmental impacts of the projects and suggest remediation as appropriate. The law did not, and does not, prohibit environmental impacts, even negative ones. It does, however, require that they be fully described and quantified for public scrutiny.

The project was under attack virtually from the beginning. Eventually, legal challenges were continually placed in the way of the project, most alleging that the EIS had not completely described the environmental impacts of the project as required by law.

There were some very real issues. Firstly, since the highway would be built using federal funds, the highway had to accommodate trucks. The West Side Highway was a parkway, and prohibited truck traffic. The generation of additional traffic in downtown Manhattan was another critical issue. The highway would attract additional traffic into what was already a highly congested area. Many feared that if Westway were built, that long-abandoned plans for cross-Manhattan expressways would be resurrected. The pros and cons of the project to the economy were debated at length. Eventually, two issues wound up being the source of multiple lawsuits:

- Air quality – air pollution estimates based upon a dizzying set of traffic forecasts were challenged for correctness; it was also claimed that generated traffic would cause multiple violations of the Clean Air Act.
- Striped Bass – environmentalists claimed that the land fill and associated Westway construction would eliminate Striped Bass from their Hudson River habitat.

In March of 1981, the Reagan Administration approved the Westway project, and designated $1.4 billion for its construction, 90% of the total since Westway was to be designated as part of the Interstate System. The amount was staggering and unprecedented. The 4.2 mile project was provided with federal funding of over $330,000,000 per mile! No highway had ever cost so much.

Despite the federal funding approval, the project remained mired in a flurry of lawsuits. As the timeline lagged, the price tag began moving higher. Finally, planners began to worry that the project would never clear all of the legal hurdles before it, and began to think about making use of a new part of the federal-aid highway act, that is, the "Interstate Trade-In." The provision allowed for urban governments to "trade in" portions of approved Interstate highway projects for alternative federal funding in like amounts, including funds for public transit.

After several years of haggling, with both the Mayor and Governor vacillating on the issue, agreement was reached in September of 1985: Westway would be abandoned in favor of a $500,000,000 surface roadway project, namely, West Avenue, and $900,000,000 of additional transit aid. The Westway trade-in provided money in addition to the $8.1 billion capital program of 1981-82, and a subsequent $8.4 billion capital program in 1986, and helped accelerate the purchase of new subway cars (and buses), and major repairs to tracks, signals, and maintenance facilities.

It is doubtful that anyone really cared about the status of Striped Bass in the Hudson River. It was not an endangered species, and the Hudson River was sufficiently polluted to discourage consumption of seafood caught in its waters. It was, however, the decisive legal issue. Air quality concerns, far more legitimate an issue, had been addressed, and were rapidly disappearing as a barrier to construction.

The Striped Bass, however, were a much trickier issue. Legal challenges focused on the failure of the EIS to properly record the impact of the project on the mating habits of the striped bass. When a judge ordered the EIS to be updated in 1985 to include such an analysis, the project was lost on a technicality, namely, the next mating season was after the federal deadline for the project to start, and the project funding would have been lost. In abandoning the project, however, New York gave up the land fill, and the housing, recreational, and commercial developments that would have come with it, all 90% paid for by the federal government. At the present time, we still don't know too much more about the mating habits of whatever Striped Bass population may still exist in the Hudson River, but no one is complaining.

15.18 Mayor David Dinkins and Renewed Fiscal Travail

Problems don't get solved forever. After leading the city's recovery from the 1975 fiscal crisis, Ed Koch left office with a $1.8 billion deficit in the budget. David Dinkins, after defeating Koch in the Democratic primary, defeated Rudolph Giuliani by a narrow margin to become the first Black mayor in the city's history on January 1, 1990.

Fig. 15.9 Mayor David Dinkins in 2007 at an award ceremony

Dinkins faced a number of daunting problems as he assumed his position. Not only was the budget in distress, but crime, fueled by the crack-cocaine epidemic and the drug wars it induced, was rising rapidly and, once again, thought to be out of control.

He attempted to close the budget gap by raising taxes. This move backfired due to an overall dip in the economy which caused the city to lose 300,000 private sector jobs. Tax revenues dipped, despite the tax increase. In 1991, he introduced a number of draconian budget cuts to balance the city budget, including the elimination of 27,000 city jobs. Education was cut $579 million, 10 homeless shelters were closed. Taxes were raised by another $1 billion. His budget cuts and tax increases were opposed by a wide spectrum of interests, but they worked: by 1992, the city had a $200 million surplus.

As the city's first Black mayor, it is ironic that Dinkins' most visible failures were in the area of race relations. Ed Koch was often considered less than fully sympathetic to the problems of minorities in the city. Dinkins had campaigned on a platform pledging racial healing and celebration of the gorgeous mosaic of New York's diverse communities. His reign, however, was marked by several major racially-polarizing events. In 1991, the Crown Heights riots, which were sparked when a seven-year old Black youth was hit and killed by a car driven by a Hasidic Jew, resulted in the murder of a Hasidic Jewish student. The Mayor was severely criticized for withholding police protection during the initial days of the disturbance. A boycott of a Korean grocery in Flatbush further intensified racial tensions. Dinkins publicly supported accused shoplifters at the grocery. Later, when a Black drug dealer was killed by police, sparking riots in Washington Heights, the Mayor used city funds to pay for his funeral.

The transit system was not immune to general climate of fiscal difficulty and burgeoning crime. To balance its budget in 1991 (a $272 million deficit was forecast), significant cutbacks in off-peak transit services were implemented. In December of 1990, the TA announced plans to eliminate seven express bus routes and 24–27 regular bus routes. Between midnight and 5 AM, several rail routes were to be shortened and replaced by buses. Up to 127 token booths were scheduled for closing or significantly reduced hours of operation. While some of the announced cuts were eliminated, most were implemented. While the fare stayed at 75 cents during his tenure, it was raised to 90 cents on January 1, 1994, just as Rudy Giuliani was walking in the door.

In 1993, David Dinkins again faced Rudy Giuliani in the mayoral election. While the vote split changed only marginally, the scales tipped to Giuliani, who won a very close election.

The defeat of Dinkens threw NYC into a new era of political irony. In the previous century, a Republican had been Mayor only twice, Fiorello LaGuardia (his first election was on a Fusion ticket) and John Lindsay (his second election was as a Liberal, and he later became a Democrat). A Republican (or, more properly, least a candidate running on the Republican ticket), however, would be Mayor for at least the next 20 years in a city with an overwhelming majority of registered Democrats.

15.19 The Giuliani Years

Perhaps no mayor in the city's history has been as controversial as Rudolph Giuliani. Widely praised for dramatic improvements in the city, and eventually for his leadership in the city's darkest hour, his personality did not evoke warm feelings for many. For eight years, it was his way or the highway. There is no dispute, however, over the dramatic upswing in the city's finances, attractiveness, and livability during his reign.

Fig. 15.10 Mayor Rudolf Giuliani

15.19.1 New Mayor, New Challenges

A December 31, 2001 *New York Times* overview of his years as mayor described the state of the city at his inauguration in the following terms (92):

> "*On a brisk January morning in 1994, Mr. Giuliani was inaugurated as the city's 107th mayor, inheriting a metropolis with a million people on welfare, an 11 percent unemployment rate, an average of five murders a day, and the widespread perception that it was a city on the skids.*" (Ref. 92, pg F4)

His signature issue was crime reduction. In 1993, there were 1927 murders in New York City, and 430,460 reported crimes. By 2001, the figures had been reduced to 671 murders and 184,111 reported crimes. He instituted a policy of virtually zero tolerance for law-breakers. He argued that the streets of New York not only had to be safe, they had to appear to be safe. He focused attention on "quality of life" crimes that had gone virtually ignored in previous administrations. "Squeegee men" and panhandlers were arrested and removed from the street, as were illegal cart vendors and others. He instituted a crime-tracking system called "Comstat" that tracked crime by type and by precinct. Individual precinct commanders were made responsible for lowering crime in their jurisdictions, and personnel were moved to meet the most serious situations as indicated by the Comstat system. A crime in Harlem or Bedford-Stuyvesant was treated the same way as a crime on Park Avenue. Eventually, he applied a form of Comstat to other city operations. Through the hard-headed application of statistics, Giuliani claimed to have transformed the mayor's office from a political position to a virtual governmental C.E.O. In over eight years, the city became safer, cleaner, and more attractive.

In another management move, Mayor Giuliani merged the city, housing, and transit police forces into a single agency, a move which is credited with making the fight against crime more efficient.

Mayor Giuliani had some help in achieving the turn-around. As he took office, thousands of additional police officers hired under the "Safe Streets, Safe City" program created by David Dinkins were coming on line. Homes built under a program called "New York, New York," started by Governor Cuomo and Mayor Dinkins, provided housing and medical care for 9000 mentally ill homeless. The last of 100,000 apartments rebuilt under a program started by Mayor Koch was completed, removing acres of blight in the South Bronx and Brooklyn. While he faced a budget deficit of $2.4 billion as he took office, he would also benefit from the booming national economy during the years of Alan Greenspan's "irrational exuberance." On the day of his first inauguration, the Dow Jones average stood at 3754; when he left, it was 10,137. The initial budget-cutting of his first year was replaced by a period in which he could invest in the city's infrastructure and services, a benefit shared by very few of his predecessors.

During his term, transit never rose to the level of a major problem needing a solution. Mayor Giuliani left the running of the transit system to the TA and the MTA. While several contract negotiations dragged on to the "wire," there were no strikes or major service disruptions. In keeping with the now-common practice, when the MTA needed more revenue, fares were raised with increasing regularity. The fare was raised to $1.00 on January 1, 1986, to $1.15 on January 1, 1990, and to $1.25 on January 1, 1992. It is no accident that the fare increases occurred with new TWU contracts every two years.

He had other notable accomplishments. The number of welfare recipients was reduced from just over 1,000,000 to 500,000, the lowest level in since 1964. Sixty new schools were built during his administration, although he lost a long and constant battle with the state legislature to remove the New York City Board of Education and place the school system under direct mayoral control. Legislative changes did allow the chancellor to appoint district superintendents, and tenure for principals was replaced with a merit system. His often harsh battles with a number of school chancellors, however, limited his effectiveness in the education area.

Rudy Giuliani was the first New York Mayor to leave office as a result of new term-limit rules established by the City Council. He was prohibited from running for a third consecutive term. As his second term entered its final months, some of the luster began to wear off the Mayor. The city's dramatic turn-around was a done deal, and was accomplished mostly during his first term. He was involved in a nasty divorce from his second wife, Donna Hanover, which he had announced at a news conference before telling her personally. They had been separated for months and clearly at odds for some time before. Donna Hanover was conspicuously absent from his reelection campaign in 1997 and participated in few formal NYC occasions. The divorce proceedings were nasty, and were played out in the media, as the opposing lawyers took almost-daily shots at Giuliani and Hanover. In a bizarre court decision, he was forced to leave Gracie Mansion, the mayor's official residence, while his wife remained. He lived in the spare bedroom of a friend's apartment, and was undergoing treatment for prostate cancer at the time.

Mayor Giuliani's most glaring failure was in race relations. While he had substantial minority support, particularly in his reelection, primarily due to the drastic improvements in crime in the city's poorest neighborhoods, his relationships with minority leaders were particularly frosty, and he rarely reached out to build solid relationships with them. His initial election over Mayor Dinkins occurred after a bitter campaign in which Dinkins' handling of several highly controversial racial events was heavily targeted. Several highly publicized police actions in which race played a major role highlighted the issue. In 1997, a police officer brutalized Abner Louima, a Haitian immigrant, in a police station bathroom. After a lengthy criminal trial, one police officer was convicted and jailed for a 30-year term. After several mistrials, another officer was convicted of complicity and sent to jail for five years. In 1999, four members of the Police Department's Street Crime Unit were pursuing a rapist. They mistakenly killed an unarmed African immigrant, Amadou Diallo, in the foyer of his apartment building. The four officers were acquitted of any wrongdoing. In 2000, an undercover police officer killed an unarmed security guard, Patrick Dorismond after a brief struggle. Dorismond had been offended when the officer inquired about buying drugs. Giuliani threw gasoline on the latter fire by releasing previously-sealed information about minor crimes Dorismond had committed as a juvenile.

Many New Yorker's praised Giuliani's efforts in turning the city around. However, most were looking forward to a new administration where some of the harder edges of his style, particularly in racial relations, would be replaced with a more conciliatory approach. In his last days in office, however, New York would be subjected to the worst day of its history, and Mayor Giuliani's image would be remade.

15.19.2 September 11, 2001

September 11, 2001 became a day that no New Yorker will ever forget. At 8:46 AM, a Boeing 767 airliner struck the north tower of the World Trade Center. At 9:03 AM, seven minutes later, a second jet struck the south tower. It would soon become clear that the strikes were part of a coordinated terrorist attack that included a third jetliner that would hit the Pentagon, and a fourth destined for the White House that was brought down by a group of courageous passengers in a Pennsylvania field.

The Mayor was at 50th Street and 5th Avenue when news of the attacks reached him. He immediately headed to the city's new two-year old Emergency Command Center, which was located in Building 7 of the World Trade Center. At 9:59 AM, the unthinkable happened, the south tower collapsed. At 10:28 AM, the north tower followed it to the ground. It was quickly determined that Building 7 was at risk, and needed to be evacuated. Thus, in the midst of the crisis, the Mayor and the rest of the Emergency Command Center team had to run for their lives to a temporary location set up at 75 Barclay Street (93). Building 7 would itself collapse from damage caused by falling debris from the twin towers at 5:20 PM.

All subway, rail, and air traffic was halted, and evacuation of all areas south of 14th Street in Manhattan was ordered. Eventually, the death toll from the attacks would reach almost 3000 people, hundreds of whom were firefighters and police involved in rescue operations, trapped in the buildings when they collapsed. Books have and will be written about the event and its aftermath. The attacks would eventually lead to wars in Afghanistan and Iraq, a new federal Department of Homeland Security, imposition of new security measures at airports at home and abroad, and new antiterrorist measures, some of which allegedly threaten some of our civil liberties.

It was during this crisis that Rudy Giuliani earned the title "America's Mayor." On December 31, 2001, the *New York Times* wrote (92):

> "*For weeks afterward, Mr. Giuliani was more than just a mayor. Day after day, his calm explanation of complicated, awful news helped to reassure a traumatized city that it would pull through, and that someone was in charge. He attended funerals, comforted survivors, urged residents to dine out and tourists to come in, all the while exuding compassion and resolve, even as the new threat of anthrax emerged. The man who seemed so finished just a few weeks earlier was now being greeted with cheers wherever he went: Rudy! Rudy! Rudy!*" (Ref. 92, pg A1)

The catastrophe humanized Rudolph Giuliani in a way all of his previous service never did. He reached out to people of all races in every part of the city to console them and give them a new confidence for the future. He worked tirelessly with renewed enthusiasm for the job he would leave in a few short weeks. For a while, there was a short term effort to extend his term for several months through state legislation, but that was quickly abandoned. In the weeks that followed 9/11, Rudy Giuliani became a national symbol of leadership and strength in one of the nation's darkest hours. He was later named Time Magazine's "Person of the Year" for 2001. Giuliani ran for President in the 2008 primaries, but dropped out early when his strategy of targeting only large states left him at the low end of the pack .

15.19.3 Impacts of 9/11 on the Subway

The September 11th attacks had a direct impact on the subway system. A section of the IRT 1 and 9 tunnels near the World Trade Center collapsed from the weight of debris falling from the towers, and tunnels near the Cortland Street Station (N and R trains) were also damaged (94). Upon inspection, there were only two completely collapsed areas of the tunnel around the Courtland Street station. The section of the IRT tunnel affected was among the oldest in the system, dating back to the original construction of the IRT in 1904. The tunnel roof was supported by a system of steel arches located every five feet, encased in a brick and concrete box no more than five feet under the street surface. The design was credited with preventing more substantial damage, and was incorporated into designs for the new replacement tunnel.

The rebuilding job was a mammoth one. Even though only 575 feet of tunnel had completely collapsed, a total of 6200 feet would have to be rebuilt because of damage. Moreover, the tunnels were filled with debris that would have to be removed before reconstruction could even begin. By mid-September of 2002, the 1 and 9 trains were running virtually through the middle of Ground Zero. The Cortland Street station, serving the N and R lines, was reopened, and the 2 and 3 trains were back on express tracks in lower Manhattan. The work in restoring the transit system was as heroic as the numerous other rebuilding projects that still go on in New York. It was finished months ahead of schedule, and $50,000,000 below budget (95)!

15.20 Mayor Michael Bloomberg and the Strange Transit Strike of 2005

Few have entered public service at so critical a time, or from the kind of background compiled by Mike Bloomberg. His election as Mayor in 2001 was his first elected position. His career had been in the investment field. After graduating from Harvard Business School with an MBA, he worked his way up to become a general partner at Salomon Bros. He made his fortune, however, in his own company, Bloomberg L.P., selling financial information terminals to Wall Street firms, and introducing a radio network dedicated to investment and business news coverage. He became, and remains, one of the wealthiest men in the world. His net worth was estimated by Forbes to be in the neighborhood of $5.5 billion at the time of his initial election.

Fig. 15.11 Mayor Michael Bloomberg in 2007

Mike Bloomberg was actually a registered Democrat for most of his life, and switched parties only shortly before his campaign for mayor started. In the Republican primary, which was delayed because of the September 11th tragedy, he defeated another former Democrat, Herman Badillo. In the election against Mark Green, Bloomberg had two critical advantages: (1) he had Rudy Giuliani's endorsement, and (2) he financed his own campaign, and was therefore not subject to campaign spending limits. He outspent Mark Green by a ratio of five to one, and won a narrow victory with 50% of the vote to Mark Green's 48%. He was reelected in 2005 by a margin of 20% over Fernando Ferrer in the biggest landslide in the history of mayoral elections in New York City.

Bloomberg had been a big supporter of term limits for many years. However, when faced with the end of his tenure as mayor, he had term limit legislation overturned by action of the City Council, with no referendum. He won a narrow victory for a third term in 2009, with many voters upset at his "change of heart" on the term limits issue.

While crime was Rudy Giuliani's signature issue, Bloomberg's main problem has been to keep up the good work. He continued most of Giuliani's management systems, and crime rates have continued to decline throughout his term.

His signature issue has been education. Where Rudy Giuliani had tried and failed to get the state legislature to abolish the Board of Education, Mike Bloomberg succeeded. In late 2000, the Board of Education and local school boards were abolished and replaced with a new mayoral agency, the Department of Education. The new agency is based in the renovated (and infamous) Tweed Courthouse near City Hall. He installed Joel Klein as Schools Chancellor. Together, they began to make significant changes. Social promotion was eliminated in favor of after-school and summer-school help for those needing assistance to get back on track. Under the new arrangement, test scores have generally risen, and the city has gotten a higher percentage of state education aid than previously. Despite many parental objections, he initiated a cell phone ban in NYC schools, labeling them a distraction. After Joel Klein stepped down, replacing him has turned out to be difficult, with two new chancellors within one year.

After the fiscal crisis caused by the September 11th attacks, Bloomberg introduced a $3 billion tax increase in 2003. While not particularly popular, the increase did restore a balanced budget, and with a recovering economy, surpluses were achieved in 2004 and 2005 at record levels.

On December 20, 2005, members of the Transport Workers Union went on strike in one of the most bizarre negotiations ever held. The TWU was led by a relatively new President, Roger Toussaint. The MTA was entering negotiations with an estimated $1 billion budget surplus, which they argued resulted from a number of one-time events related to capital construction projects and bonds.

Before the full strike, a more limited strike was staged on December 19th by workers at two private bus lines, that is, Jamaica Buses Incorporated and Triboro Coach Corporation. This was done to avoid provisions of the Taylor Law. What was not clear was whether these workers were even part of the negotiations with MTA at the time. By January 9th, the city absorbed these bus companies into MTA Buses, and the workers fell under the Taylor Law at that time.

The primary sticking point in the negotiations involved pensions. The original MTA proposal sought to increase the retirement age from 55 years of age to 62, and to reduce benefits through the creation of a new "tier" of workers. The MTA, however, dropped this request in return for increased pension contributions of 6% from new employees for their first ten years of employment. One major sticking point was that the MTA insisted that the pension benefit be negotiated as part of the contract negotiations, a clear violation of the Taylor Law, which prohibits such bargaining.

The MTA eventually converted the pension contribution to a contribution towards the cost of health benefits. On December 8th, MTA offered a two-year contract with raises of 3% in Year 1 and 2% in Year 2. Workers would have to contribute 2% of their earnings towards health-care premiums (96).

On the issue of salary, the TWU demanded a 6% raise during each year of a three-year contract, and reduction of the retirement age to 50. At the strike deadline, the MTA offered two 3% annual wages increases, but demanded concessions on health-care costs, pension costs, or productivity (97). The TWU rejected the offer, and started the system-wide strike, in violation of the Taylor Law. Further complicating the issue was that the International TWU disavowed the strike, and claimed that sufficient progress in negotiations had been made to avoid it (98). The union itself was in turmoil, and a major side-issue was the large number of disciplinary proceedings that were being brought against union workers.

On December 22, the third day of the strike, with the assistance of a state mediator, the strike was ended without a contract settlement. The sides agreed to continue negotiations after a return to work (99). Agreement was tentatively reached on December 27, 2005. Salary increases of 3%, 4%, and 3.5% were awarded, while the MTA withdrew its insistence on increased pension contributions. Workers agreed to contribute 1.5% of their salaries for medical benefits. Martin Luther King Day was added as a holiday, and the union won a one-time refund of some prior employee pension contributions.

At this point, bitter bickering amongst union members began to affect the process. Toussaint was accused by some members of not properly informing workers about contract provisions, and some believed that he had "caved in" to the MTA. On January 20, 2006, members of the TWU rejected the contract by seven votes out of over 22,000 cast (100). After much internal squabbling, the union held a re-vote in April 2006, approving the contract by an almost three to one majority (101). At this point, the MTA declared that the contract offer was no longer on the table, and that an impasse had been reached. They called for binding arbitration under the Taylor Law.

In June 2006, court-ordered Taylor Law penalties were deducted from striking workers' checks amounting to two days' wages for every day of the three-day strike. The TWU was also forced to pay over $300,000 per month towards strike-related penalties that had been assessed at $1,000,000 for each day of the strike. Automatic withholding of union dues was suspended until early 2009.

After nearly a year of court challenges and counter-challenges, the arbitration panel mandated a settlement that was nearly identical to the one originally rejected and later accepted by the TWU (102).

Shortly after the arbitration announcement, Roger Toussaint narrowly won re-election as union President over four other challengers. He amassed only 45% of the vote, and presided over a union that was sharply divided over his handling of the 2005 negotiations and strike (103). In 2009, Toussaint declined to run again. His hand-picked replacement, however, was narrowly defeated. John Samuelson, a former ally, was elected

15.21 Fare Wars Go On and On and…

The five-cent fare has now been gone for over 60 years. Its demise in 1948 has been followed by additional fare increases at an ever-increasing rate, and at more frequent intervals, as shown in Figure 15.12.

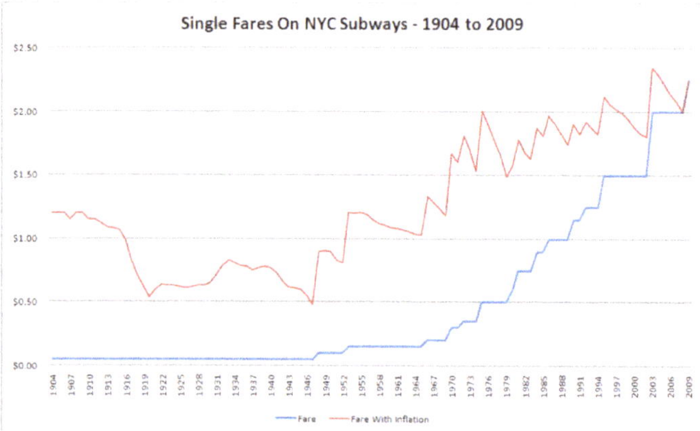

Fig. 15.12 History of the NYC Subway fare

Fare increases have come with regularity, often resulting from new labor agreements. Each increase has come with public opposition, and with questions and accusations over whether or not MTA financial data used to justify the increases are accurate.

Over the years, the method of fare collection has changed radically. Initially, riders bought a ticket which was manually collected by a ticket agent as they entered the station platform on subways and elevateds. In the 1920s and 1930s, rising costs and reduced revenues forced private operators to install automated turnstiles, which accepted coins in payment of the fare. When the fare was increased to 15 cents, the TA began to use tokens, purchased from token agents in stations, in the turnstiles. The token, it was argued, could be used irrespective of the fare. With a fare increase, the cost of a token would rise, but all system turnstiles would not have to be refit to handle a different set of coins. This practice, good in theory, never really worked in practice. While the same token was retained during some fare increases, different tokens were often minted to avoid riders hoarding cheaper tokens prior to a fare increase. New tokens, of course, had to be manufactured, and all turnstiles would have to be refit to accept them.

In 2003, the NYC transit system entered the electronic age with the elimination of tokens, and introduction of electronically coded "metrocards." Such cards have been used in other transit systems for many years, including the Bay Area Rapid Transit System (BART), and the Washington Metro.

The use of cards has allowed new modes of buying service. Single-fare cards are available, but seldom used. Most popular are multitrip tickets, which are available for almost any amount. Depending upon how many trips are purchased, tickets are encoded with extra "free" rides. Unlimited ride tickets for a day, a week, or a month are also available, as are senior citizen discounts. Fare increases are no problem: the ticket has an amount encoded, not a number of trips. When processed through electronic turnstiles, the fare in effect at the time is deducted. The fare was increased from $1.50 to $2.00 on May 4, 2003, and was subsequently raised to $2.25. It was the largest single percentage increase in the history of the transit system, and the largest in dollar amount. In December of 2010, the single-ride fare was raised to $2.50, but the base fare remained at $2.25.

The use of the metrocard allows for even more radical changes in the future. Fares could vary based upon distance, as they do in the Washington Metro. Fares could vary by time of day, encouraging off-peak usage by offering lower fares. Transfers could be free, or could be made at a reduced fare, perhaps depending upon the time-lag between leaving one service and entering another. The possibilities are endless, but the politics involved in any one of these ideas could be very difficult indeed.

Urban mass transit has never really broken through the barrier created by its roots in private enterprise, the concept that it should be a self-supporting system. Necessary subsidies are provided with great trepidation and dismay. This will doubtless continue, and mirrors the national view of public transportation across the country. Most European countries, and many Asian nations, have significant national railway and air networks, many of which are subsidized on a regular basis. The passenger rail system in the U.S. is in shambles, run by a public corporation, that is, AMTRAK, that Congress regularly threatens with extinction. In our urban centers, public transportation has never been seen as a vital service as important as police and fire protection, sanitation, education, hospitals, or others that are directly supported from general taxes. In New York, the never-ending fight to keep the fare low has compromised the ability of the system to keep up with the needs of the population, and clearly paid a major role in defeating several major planned expansions of the subways.

There will doubtless be more fare increases in our future. Hopefully, at some point there will be a real discussion of the issue of how much subsidy should transit receive, and from what sources. Unfortunately, every time the MTA and/or NYCTA faces a potential deficit, the same arguments occur over and over again, with no comprehensive solution ever reached. The fact that transit is critical to New York City's well-being has been proven every time there is a strike, even a short one. The city's commerce cannot be sustained without the transit system. Access to and from Manhattan, and many other parts of the city, cannot be handled by the region's highway systems. If the system is a vital public service, then direct public subsidies are not just the unhappy results of operating deficits, as they are an appropriate way to pay for at least part of the cost of a system without which the city could not function.

15.22 The Arts for Transit Program

One of the most successful ongoing programs in the New York City subways is the "Arts for Transit Program." Initiated in 1985, the Arts for Transit program was established to oversee art commissioned as part of station rehabilitation projects. Design criteria were established that would respect the history of the transit system. Stations rebuilt in the IRT division would be in keeping with original designs and style of the 1904 IRT. BMT renovations would adhere to the style of BRT stations of the early 1900s. IND stations would reflect the modernism that dominated that system, built between 1932 and 1940. The program provided a sense of history, and was meant to create an underground museum that reflected the transit system's unique and varied past.

Art was also to reflect the character of the neighborhood served by the station, thereby integrating the station into its surroundings. The program continues today, and has been responsible for some of the most interesting art in the city.

The program celebrated its 20th anniversary in 2005, a marked the occasion with new artworks and exhibitions of artwork produced through the program since its inception.

15.23 New Subways ????

On April 12, 2007, a ceremonial ground-breaking ceremony took place for the beginning of the construction of, believe it or not, the Second Avenue Subway (104). Actual work commenced on April 23, 2007. The latest reincarnation of the Second Avenue Subway is actually under construction. Federal funding for the first phase of a four-phase project is secured, and a construction contract in place.

Phase I of the project will involve tunnels from 105th Street and 2nd Avenue through 96th Street and Second Avenue, with a connection to the current Broadway Line at 63rd Street and Third Avenue. Three stations, at 72nd Street, 86th Street, and 96th Street will be built. The first phase is expected to be in service by December of 2016, and would be an extension of the existing Q line which services the Broadway subway. Additional phases are planned to extend south on 2nd Avenue to Hanover Square, with a station at Chatham Square, the historic confluence of the 2nd and 3rd Avenue Els. A northern extension would terminate at 125th Street.

The entire project was originally planned for completion in 2020, although none of the funding is in place for Phases 2, 3, and 4. While this deadline will not be met, no new projection of full completion has been established. Phase 2, that is, the northern extension to 125th Street, is thought to be relatively probable, as it will make use of 15 blocks of tunnels built in the 1970s attempt at a Second Avenue Subway. These have been kept in usable condition by the MTA since their abandonment.

As part of a redevelopment plan for Manhattan's westside, including the Hudson Yards, the No. 7 Flushing line is to be extended to a station at 11th Avenue and 34th Street. The centerpiece of the original redevelopment plan was a stadium for the football Jets, to be located over Hudson Yards. Plans for the stadium were abandoned when local opposition and legal challenges became too costly to overcome. Despite this, the No. 7 extension is under construction, as it would also serve an expanded Javits Convention Center. In November of 2006, the city began issuing bonds to fund the extension project. In September of 2007, the city and the MTA signed a deal in which the MTA would be in charge of design and construction, with the city providing $2.1 billion in support. The original plan includes service tracks that extend to 23rd Street for train storage.

There is currently some consideration of further extending the No. 7 train under the Hudson River to New Jersey.

15.24 Into the Future

New York City and its transit system share an integrated history for over two centuries. From the days of horse-drawn omnibuses to the latest subway plans, the city is very much shaped by the access provided by its rapid transit system.

As we look to the future, there are plans for new expansions of the system, some of which look promising, others of which are more doubtful. The fate of these and future projects will hinge on the details of finance, and the general economic background that can and does have a monumental impact on what the city can afford.

New York has the most comprehensive transit system in the world. Hopefully, it will continue to grow and to serve the city's inhabitants as it has in the past. It is still possible to live in Queens, work in Brooklyn, catch a matinee on Broadway, and a Yankee game at night in the Bronx (or a Met game in Queens), that is, all in one day, and all without driving a car. In the early 1900s, this ability was new and marvelous, and allowed the city to grow into one of the great cultural centers of the nation and the world. We take it for granted these days, but it allows New Yorkers mobility and accessibility that is virtually unduplicated anywhere else. The nickel fare is long gone, but a $2.50 ride from Coney Island to Yankee Stadium is still a bargain.

References

1. Abraham Beame is Dead at 94; Mayor During 70's Fiscal Crisis. New York Times, 1 (February 11, 2001)
2. Recalling New York at the Brink of Bankruptcy. New York Times, B3 (December 5, 2002)
3. Beame Angrily Vows No More Cuts in Service. Chicago Tribune, 5 (August 22, 1975)
4. Carey: State Loan to City Possibility. The Washington Post, D7 (August 22, 1975)
5. Unions Funds Enable N.Y. to Meet Its Payroll. Los Angeles Times, B1 (September 5, 1975)
6. Teachers Strike as City Gets Reprieve. The Irish Times, 7 (September 10, 1975)

7. New York City and State Downgraded by Moody's. Los Angeles Times, A6 (October 3, 1975)
8. N .Y. on the Brink of Default. Los Angeles Times, A1 (October 17,1975)
9. N.Y. Avoids Default as Teachers Pledge Funds. Los Angeles Times, A1 (October 18, 1975)
10. Ford's $2.3 Billion New York Aid Plan Is Strict but May Face Legislative Hurdles. Wall Street Journal, 3 (November 28, 1975)
11. $724 Million to Be Cut From N.Y. Budget in 3 Years. Los Angeles Times, B1 (October 21, 1975)
12. The Mayoralty of Abraham Beame. New York Times, A22 (February 12, 2001)
13. Routes Outlined for New Subways. New York Times, 1 (January 2, 1968)
14. $2.9 Billion Transit Plan For New York Subways, Rails, Airports. New York Times, 1 (February 29, 1968)
15. Highlights of Program For Subway, Rail, and Air. New York Times, 26 (February 29, 1968)
16. Subway Expansion is Backed in Report. New York Times, 35 (July 31, 1968)
17. Rockefeller and Lindsay Break Ground for 2nd Avenue Subway. New York Times, 35 (October 28, 1972)
18. Mayor and Governor Unite to Start Transit Tube. New York Times, 49 (November 25, 1969)
19. Rockefeller Asks $2.5-Billion Bonds to Aid Transport. New York Times, 1 (March 21, 1971)
20. The Vote. New York Times, E1 (November 7, 1971)
21. Third Ave. El Reaches the End of Its Long, Noisy, Blighted, Nostalgic Line. New York Times, 24 (April 29, 1973)
22. Brooklyn's Culver Shuttle Makes Festive Final Run. New York Times, 20 (May 12, 1975)
23. M.T.A. Expected to Save Franklin Avenue Shuttle, Once an Austerity Target. New York Times, 27 (February 10, 1977)
24. Tips on Trains. New York Times, 61 (September 9, 1977)
25. Feinman, M.: New York City Transit Authority in the '70's (2002), on-line manuscript available at `http://www.nycsubway.org`
26. Patrolling the Muggers Express. New York Amsterdam News, 45 (April 28, 1979)
27. Guardian Angels Fighting Against Crime. New York Amsterdam News, 26 (September 29, 1979)
28. In New York, Good Guys Wear Red Hats. Chicago Tribune, 3 (March 23, 1980)
29. American Graffiti Today. The Irish Times, 10 (January 12, 1968)
30. New Graffiti on the Old IRT. New York Times, 33 (May 1, 1971)
31. Subway Graffiti Here Called Epidemic. New York Times, 39 (February 11, 1972)
32. Graffiti Epidemic Strikes 7,000 Cars In Subway System. New York Times, 37 (October 21, 1972)
33. 1,562 Youths Seized in '72 For Their Graffiti Work. New York Times, 14 (January 14, 1973)
34. Fight Against Subway Graffiti Progresses From the Frying Pan to Fire. New York Times, 39 (January 26, 1973)
35. At $10-Million, City Calls It a Losing Graffiti Fight. New York Times, 51 (March 28, 1973)
36. Graffiti Goes Legit – But the 'Show-Off' Ebullience' Remains. New York Times, 147 (September 16, 1973)

37. Norman Mailer on the Urban Scrawl. Los Angeles Times, D14 (March 7, 1974)
38. M.T.A. to Use Dogs in Its Battle on Graffiti. New York Times, 35 (July 30, 1974)
39. Subway Graffiti Campaign Given Lower Priority. New York Times, 24 (August 7, 1975)
40. New Subway Cars Decried by Levitt. New York Times, 126 (April 21, 1974)
41. Head of Council's Transit Panel Calls New Subway Cars Unsafe. New York Times, 43 (December 5, 1975)
42. Pullman Bids Lowest on 900 Subway Cars. New York Times, 31 (April 8, 1972)
43. 752 Air-Conditioned Cars Ordered for City Subways. New York Times, 1 (September 6, 1972)
44. Finding of Flaw Said to Predate Subway Mishap. New York Times, 131 (June 20, 1979)
45. Subway Cars Cracked, NY Sues for New Fleet. Chicago Tribune, E1 (June 24, 1979)
46. New York Is Countersued By Makers of 754 Subway Cars. New York Times, 19 (July 7, 1979)
47. Replacement Voted for Undercarriages on 754 Subway Cars. New York Times, 1 (September 15, 1979)
48. U.S. Cites Flaws in R-46 Cars; Transit Authority to Cut Their Use 47%. New York Times, B1 (April 14, 1980)
49. $72 Million Is Won By City On Defects In New Subway Cars. New York Times, A1 (December 24, 1980)
50. Transportation Bonds Pass; Job and Tax Plan Lose. New York Times, B4 (November 7, 1979)
51. Feinman, M.: New York City Transit Authority in the '80's, online manuscript (2005), `http://www.nycsubway.org`
52. Transit Union Explains Its Case for 30% Pay Increase. New York Times, B3 (March 19, 1980)
53. Prospects 'Very Grim' For Averting Transit Walkout in New York. Wall Street Journal, 46 (March 19, 1980)
54. N.Y. Judge Bars Transit Workers From a Walkout. The Washington Post, A8 (March 29, 1980)
55. New York is Hit by Transit Strike. Chicago Tribune, 1 (April 1, 1980)
56. Discord in Union Is Seen as Posing Problem in Talks. New York Times, B1 (April 7, 1980)
57. New York's Transit Strike Ends in Disputed Settlement. Chicago Tribune, W3 (April 12, 1980)
58. City's Transit Pact Is Backed Strongly in Vote By 2 Unions. New York Times, A1 (May 13, 1980)
59. Judge Fines Striking N.Y. Transit Unions $1 Million. Los Angeles Times, A2 (April 8, 1980)
60. City to Buy 837 Buses With Low Fuel Efficiency. New York Times, 23 (February 9, 1980)
61. First Eight Buses Of a Fleet of 837 Arrive in the City. New York Times, 28 (June 14, 1980)
62. Defect Puts 57 of 217 New City Buses Out of Service. New York Times, B14 (August 14, 1980)
63. Cracks Are Found in 13 City Buses; Grumman Accepts Responsibility. New York Times, 27 (November 22, 1980)

64. Cracks Found in Frames of RTD and OCTD Buses. Los Angeles Times, OC-B9 (December 6, 1980)
65. All Grumman Buses Out Pending Repair. New York Times, 1 (December 14, 1980)
66. Grumman-Made Buses Removed From Service In 2 Metropolitan Areas. Wall Street Journal, 21 (December 15, 1980)
67. Builder of Buses Says City Roads Cracked Frames. New York Times, A1 (December 16, 1980)
68. Grumman to Fix Cracks in Flxible Buses Removed From Service in Several Cities. Wall Street Journal, 18 (December 16, 1980)
69. Chicago Finds Cracks In Some of Its Flxibles. New York Times, B4 (December 26, 1980)
70. New York Getting Washington Buses. New York Times, A1 (December 18, 1980)
71. M.T.A. Will Pay Grumman in Full For 837 Buses In Return for Repair. New York Times, A1 (January 6, 1981)
72. New Defect to Delay Flxible Buses' Return. Chicago Tribune, 3 (January 25, 1981)
73. Grumman Is Returning Fixed Buses to New York. Wall Street Journal, 15 (July 17, 1981)
74. Engine Problem Sidelines N.Y. Buses. Los Angeles Times, OC2 (August 9, 1981)
75. New Problem Idles All of City's 850 Grumman Buses. New York Times, 27 (January 7, 1984)
76. All Grumman Buses To Be Put Off Streets In City Permanently. New York Times, A1 (February 8, 1984)
77. M.T.A. to Buy 400 New Buses For $68 Million. New York Times, B4 (March 30, 1984)
78. U.S. Says City Owes $56 Million For Discarding Grumman Buses. New York Times, A1 (June 1, 1984)
79. City Is To Sell Flxible Buses, Some to Go to N.J. Transit. New York Times, B3 (March 21, 1986)
80. Subway Shooting Splits N.Y. Chicago Tribune, 5 (December 27, 1984)
81. Man Tells Police He Shot Youths In Subway Train. New York Times, 1 (January 1, 1985)
82. 2nd Grand Jury Indicts Goetz in 4 Subway Shootings. Los Angeles Times, A1 (March 27, 1985)
83. 4 Youths Shot By Goetz Faced Criminal Counts. New York Times, B3 (January 10, 1985)
84. Goetz Shooting Victim Seized in Bronx Rape. New York Times, 30 (June 29, 1985)
85. Jury Exonerates Goetz in 4 Subway Shootings. The Washington Post, A1 (June 17, 1987)
86. Goetz Receives 6-Month Jail Term, $5,000 Fine. Los Angeles Times, C19 (October 20, 1987)
87. Goetz ad the Common Man. Wall Street Journal, 30 (June 18, 1987)
88. Kiley Appoints Philadelphian to Transit Post. New York Times, A1 (June 12, 1984)
89. City Transit: Gunn Looks, Leaps Away. New York Times, E7 (January 15, 1984)
90. Heller, V.: The City Beneath Us, New York City Transit Museum. W. W. Norton and Company, New York (2004)
91. New York Buried Its White Elephant. The Washington Post, F6 (August 25, 1985)
92. A Man Who Became More Than a Mayor. New York Times, A1 (December 31, 2001)

93. Trying to Command An Emergency When the Emergency Command Center Is Gone. New York Times, A7 (September 12, 2001)
94. Part of Subway Tunnel May Have Collapsed Under Weight of Debris, Officials Fear. New York Times, A13 (September 13, 2001)
95. With Station's Reopening, Even Commuters Smile. New York Times, B3 (September 17, 2002)
96. M.T.A. Offers 2-Year Pact; Union Balks. New York Times, B1 (December 8, 2005)
97. Transit Talks Pass Deadline For a Strike. New York Times, B1 (December 16, 2005)
98. A Transit Local Riven With Dissent, and Lacking Support of its Parent Union. New York Times, B8 (December 21, 2005)
99. Transit Strike Ends On 3rd Day; Framework For a Deal Reached. New York Times, A1 (December 23, 2005)
100. Transit Workers Reject Contract By 7-Vote Margin. New York Times, A1 (January 21, 2006)
101. Transit Union Approves Contract That It Rejected Before. New York Times, B1 (April 19, 2006)
102. Arbitrators Rule on Deal for Transit. New York Times, B1 (December 16, 2006)
103. A Transit Union Vote So Close They Counted It Five Times. New York Times, B3 (December 27, 2006)
104. Was There a Ghost, or Just a Tunnel At the Latest Subway Groundbreaking? New York Times, B3 (April 13, 2007)

Figure Sources

15.1: *Courtesy of* New York Historical Society.
15.2: *Courtesy of* New York Transit Museum.
15.3: *Courtesy of* Wikimedia Commons.
15.4: *Courtesy of* New York Transit Museum.
15.5: *Courtesy of* MTA New York City Transit.
15.6: *Courtesy of* MTA New York City Transit.
15.7: *Courtesy of* Wikimedia Commons. (a) Photo by David Shankbone, used under Creative Commons Attribution 2.5 Generic license, (b) Photo by Daniel Schwen, used under Creative Commons Share-Alike 2.5 Generic license.
15.8: *Courtesy of* Wikimedia Commons. (a) Photo by Pacific Coast Highway, (b) Photo by David Shankbone, used under Creative Commons Attribution 2.5 Generic license.
15.9: *Courtesy of* Wikimedia Commons.
15.10: *Courtesy of* Wikimedia Commons, Jason Bedrick, photographer.
15.11: *Courtesy of* Wikimedia Commons, Rubenstein, photographer.
15.12: *Courtesy of* Wikimedia Commons.

Chapter 16
Rolling Thunder

A modern rail rapid transit system is a complex combination of many technologies. Systems involve tunnels and elevated structures, tracks and switches, complex maintenance facilities, stations, security systems, fare collection systems, train control systems, and rolling stock.

The history of the development of subway cars is as interesting as the history of the transit system on which they travel. From the earliest days of rudimentary cars hauled by cables, and later by steam engines, to the most recent subway cars with a new generation of technology, the subway car is the place where transit patrons spend most of their time. It is their primary interface with the system. Thus, the design and operation of subway cars is critical to the user's riding experience, and critical to the proper operation of the system.

16.1 Introduction

This chapter is devoted to providing a abbreviated overview of the development of NYC rapid transit cars, emphasizing some of the important technologies and characteristics that shaped the operation of the system. Most of the information in this chapter is drawn from a single source, namely, *New York Subways: An Illustrated History of New York City's Transit Cars*, Centennial Edition, by Gene Sansone, 1997. Mr. Sansone, recently retired, was the Chief Mechanical Officer of MTA New York City Transit, and is an Adjunct Professor at Polytechnic Institute of NYU, where he regularly teaches graduate courses related to transit management. His work is the most comprehensive presentation of NYCTA rolling stock, and the technologies that accompany it available, and is the result of over thirty years of comprehensive research and documentation (1).

16.2 Why Aren't They the Same Size?

The NYC rapid transit system is separated into two divisions: Division A and Division B. The primary reason for the distinction is that the railcars for the two divisions have distinctly different dimensions, although they run on the same standard-gauge track. It is logical to ask why and how this inefficiency came about.

The difference came about in the very beginnings of the subway system, in fundamental decisions made by the Interborough Rapid Transit (IRT) and the Brooklyn Rapid Transit (BRT). Surprisingly, there was little difference in the typical sizes of presubway elevated cars operated by both the IRT and BMT. In both

R.P. Roess and G. Sansone: The Wheels That Drove New York, STTT 1, pp. 377–410.
springerlink.com

cases, relatively small cars were used with an approximate width of 9 ft and a length of between 47 and 48 ft. These dimensions accommodated the relatively tight geometry of the city streets over which the elevateds had to navigate, particularly in Manhattan, where sharp curves and narrow clearances dictated the use of smaller, more maneuverable cars.

When the IRT began to build the first subway in 1902, they made a fundamental decision to buy railcars that could operate over both the subway and elevated system. This led to the use of cars very similar to original elevated cars in Manhattan and the Bronx. It also allowed the IRT to build fairly sharp curves into the subway, particularly in downtown Manhattan.

When the BRT (later the BMT) began to build its subways under the Dual Contracts, they made a fundamentally different decision. They wanted to provide more capacity with fewer cars, which meant making them larger. Much of the BRT elevated system could be easily reconfigured for the larger cars, as the curvature (the real critical factor) was less severe than in the IRT elevateds. Stations platforms would have to be "shaved" to accommodate the larger cars, but this was not thought to be a difficult or very costly process.

By the time the Independent Subway was being built, the benefits of the larger BRT/BMT cars had been generally established, so the Independent selected a similar size for use. Further, the construction of the Independent sought to avoid sharp curves, looking to maximize speed through the system.

In today's subways, the Division B rolling stock is generally around 10 ft wide and 60 ft long. Thus, while IRT rolling stock can roll over any track in the transit system, BMT/IND rolling stock cannot roll over most Division A trackage, either due to the narrower station openings or tighter curvature in that part of the system.

Table 16.1 Comparison of typical rolling stock dimensions

Type of Car	Width	Length	Height
	Interborough Rapid Transit		
Manhattan Elevated Car	8 ft 10 7/8 in	47 ft 4 in	12 ft 20.5 in
Original Composite Cars	8 ft 11-3/8 in	51.5 ft	12 ft 0.75 in
Original All-Steel Cars	8 ft 8 in	51.5 ft	12 ft
Typical Modern Car	8 ft 10 7/16 in	51.5 ft	12 ft 10 in
	Brooklyn Rapid Transit/Brooklyn Manhattan Transit		
BU Elevated Car	8 ft 9 7/8 in	48 ft 11 in	12 ft 10 in
QX Car	9 ft 10 in	49.3 ft	12 ft 5 in
BMT Standard Car	10 ft	67.3 ft	12 ft, 1 11/16
	Independent Subway		
R-1 through R-9	10 ft	60 ft 2.5 in	12 ft 1 15/16 in
	Typical Division B Car		
Typical	9 ft 9 7/16 in	60 ft 2.5 in	12 ft 1 ¾ in

Table 16.1 shows a summary of some key parameters concerning the size of rolling stock in the New York City transit system. Between the beginnings of the rapid transit system and the current time, a number of different railcar sizes have been tried. The standards widths, however, are fixed by the dimensions of the subway itself, approximately 9 ft for Division A and 10 ft for Division B.

16.3 Rolling Stock in the Steam Era

Except for the brief operation of Harvey's Ninth Avenue El using a cable powered by stationary steam engines, the elevateds before 1900 used steam engines hauling unpowered passenger cars. The first steam engines used on the Els were referred to as "dummies" because they were disguised to look like a regular railcar. Dummy engines shared a common 0-4-0 configuration, that is, no pilot wheels, four driving wheels on two axles, and no trailing truck, and were manufactured by a variety of companies. The first such engine, for the Ninth Avenue Elevated, was named the Pioneer, and was built by Handren & Ripley.

(a) Pioneer, built by Handren & Ripley

(b) Spuyten Duvvel

Fig. 16.1 Dummy engines used on the early Ninth Avenue El

The first group of engines each bore the name of a prominent place in New York or its environs, until about 1877, when too many engines were in service to continue the practice.

Over time, dummy engines were replaced by a variety of more typical steam engines. In general, they provided better accommodations for engineers, a larger fuel carrying capacity, more horsepower, and greater efficiency.

After 1881, most of the steam engines used were referred to as "Forney's." Developed by Matthew Forney, the engines had several attractive features. They had a 0-4-4 wheel arrangement, and could negotiate sharp curves. The weight of the boiler was directly over the driving wheels. This provided for excellent traction, and good acceleration. Because the tender was built into the engine, it could be operated in either direction with equal ease.

The first coach cars used on the Ninth Avenue El when steam was adopted were converted cable cars. Starting in 1872, these were rapidly replaced by cars that became known as "Shad Belly" or "Drop Center" cars. The floor of the car

Fig. 16.2 Typical Forney engine in use 1881–1894

between the wheel trucks was lowered to give them greater stability, and to give riders more confidence about their safety. When later coaches were designed with flat floors, the floors of the Shad Belly cars were raised for conformity. Access and egress from these cars was through open-end platforms.

Fig. 16.3 "Shad Belly" cars for the early New York Elevated RR

Over time, coach designs evolved. Some were equipped with a center door in addition to the end platforms for faster access and egress. Most coaches were of the "14 window" design, without a center door, with fourteen windows on each side of the car. Depending upon the manufacturer, the shapes and arrangements of the windows would vary.

Fig. 16.4 Forney engine pulling fourteen window coaches on the Third Avenue El, 1885

Access and egress through open-end platforms required a conductor to be located between every two cars to manually open the gates separating the train platform from the station platform. Eventually, cars requiring this became known as "Gate Cars."

16.4 Electrification of the Elevateds

The initial electrification of the elevated lines occurred in Brooklyn in 1898. Manhattan elevateds began the process of electrification in 1901. In Manhattan, the Els had been leased by the IRT. The first electric cars were lettered *Manhattan* on the side, but this was quickly replaced by *Interborough*. However, all gate cars manufactured for the system became known as "Manhattan Standards" until multiunit door control (MUDC) was introduced.

Fig. 16.5 Exterior and interior of the "Manhattan Standard" elevated car

In Brooklyn, elevated lines of three railroads, that is, the Brooklyn Elevated Railroad, the Union Elevated Railroad, and the Seaside and Brooklyn Bridge Railroad, were acquired by the BRT, which was already operating streetcars and trolleys in Brooklyn. After acquisition, the Brooklyn Els were operated under the Brooklyn Union Elevated Railroad Company. Gate cars manufactured for the Brooklyn elevateds became known as "BU" cars. There were a fairly large number of different cars under this catch-all category over the years.

When electric operation commenced, trains consisted of motorized cars and nonmotorized trailer units. Motorized units had a motorman's cab, initially at both ends of the car. Trailer units had no motorman's cabs. The motorman's cab was within the covered portion of the car, that is, the gated car platforms at the end of the car were in front of the motorman's location. A typical six-car train in Manhattan consisted of four motorized cars and two trailers, connected in an alternating pattern: A:B:A:A:B:A (A=motorized; B=trailer). Three-car trains were arranged as A:B:A, and four-car trains as A:B:B:A. Motorized cars generally had a motor (or motors) on only one truck of the car.

Manhattan Standards were built by several companies, including American Car and Foundry (ACF), Barney and Smith, Cincinnati Car Company, Jewitt Car Company, St. Louis Car Company, and Wason.

On the BRT-owned lines, the conversion to electric operation was a bit more difficult. Many of the steam-era coaches were converted to electric operation, and new cars were ordered. Because the BRT operated some sections of track that were on the surface, adjacent to or in the street, a third rail could not be used. The cars were equipped with third rail shoes, as well as trolley poles. When operating on the surface, the trolley pole would have to be attached to the overhead catenaries, from which power would be drawn during surface operations.

(a) Early BU Car exterior built by Pullman

(b) Early BU Car interior built by Pullman

(c) First electrified cars manufactured for the Brooklyn Elevated, with center doors for Brooklyn Bridge operation

(d) Interior of first electrified cars showing center door blocked by seating when not in use

(e) Brooklyn Elevated convertible car, with removable side panels for summer use

(f) Interior of convertible car

Fig. 16.6 Some samples of "BU" Cars

Another unique feature of the Brooklyn elevated cars was a hinged platform. In addition to the standard gate separating the train end platform from the station platform, the train platform included a hinged piece that would have to be lowered to the station platform before anyone could get on or off the train. Since the BRT also served areas with low-level platforms, the cars also included a set of steps on each end platform which tucked under the platform during operation. All of this, of course, was labor intensive and time-consuming during a normal run.

BU cars continued to be made by a variety of manufacturers until approximately 1908. Between 1893 and 1908, a number of cars known as C-Type cars were also used on the Brooklyn Elevateds. Originally designed as gate cars, they came in three-car A:B:A units, and were used to test new technologies for future designs. They were the first to operate under multiunit door control, and were originally designed using articulated trucks, a concept that was not actually built. Eventually, all of the C-Types had their car platforms enclosed, and a number of these cars successfully operated until the 1950s.

Fig. 16.7 Exterior and interior of typical C-Car used on BRT Elevateds

Some of the early BU cars were rebuilt into Q-Cars for the 1939 World's Fair. Open car platforms were enclosed as steel vestibules, and the motorman's cab was moved forward. After the World's Fair, they were used on the Astoria and Flushing Els, and eventually ran on the Myrtle Avenue El until it closed in 1969. They were the last of the wooden transit cars to operate in New York.

Fig. 16.8 Exterior and interior of renovated Q-Cars

16.5 Early Rolling Stock for the IRT Subway

As the IRT Contract #1 subway was being built, a feverish effort to design and build the first electric-traction subway car for underground operation took place. The IRT was the first electrified heavy rail operation underground, as earlier subways were trolley operations that were placed underground. Engineers wanted an all-steel car to minimize the impact of a fire in a tunnel. At the time, however, an all-steel car was a radical idea. The only all-steel cars had been built by the Pennsylvania Railroad in their Altoona shops, and they were motorless coaches.

When initial orders were placed, they were rejected by every manufacturer. Most were busy filling railroad orders, and none was willing to undertake the reconfiguration of their factories that would be necessary to build all-steel cars.

The IRT reconsidered its desire for all-steel cars, and developed a design for a composite car, that is, wooden cars with enough metal and fireproofing material to protect them in accidents and fires. To get a sufficient number of cars to open the subway on time, orders were placed among four manufacturers, namely, Jewett, St. Louis, Wason, and John Stephenson Car Companies. At the time, the cars were believed to be the strongest and safest electric railcars in existence. Metal cladding and extensive use of asbestos insulation material were incorporated into the design to provide strength and fire protection. Two thirds of the fleet was ordered as motorcars, and one third as trailer units. In a unique arrangement, car builders supplied only the car bodies and trucks. All under-car equipment was installed in the IRT shops. As IRT engineers had decided that the ratio between motorized and trailer units should be 3:1 (not the 2:1 pattern in which they were ordered), a number of cars were converted from trailers to motor cars in the shops as well. Test composite trains were tested on the Second and Third Avenue Els in 1903 and 1904.

Motorized composite cars were all initially designed to operate as high-voltage (or HV) cars. In this design, 600 volts of direct current power from the third rail fed directly into the master controller and then to relays that operated the motors. The IRT later converted these to the more efficient low-voltage (LV) operation, in which battery power was used in the master controller unit, with high-voltage current directly connected to the traction motors.

The first two composite cars to be delivered and tested were the *August Belmont* (Car No. 1) and the *John MacDonald* (Car No. 2). The *Belmont* was lavishly decorated and equipped, and was to be a model for a possible "first class" car for the subway. The first class idea was never put into service, although August Belmont had his own personal car, the *Mineola*, that was said to cost more than the rest of the initial car order combined.

The first all-steel sample car was designed by George Gibbs and manufactured by the Pennsylvania Railroad in its Altoona shops. The car proved to be too heavy for operation in passenger service, resulting in a number of design changes targeted at weight reduction. The car was converted to a "pay car," and operated in this capacity until 1956.

The first redesigned steel cars, known as "Gibbs Cars" in recognition of the designer, were built in 1904 and 1905 by American Car Foundry (ACF). They were equipped with high-voltage controllers. The first 63 cars had manually operated doors; subsequent cars were equipped with multiunit door control (MUDC). Like the composites, they had two doors at the end of the cars, and no center door. It was soon decided that two doors were insufficient to load and unload passengers during the rush hour, and, beginning in 1912, Gibbs Cars were modified to have a center door. All subsequent car orders would also have three doors.

In 1908, a small number of cars, namely, called the "Gibbs Modified" cars, were built with four doors, that is, two entry doors at the car ends, and two separate exit doors located several feet to the interior of the car.

(a) Gibbs Car with center door

(b) Experimental Gibbs Car with separate access and egress doors

Fig. 16.9 Gibbs Cars on the early IRT

Subsequent IRT rolling stock would retain a common appearance until new World's Fair cars were ordered for 1939. Technological changes, however, were gradually introduced. Eventually, all controllers would be of the low-voltage type, and all cars would be equipped for MUDC. The three-door design, with two end doors in enclosed vestibules and one center door, would also be retained.

The first low-voltage cars to be ordered were the "Steinway" cars. These were placed in service in 1915, when the IRT took over operation of the Steinway Tunnel. Because of the steep grades, the existing fleet of cars was not considered appropriate for use. The Steinways were designed with a lower weight and a better gear ratio for operation in the Steinway tunnel, which had steeper grades than elsewhere in the system.

By the1920s, the IRT was already experiencing difficult time financially. During this period, a number of modifications to existing cars were undertaken as an alternative to buying new cars. Many of the modifications were related to the need to eliminate composite cars, or at least restrict their operation to elevateds. As composites were taken out of service, their motors and trucks were used to convert trailer cars of other types to motor cars. The last of the traditional Lo-V cars to be ordered was in 1925.

(a) Lo-V car manufactured by Pullman in 1916

(b) Lo-V car manufactured by ACF in 1925

(c) Interior of a typical Lo-V IRT car

Fig. 16.10 Typical low voltage IRT cars

In 1938, with the 1939 World's Fair on the immediate horizon, the IRT ordered 50 modified Steinway cars from the St. Louis Car Company. The cars were jointly designed by the IRT and the Board of Transportation, and they represented a mix of the design practices of the IRT and the new IND system. End vestibules were eliminated, and new placements of the three doors were introduced. While all cars were motorized, each had only one motorman's cab. The cars were coupled in pairs, with the "no cab" ends joined directly to each other. Thus, each two-car unit had a motorman's cab at each end. These were the last cars ordered by the IRT before unification of the subways under New York City control.

Fig. 16.11 1939 World's Fair cars for the IRT

16.6 The BMT Standard

Until the Dual Contracts of 1913, the BRT did not operate any subways. All of its lines were elevated, on the surface, or in open cuts. As early as 1910, with subway expansion already an active discussion, the BRT began to design a new type of car for subway operation. Given the rapid growth of ridership and the resulting crowding on the IRT, the BRT designed a car that was longer and wider and, therefore, carried more passengers than a standard IRT car. The design was for a 67 ft long, 10 ft wide car, which became known as the BMT "Standards." A full scale model was constructed in the BRT's 39th Street shops, and an initial order for 500 cars was placed with ACF. The cars were delivered between January 1915 and 1918. The first order did not have ceiling fans, and featured a single row of ceiling lights down the center of the car. In subsequent orders, a staggered pattern of 20 lights was used, four ceiling paddle fans were added, and hand grips to assist a larger-than-expected number of standees were added on seatbacks. Eventually, the original 500 cars were retrofitted to include these modifications.

The BMT Standards represented a number of technological advances. The cars were all steel, and had large windows, providing riders with an airy feel and excellent visibility. Equipped with MUDC, the conductor's position was inside the car, with all operating controls located on a wide column separating the two middle-door panels. The cars had 78 seats in both transverse and longitudinal positions. Because of the length of the cars, end doors were kept closed during operation. On curves, the separation between adjacent cars was too large to allow riders to move between cars. Initially, all cars were independent motor cars, with motorman's cabs at each end of every car. Later, for operational efficiency, the cars were joined into three-car A:B:A units, with the interior motorman's controls removed. Even though all cars were placed into three-car units, some independent motor cars continued to be ordered to provide for maintenance flexibility, and the ability to run seven and eight-car trains.

Most of the "BMT Standards" were ordered by the BRT. Only the last order, placed in 1924, was placed after the reorganization of the BRT into the BMT. All fifty cars ordered were trailer units in the interests of economy.

(a) Three-car BMT Standard A-B-A unit (b) Interior of typical BMT Standard

Fig. 16.12 Exterior and interior of the BMT Standard Subway Car

16.7 The BMT Experimental Fleet

Through the late 1920s and 1930s, the BMT was the most innovative of the transit operating companies in terms of rolling stock. Over this period, they invested in rolling stock using new materials, technologies, and design approaches in attempts to create a more efficient, more comfortable, and faster subway car.

The one common characteristic of all of these cars were that they were "articulated." In articulated units, a single truck carries both the lead and lagging cars. Thus, a three-car articulated unit would consist of three cars and four trucks, that is, one at each end of the articulated unit, and two at the interior connections, each supporting two cars. Similarly, a five-car articulated unit would have five cars and six trucks. Articulation allowed passengers to move from car to car through an enclosed vestibule, as there was no displacement between cars (of a unit) when the train executed a curved section of track. It also allowed for some efficiencies in locating motors and other truck-mounted equipment. Articulation also meant that the multicar unit was more-or-less permanently connected.

Fig. 6.13 An articulated truck supporting two cars

16.7.1 The Triplex Units

The first of these innovative cars was the D-Type or Triplex cars, which were ordered between 1925 and 1929. The Triplex cars came in units of three "articulated" cars. Despite a reduced number of trucks, the cars were the heaviest manufactured to that time. They were, however, also the fastest cars made to date, reaching speeds of over 50 mph on the Sea Beach Line. Because of their weight, they could not be used on older elevated sections.

(a) Two Triplex units on the Brighton Line in 1946 (b) Interior of a Triplex

Fig. 16.14 BMT Triplex units

16.7.2 The Green Hornet and the Zephyr

When the BRT began to operate subway lines, it continued to use older wooden El cars in subway service. After the Malbone Street wreck on November 1, 1918, wooden cars were removed from subway operation. Because the BMT operated many elevated lines, it continued to experiment to develop a train that would be sturdy enough for subway use, and light enough for use on even the oldest elevateds. The BMT Standards and Triplex Units were both too heavy for use on many of the older elevateds in the system.

In 1934, two unique multisection articulated car units were built. They were both built in five-car articulated units, and at 179 ft for each unit, they were the longest to ever operate in the NYC transit system. Both were relatively light in comparison to the Triplex Units, and were capable of fast acceleration and speeds of 55 mph.

The "Green Hornet," as manufactured by Pullman Standard, was made of aluminum. The cars were named after a popular radio character of the day, and because of its green exterior. Each car of the multiunit had one side door, located in the middle of the car. The car had a number of new features, including chimes that sounded when the doors were closed, light sensors that automatically turned on interior lights in tunnels and after dark, and route and destination signs that could be centrally controlled automatically. The Green Hornet was designed to be operated by one person. BMT purchased a total of 16 units from Pullman Standard.

The Green Hornets, however, were plagued with many problems, and saw little service after 1938. In 1942, the United States War Scrap Administration ordered the cars dismantled for their much-needed aluminum.

Fig. 16.15 Exterior and interior of the Green Hornet

The "Zephyr" was originally built by Budd. They were also five-car articulated units, and were the first stainless steel cars to operate on the NYC transit system. It shared the characteristics of the Green Hornet in terms of light weight, rapid acceleration, and high speeds (55 mph was attainable). The Zephyr had two doors in each car section, and most of the high-tech features of the Green Hornet. BMT eventually took delivery of 11 units.

The Zephyr proved to be much more reliable than the Green Hornet in service. It continued to run on the Franklin Shuttle until 1954.

Fig. 16.16 Exterior and interior of the Zephyr

16.7.3 The Multis

Because of the relative popularity of the Green Hornet and Zephyr, the BMT ordered another 25 lightweight five-car articulated units from St. Louis Car Company and Pullman. Simply known as the "Multis," these units retained the major features of the more experimental Green Hornet and Zephyr. They were relatively lightweight, although not containing aluminum, and had rapid acceleration and a top speed of 58 mph. The cars were delivered in 1936, but were quickly removed from service to modify the trucks. They resumed regular operation in 1937, and continued to make revenue runs until 1961.

Fig. 16.17 Exterior and interior of a BMT Multi

16.7.4 The Bluebirds

The last preunification car design was perhaps the most innovative of all. Built in 1938 and 1940, the "Compartment Cars" or "Bluebirds" were manufactured by Clark Equipment Company in three-car articulated units. Like the Green Hornet, each car had a single set of passenger doors in the center of each car. The units were extremely lightweight at only 38 tons. They could accelerate very quickly, but their top speed was only 39 mph.

Clark Equipment was essentially a trolley-car and motor truck manufacturer. When buses started to seriously cut into to trolley ridership, a national committee, namely, the Presidents' Conference Committee, was formed to design a common, uniform, state of the art trolley car. The "Presidents" referred to were presidents of trolley operators throughout the country. The cars resulting from the common design were referred to as "PCC" cars.

The Bluebirds were basically manufactured to PCC standards, and used a truck unit common to trolleys. The Bluebirds also adopted the more luxurious interior of the PCC design, including pale blue colors, mohair-covered seats, PCC lighting, and a Westinghouse Air Brake controller that used one single motorman's handle for both acceleration and braking. A total of six units were ordered. The BMT was about to order another 40 units when unification occurred. The city cancelled the order. The six units remained in service on the Canarsie Line until their retirement in 1956.

Fig. 16.18 Exterior and interior of a Bluebird

16.8 The First Independent Cars: R1–R9

As the construction of the Independent Subway commenced, New York City entered the transit business with full operational control for the first time. The Independent was able to capitalize on past experience in its design and layout.

While relatively plain and Spartan in appearance compared to the IRT and BMT, the IND was built to avoid sharp curves that slowed IRT and BMT trains and limited their capacity. Stations were more methodically placed based upon demand, and average station spacing was longer than those of the other two subways. High-speed switches were used, which allowed trains to change tracks at critical junction points at significantly higher speeds.

Because of this, the nominal capacity of the IND system was said to be 90,000 people per hour, per track, compared to approximately 74,000 for the BMT and 60,000 for the IRT. It should be noted that the IND never actually reached the nominal capacity, but it does carry generally larger hourly flows than either the BMT or the IRT to this day.

In developing a design for the city's first municipally-operated railcars, a committee was formed by the Board of Transportation that included representatives from the IRT and BMT. The committee also sought input from car manufacturers throughout the country. With unification already in the air, the committee sought to develop a single standard for such things as car length, car width, and configuration that could be used on any subway or elevated in the transit system. Because IRT cars were narrower than BMT cars, station platforms on the IRT system would not accommodate wider BMT cars. After much discussion, the committee abandoned the idea of a single car design. It turned out that a good deal of structural work would have been needed to bring the IRT into conformity. The committee proceeded to develop a car for the IND that would be compatible with the BMT system. This meant, essentially, that the IND cars would be 10 ft wide.

The committee also introduced a new nomenclature that would be used to identify subway car models. Every car order bore a number in the form R-#. The "R" stood for "railcar," and each contact was sequentially numbered. After unification in 1940, car orders for the entire system were named in this fashion, although some of cars developed "nicknames" as well.

The committee originally considered a number of alternatives, including a 54 ft car with three doors on each side, a 54 ft, six in car with four doors on each side, a 67 ft car along the lines of the BMT Standard, and three-car articulated units similar to the BMT Triplex units. Eventually, a compromise solution was adopted: a 60 ft, six inch car with four doors on each side. The doors were widened from the previous standard of 32 to 48 inches to speed the boarding and alighting of passengers. A mix of transverse and longitudinal seating was also adopted. The cars were all built as individual motor units with a motorman's cab at each end.

The first nine car contracts were for basically the same car design, with minor modifications occasionally made. Referred to as the R 1–9 fleet, these cars were somewhat less technologically advanced as the BMT Standards or Triplex units, but served reliably for over 40 years.

Fig. 16.19 Exterior and interior of an R-1 Car

Figure 6.19 shows car number 100, the first R-1 car delivered for the Independent Subway in 1930.

16.9 The R-Cars: 1948 to 1970

16.9.1 The R-10

The R-10 was designed as the first IND car to differ from the basic R1-9 look. First appearing in 1948, the cars were noisy, dimly lit (despite introducing fluorescent lighting for the first time), and hot during the summer. It was the first car to have small pole-mounted rotary fans instead of overhead ceiling fans.

Several major technological changes were, however, introduced with the R-10s. They were the first cars that could accelerate at 2.5 mi/h/s (compared to the 1.75 mi/h/s of the R1-9s and previous cars). The cars were powered by a 100 hp traction motor on each of the four axles of the trucks. Older cars used two large traction motors, either on one of the two trucks, or on one axle of each truck. The braking system was a major improvement over previous designs. The cars used both pneumatic (air) and dynamic braking, and had an easier-to-use brake valve. Dynamic braking provides for some feedback of amperage by using the traction motors as generators. Energy is dissipated through electrical resistors instead of as heat generated by the friction of brake shoes on wheels. This system provides major decreases in brake wear, and greatly reduces the presence of brake shoe dust in the system.

The R-10s, never popular with riders, were steady performers. Many were overhauled in 1984, a continued in service until 1989. All R-10s were manufactured by ACF.

Fig. 16.20 Exterior and interior of an R-10 Car

16.9.2 The R-11

In 1949, Budd built a prototype 10-car train known as the R-11. It was actually intended to be a prototype for cars on a new Second Avenue Subway that would, of course, not be built. The cars featured stainless steel construction, a rounded-roof profile, a PA system, germicidal lamps, forced air ventilation, and crank-operated windows. Doors were operated by electric door-openers rather than by air engines, and disc brakes were used. A production order was never placed, and the cars were relegated to part-time use, or as part of consists of the later R-16 cars. They were rebuilt in 1964, and became the only rebuilt car to assume a new designation, the R-34, and operated until 1976. After rebuilding as R-34s, the side and front/back doors were painted blue.

Fig. 16.21 Exterior and interior of an R-11 Car

16.9.3 The R-12 and R-14

The R-12 and R-14 series, which were virtually the same, were the IRT versions of the IND R-10s. They shared the same general look, and had all of the same equipment, except that they were smaller, that is, shorter and narrower, and had three doors per car (on each side) instead of four. They were manufactured by ACF, as the R-10s had been, and were placed into service starting in 1948. As the R-10 was the first different looking car on the IND, so was the R-12 on the IRT. The shape and color of the earlier Hi-V and Lo-V fleet were replaced. The conductor still had to stand on steps between the cars to operate doors, but these would be the last IRT cars to require that.

Fig. 16.22 Exterior and interior of a typical R-12/R-14 Car

As the R-12/14 was the first IRT car to have front destination/route signs, they introduced the practice of identifying IRT lines by number.

16.9.4 The R-15

R-15 cars were introduced on the Flushing Line in March of 1950. Like the R-11s for the BMT, the R-15s represented a substantial change in look. Like the R-11s, they featured porthole style round windows on all doors, but they were not stainless steel. The roof was rounded, and the exterior color scheme was a dark maroon separated by distinctive tan bands above and below the windows. They incorporated some brake improvements, and the conductor's station was moved inside. Like all of the cars during this era, every car was motorized, and had a motorman's cab on both sides of the car. Each car could, therefore, be operated independently, and there were no restrictions on train length. The first experiment in retrofitting existing cars with air conditioning was an R-15, in 1955. Initially, all R-15s were used on the Flushing Line. After reconditioning, however, they were used throughout the system, usually mixed with other car types. They were first used as part of an experimental "super express" service on the Flushing Line, in which a selection of trains ran virtually nonstop between Flushing and Times Square.

Fig. 16.23 Exterior and interior of an R-15 Car

16.9.5 The R-16

The R-16 was built for BMT/IND service, and was the first in a series of cars for all divisions that had the same general look and feel. The R-16 was the heaviest of all the "R" cars, except for the 75-footers that came later. Except for the BMT Standards, virtually all previous BMT/IND cars required the conductor to mount the outside step-holds between cars to operate the doors. The R-16 was the first BMT/IND car to return the conductor door controls to the interior of the car. The R-16s had a poor maintenance record due to a GE dynamic brake system that generated blasts of warm air between the car and station platforms, and collected a lot of dirt and debris which caused occasional fires. GE later redesigned some of the propulsion and braking components, which proved to be very reliable.

Fig. 16.24 Exterior and interior of an R-16 Car

The R-16s also started the sequence of cars that used a deep red exterior color. Later, all of these cars, and rebuilt cars that used the same basic color, would become generally known as "redbirds," although the term was usually applied to IRT cars, as opposed to similar BMT/IND cars. They were also the first BMT/IND cars to have large side destination roll signs since the R1-R9 series. The R-16 featured a porthole window on the front and rear doors, and traditional rectangular windows in the side doors. They began service in 1955 on the Jamaica Line.

16.9.6 The R-17, R-21, and R-22

The R-17, R-21, and R-22 series were all similar, and were the IRT versions of the R-16. In all, 1100 of these cars were manufactured by St. Louis Car Company. The R-14 and R-15 models had been introduced exclusively on the Flushing Line. The R-17/R-21/R-22 cars were introduced on the Manhattan mainlines (Lexington Ave. and 7th Ave.) and their Brooklyn and Bronx tributaries, replacing the last of the Gibbs Hi-V cars on those services, beginning in 1955. They were the first new cars introduced on the Manhattan mainlines since the last Lo-Vs were ordered in 1925. The R-17s were maroon and had porthole windows on front and back doors. The R-21's and R-22s were kale-green, and featured a square drop-sash window on the front and back doors. Ten R-17s were delivered with air conditioning, but they did not perform well, and were retrofitted with axiflow fans. The R-22s were the first cars to have sealed-beam headlights.

Fig. 16.25 Exterior and interior of an R-21 Car

The R-22s were the last single cars ordered for 26 years, except for a special order of forty R-33s ordered to be the 11th car on Flushing Line trains, the only line to run such a consist. Three of the cars participated in a unique experiment in automatic train operation (ATO). The NYCTA had demonstrated the feasibility of running a fully automated, crewless, three-car train on a section of the Sea Beach Line in October of 1960. It began regular automated operation on the Grand Central-Times Square Shuttle in 1961. The experiment, which worked without significant incident, ended when a mysterious fire destroyed the on-board equipment on the cars.

16.9.7 The R-26 and R-28

Between 1959 and 1961, ACF manufactured 110 R-26 cars and 100 R-28 cars for the IRT, which were virtually identical. They were the last cars built by ACF for the NYC system. They were relatively light, and were manufactured in married pairs. One car had a motor generator and battery set to supply low voltage for control equipment for propulsion, doors, and the like, while the second car of each pair had an air compressor to supply the air for the braking system. The pairs, however, could be relatively easily separated, and they were referred to as semi-permanent pairs. In style, these cars returned to the "redbird" maroon color of the R-17s, but had a sealed rectangular window on the front and rear doors. The cars were all retrofitted with air conditioning as part of a NYCTA program between 1975 and 1982.

16.9.8 The R-27, R-30, and R-30A

The BMT/IND versions of the R-26/R-28 class were designated as R-27, R-30, and R-30A. They were manufactured by the St. Louis Car Company, who delivered 550 cars. The cars looked the same as their IRT counterparts, except that they were wider and longer, and had four side doors instead of three. Like the IRT versions, these cars were delivered in married pairs that could be relatively easily detached.

Fig. 16.26 Exterior and interior of an R-27 Car

This generation of railcars inspired some amusing designations created by rail fan groups. Since the cars were "married" but could be easily decoupled, they were referred to as "Protestant" marriages. Later pairs would be more permanently attached, and were referred to as "Catholic" marriages. The IRT division did not separate their Protestant cars, even during maintenance, while the BMT/IND division did so regularly, reshuffling the pairs frequently. Rail fans dubbed this practice "bigamy on the BMT."

16.9.9 The R-29

In 1962, St. Louis Car Company built 236 cars for the IRT division in the R-29 series, but these were a virtual copy of the R-26 and R-28 models. The major difference in the R-29 series was that the married pairs were built with link bars, making the marriages "Catholic." The red color used for the exterior was brighter than on previous cars.

This and subsequent NYCTA orders helped make St. Louis Car the largest manufacturer of transit passenger cars in the world. Unfortunately, it would be a NYCTA order that forced it out of business just a few years later.

16.9.10 The R-32 and R-32A

The R-32/R-32A series marked the introduction of the NYCTA into the era of stainless steel cars. While stainless steel construction was more expensive, it produced a much lighter car (a weight advantage of about 4200 lbs per car over conventional steel construction). This, in turn, would lead to savings in power requirements. Because stainless steel did not have to be repainted, it would also lead to maintenance savings. The 533 cars of this series were manufactured by Budd for the IND/BMT division, and featured a distinctive "fluted" design for the stainless steel car sides. All were delivered as permanent married pairs. The only reason for the double car numbers was that they were purchased under two different subcontracts, with the R-32 being paid from the proceeds of a revenue bond issue, and the R-32A from the 1963-1964 New York City budget.

The cars experienced some initial growing pains. The cars initially sat too high to safely clear some subway tunnels. No one had redesigned the suspension for the lighter weight of the cars. This and other bugs would eventually be worked out, and the cars became extremely reliable. Operating personnel generally loved these cars, as they had extremely high acceleration and were fast. This, of course, was because they had the same motors as previous cars, but (in a typical consist) had a 10,000 lb weight advantage over a train of steel cars. After reconditioning, some of the R-32s sported blue side and end doors.

Fig. 16.27 Exterior and interior of an R-32 Car (after renovation)

While many of these cars have been replaced by the R-160/R-160A cars, as of November 2011, there were still 222 in operation. Their retirement was postponed, when additional problems with the R-44s caused their retirement schedule to be advanced.

16.9.11 The R-36

The No. 7 Flushing line, as noted previously, is the only one to run eleven car trains. As the 1964 World's Fair approached, the line's entire fleet was replaced with new cars sporting a unique and distinctive exterior look. The cars were made up of R-36 cars in married pairs, and a special order of forty single R-33 cars that comprised the "11th" car on trains. The cars were manufactured by the St. Louis Car Company. The cars were all air conditioned, except for the forty R-33s, which were not run during the summer. With the arrival of this special fleet, all of the IRT's pre-WWII fleet was retired, except for a small number of LVs and Steinways operating on the remaining Bronx portion of the Third Avenue El at the time.

The cars were maintained at the Corona Shop, which serves only Flushing Line cars, and were quite reliable. In 1982, the cars were the first to enter the TA's general overhaul program, where they transitioned to the "redbird" exterior.

Fig. 16.28 Exterior and interior of an R-36 Car

16.9.12 The R-38

The R-38 cars were manufactured by St. Louis Car Company for the BMT/IND divisions. They were very much like the R-32s, and were clad in stainless steel. Unlike the R-32s, the R-38 stainless steel cladding was fluted only halfway to the bottom of the windows, then smooth to the top of the car. The last ten cars delivered were air conditioned, an experiment that worked very well, leading to a decision to air condition all future orders, and to retrofit older cars as appropriate. The R-38 fleet has been recently retired.

Fig. 16.29 Exterior and interior of an R-38 car

16.9.13 The R-40 and R-42

The R-40/R-42 class cars for the IND/BMT division closed out the 1960s. These cars were all manufactured by the St. Louis Car Company, and were known for their lack of uniformity. The initial fleet of R-40s had a distinctive sloped end at the No.1 end of each car of a married pair, and were stainless steel. It is rumored that the slant ends were encouraged by Mayor John Lindsay, who wanted a more stylish look for subway cars. The first 200 cars were not air conditioned; the last 100 R-40s were air conditioned.

The sloped ends, however, created a wide gap for passengers passing between cars, so some hardware was added for their protection. Of course, the hardware literally destroyed the aesthetics of the cars. Later, a modified version of the R-40 with straight ends was delivered with air conditioning. The modified R-40s were virtually identical in appearance to the subsequent R-42s, all of which were air conditioned.

The R-40s also introduced semiautomatic door control. On previous rolling stock, individual zones on each train had separate door control switches. On the R-40s, a single control operated all doors on the train. They were zoned through a series of push-button relays, allowing the conductor to set up door operation on the entire consist in a timely fashion.

Thirty-two of the R-40s had package brakes with composition brake shoes, which later became standard. Composition brake shoes created less heat, and less iron dust. They also stopped trains more quickly in the emergency braking mode, and reduced the occurrence of "flat wheels."

The R-42 class was the first to be fully air conditioned, and used a solid-state converter to provide Low-V power supply, and a solid-state inverter to provide interior lighting. Most of the R-40s and R-42s have been retired. As of November of 2011, 48 R-42's remained in service.

The R-40/R-42 series brought the 1960s to a close. In the next decade, the NYCTA would make its first attempt at "high-tech" subway cars, with very mixed results.

(a) Exterior of an R-40 "Slant Roof" car

(b) Interior of an R-40 car

(c) Exterior of an R-42 car

(d) Interior of an R-42 car

Fig. 16.30 The R-40 and R-42 subway cars

16.10 R-Cars of the 1970s and 1980s

16.10.1 The R-44 and R-46

The R-44s represented a series of firsts for the NYCTA. Sleek and silvery with a polished stainless steel exterior, dark blue belly band, and pastel-colored interior, the R-44s were the first subway car to sport the MTA's "corporate look."

The handsome air-conditioned cars were originally slated for the planned "new routes" of 1968, which included a Second Avenue Subway and Queens super-express along the LIRR mainline right-of-way. Because the new routes would have longer distances between stations, the cars were designed for 70 mph operation. In order to validate this new high-speed equipment (no existing NYCTA tracks could accommodate more than 55 mph operation safely), an eight-car test train of R-44s was transferred to the LIRR and tested on a 5.9 mile long track between Jamaica and Woodside in Queens. On January 31, 1972, on the very first run, the train set a world record for subway cars of 87.75 mph at full power. A second run was made with the motors cut out on two cars to simulate a fully-loaded train; a speed of 77 mph was attained. The results were gratifying, but an inspection of the equipment a few days later revealed that 8 out of the 32 DC traction motors were flashed due to the extremely high dynamic currents during dynamic braking. A design modification was performed, and a second high-speed test was successfully performed without any component failures.

With a length of 75 ft, the R-44s were a full 15 ft longer than the standard BMT/IND length of 60 ft, but still 10 ft shy of the 85-foot length of the LIRR commuter cars they so closely resembled. Engineers at that time assumed that with fewer cars and fewer components in an eight-car train of R-44s as compared to the same train length with ten 60 ft standard cars, both capital and maintenance costs would be reduced. However, an eight-car train of R-44s has only 32 door openings, eight less than a 10-car train of 60-foot cars, which slows passenger access and egress and increases dwell times at stations. NYCTA engineers did explore adding a fifth door opening on each side of the R-44, but this would have made the car too heavy for the right-of-way structures.

The cars were delivered in four-car A-B-B-A units, with one motorman's cab at each end of the units. While they could be decoupled, doing so was a time and space-consuming task, and it was rarely done. The cars were also equipped with cab signals and ATO (automatic train operation) equipment, neither of which was ever used. The brakes were controlled by a trainline electrical analog P-wire. The cars were subject to a number of electronic equipment problems, and had a generally poor maintenance record. The softer ride was intended to be coordinated with installation of welded rail, which would have been the practice on the new routes, if any of them had ever been built. Running primarily on standard section length rails caused a good deal of excess wear and tear on the suspension.

The cars were built by St. Louis Car Company. The problems with the car were so great that St. Louis Car Company went out of business. The R-44s were the last cars they manufactured. A total of 352 cars were ordered, 300 for the NYCTA, and 52 for the Staten Island Rapid Transit with minor changes to comply with ICC/FRA rules. After their general overhaul, and additional 12 NYCTA cars were transferred to the SIRT.

The R-46 looked virtually the same as the R-44, but they were manufactured by Pullman Standard, as the St. Louis Car Company had already gone out of business. The cars had a number of improvements based upon the R-44 operating experience, and (except for one major flaw) performed better than the R-44s.

The "one major flaw" was an understatement. The R-46 used a Rockwell articulated truck that had been untested on any NYCTA property. On March 27, 1977, the first structural crack in a Rockwell truck was reported. It was the first of many, leading to massive downtime for the fleet. By 1983, every one of the Rockwell trucks (1548 of them) was replaced with new truck castings of conventional design. Eventually, the NYCTA won a financial settlement from Rockwell in the courts. While the R-46 did not put Pullman Standard out of business, the R-46 was the last passenger rail car they manufactured, limiting their subsequent business to freight cars. Rockwell, however, did go bankrupt. All R-46 cars are still in service as of November 2011.

Fig. 16.31 Exterior view of an R-46 car

The TA learned two things from the R-44/R-46 experience: (1) never buy technology in bulk before testing it on the New York system, and (2) don't invest in technology that cannot be used within the expected life of the car.

As the 1980s dawned, no new cars for the IRT division had been purchased since the R-36 cars for the 1964 World's Fair. In general, the condition of the subway fleet was fairly poor, with an MDBF of less than 10,000 miles, and an average of one-third of the entire fleet out of service at any given time.

16.10.2 The R-62 and R-62A

The R-62 was the first stainless steel IRT car. Because of the poor performance of the R-44s and R-46s, virtually all of the "high-tech" features of those trains were eliminated, relying instead, on the tried and true technology of the R-36 cars, at the time, twenty years old. The TA also abandoned its long-standing practice of ordering cars as married pairs, and the R-62 was delivered as single cars with two motorman's stations in each. In 1989, after great improvements in general rolling stock performance, and new maintenance facilities on line, the cars were modified to run in five-car units with full-width motorman's cabs at each end of the five-car unit. The five cars were joined by link bars.

The R-62s were the first cars to be manufactured by a foreign company, Kawasaki of Japan. When St. Louis Car Company went out of business, and Pullman Standard ceased manufacturing passenger rail cars, only one American manufacturer of rail cars was left, that is, Budd. The TA selected Kawasaki because the financial arrangements were more favorable and flexible than Budd was willing to offer. As Budd held several patents on stainless steel car manufacturing, Kawasaki had to purchase a license and technology from Budd.

When a subsequent order for R-62s was issued, Kawasaki did not bid. The contract (referred to as R-62A) went to Canadian Bombardier, who agreed to use the Kawasaki design. The only difference remaining between the R-62 and the R-62A is that while all R-62s are in five-car units, there are still about 250 individual R-62A cars left. The full fleet of R-62 cars is still in service as of November 2011.

Fig. 16.32 Exterior and interior of an R-62 car

16.10.3 The R-68 and R-68A

The R-68 was basically the IND/BMT equivalent of the R-62. Stainless steel, fully air conditioned, with comfortable seating and bright interior colors, the R-68 fleet has not performed well. The cars retained the 75 ft length of their R-44 and R-46 predecessors, and were extremely heavy. They were built by a French consortium led by Bombardier (of France). After some modifications, and creation of four-car units, the maintenance record of these cars has improved considerably. A subsequent order (R-68A) was built by Kawasaki. All of the R-68 cars are still in use as of November 2011.

Fig. 16.33 Exterior and interior of an R-68 car

16.11 R-Cars from 1990 to the Present

After a decade of building new cars to older, well-established standards, as the 1990s approached, the NYCTA was ready to jump back into the "high-tech" business again.

16.11.1 The R-110A and R-110B

In 1992, two specially-designed experimental trains were delivered for extensive testing in mainline use. The cars were part of the "New Technology Tomorrow Trains" program (NTTT). The R-110A was ordered for the IRT division (now called the A Division), and the R-110B for the IND/BMT division (now called the B Division). Each train represented a $20 million investment from the TA, and another $20 million from various suppliers to introduce and qualify new technologies for the future. Some of the innovative features on these cars included:

- AC propulsion motors with regeneration
- Microprocessor-controlled doors and brakes
- Roof-mounted hermetic A/C units
- Fabricated trucks with air bag suspension
- Doors that automatically reopen when something is stuck between them
- Passenger emergency intercoms
- Passenger alarm strips
- Improved lighting
- Computerized announcements

The R-110A cars were manufactured by Kawasaki, and included many high-tech improvements in the motorman's cab. The cab included a keypad for programming automatic signage and prerecorded announcements, a built-in radio, a microprocessor screen with selection buttons on each side, and a single-handle controller and communications handset. The seat is adjustable and ergonomically designed.

To improve passenger flow, the doors on the R-110A were increased from 50 to 64 inches, and cross-sectional seats were introduced on the IRT for the first time in many years. These aspects, however, resulted in fewer seats (but greater capacity for standees), and were modified for later production runs.

The R-110B was manufactured by Bombardier of Canada. A nine-car train consists of the triplex units in which each car is 67 ft long, a length not used since the BMT Standards. The R-110B cars included electronic strip maps showing each station on the A-Line, where the cars operated.

16.11.2 The R-142 and R-142B

The R-142 was the production version of the IRT R-110A. The major modification was that the doors were reduced to 54 inches in width, and the cross-sectional seating was eliminated. These changes increased the seating capacity of the cars. For the first time in memory, fierce competition for the R-142 order drove priced down, and allowed the TA to increase the number of cars it ordered. The cars were made by two manufacturers, both of whom agreed to build cars that were fully compatible and looked exactly the same. Bombardier of Canada was eventually contracted to produce 1030 cars, while Kawasaki built 600 cars. Virtually all of the technologies tested on the R-110A were incorporated into the production runs of R-142s. After some initial reliability problems, the cars have performed extremely well, and are setting new MDBF records.

The cars are manufactured to run in five-car A-B-B-B-A units. The "A" cars have a motorman's cab at one end, and two motored trucks. The "B" cars have no motorman's cab, and one motored truck. The consist is relatively flexible, allowing trains of various lengths to be run.

Fig. 16.34 Exterior and interior of an R-142 car

16.11.3 The R-143 and R-160

The production model of the R-110B for the IND/BMT division was the R-143. It included virtually all of the new technologies of the R-110B, with one critical addition: these were the first cars equipped to run on a Communications Based Control System (CBCS). This new technology relies on the use of microprocessors on trains and in wire loops on the wayside. Also known as "moving block" control, it allows the signal block, that is, traditionally fixed distances set by wayside signals, to move with the train. This allows trains to move closer together, and will increase the capacity of each track. The technology is now being tested on the Canarsie L-Line. The R-143 was originally built by Kawasaki, but large orders were subsequently placed with both Kawasaki and the ALSTROM consortium of France. Initial orders were placed in 1998 and 1999. Subsequently, an order for R-160 cars (essentially the same car) was issued in 2002.

In another key difference between the R-143/R-160 cars and the R-110B, the length of the car was returned to the typical 60 ft length used on most cars in the IND/BMT division.

Fig. 16.35 Exterior and interior of an R-143 car

As the R-160s continued to arrive, the NYC subway fleet was in better shape than it had been in years. New technologies are being introduced and are proving to be effective and reliable. Records are being set for MDBF. Semiautomated control is being introduced and tested in ways that will lead to increased track capacities throughout the system.

16.12 Into the Future

As this is being written, new orders for subway cars are in progress. Several hundred R-179 cars are to be ordered. They were originally intended to replace the last R-44 cars in service. Due to new structural issues concerning the R-44s, however, the R-44s have been retired prior to taking delivery on any R-179s. They will now replace remaining R-32s and R42s still in service on the BMT/IND Division B.

A fleet of R-188s will be ordered to replace the R-62s and R-62As that now service the Flushing Line. The R-62s will be moved back to the mainline service. The R-188s will incorporate Communications Based Train Control (CBTC) that is now being tested on the Canarsie Line. The R-188 fleet will consist of new cars and R-142s that will be modified to similar standards.

A fleet of R-211s is contemplated to replace the R-46 cars in 2015. No award data has yet been scheduled, and cars would be expected within 3–5 years of an award being made.

All capital programs of the MTA, including the purchases of new rolling stock, have been affected by the recent economic difficulties in the nation and the city, which have constrained the MTAs budget flexibility. Despite this, the program of regular car renewals is moving forward, albeit at a somewhat slowed pace.

New York City's rolling stock has moved countless billions of passengers to the far reaches of the city for over 100 years. Our rail cars have become part of our society, and reflect the complex technologies that develop over time. From the hand-operated gate cars to the newest trains with CBTC, they have truly been the "wheels that drove New York."

Reference

1. Sansone, G.: New York Subways: An Illustrated History of New York City City's Transit Cars. Centennial Edition. The Johns Hopkins University Press, Baltimore (1997)

Figure Sources

16.1 *Courtesy of* MTA New York City Transit, from Sansone, G., *New York Subways: An Illustrated history of New York City's Transit Cars*, Johns Hopkins University Press, Baltimore MD, 2000, pg 15 and 17.
16.2 *Courtesy of* MTA New York City Transit, from Sansone, G., *New York Subways: An Illustrated history of New York City's Transit Cars*, Johns Hopkins University Press, Baltimore MD, 2000, pg 18.
16.3 *Courtesy of* MTA New York City Transit, from Sansone, G., *New York Subways: An Illustrated history of New York City's Transit Cars*, Johns Hopkins University Press, Baltimore MD, 2000, pg 32.
16.4 *Courtesy of* MTA New York City Transit, from Sansone, G., *New York Subways: An Illustrated history of New York City's Transit Cars*, Johns Hopkins University Press, Baltimore MD, 2000, pg 36.
16.5 *Courtesy of* MTA New York City Transit, from Sansone, G., *New York Subways: An Illustrated history of New York City's Transit Cars*, Johns Hopkins University Press, Baltimore MD, 2000, pg 45 and 47.
16.6 *Courtesy of* MTA New York City Transit, from Sansone, G., *New York Subways: An Illustrated history of New York City's Transit Cars*, Johns Hopkins University Press, Baltimore MD, 2000, pg 135, 138, and 139.
16.7 *Courtesy of* MTA New York City Transit, from Sansone, G., *New York Subways: An Illustrated history of New York City's Transit Cars*, Johns Hopkins University Press, Baltimore MD, 2000, pg 148.
16.8 *Courtesy of* New York Transit Museum.
16.9 (a) *Courtesy of* New York Transit Museum. (b) *Courtesy of* MTA New York City Transit, from Sansone, G., *New York Subways: An Illustrated history of New York City's Transit Cars*, Johns Hopkins University Press, Baltimore MD, 2000, pg 67.
16.10 *Courtesy of* New York Transit Museum.
16.11 *Courtesy of* New York Transit Museum.
16.12 *Courtesy of* New York Transit Museum.
16.13 *Courtesy of* New York Transit Museum.
16.14 *Courtesy of* New York Transit Museum.
16.15 *Courtesy of* New York Transit Museum.
16.16 *Courtesy of* New York Transit Museum.
16.17 *Courtesy of* New York Transit Museum.
16.18 *Courtesy of* New York Transit Museum.
16.19 *Courtesy of* New York Transit Museum.
16.20 *Courtesy of* New York Transit Museum.
16.21 *Courtesy of* New York Transit Museum.

16.22 *Courtesy of* New York Transit Museum.
16.23 *Courtesy of* New York Transit Museum.
16.24 *Courtesy of* New York Transit Museum.
16.25 *Courtesy of* New York Transit Museum.
16.26 *Courtesy of* New York Transit Museum.
16.27 *Courtesy of* New York Transit Museum.
16.28 *Courtesy of* New York Transit Museum.
16.29 *Courtesy of* New York Transit Museum.
16.30 *Courtesy of* MTA New York City Transit, from Sansone, G., *New York Subways: An Illustrated history of New York City's Transit Cars*, Johns Hopkins University Press, Baltimore MD, 2000, pg 242, 243, 250, and 251.
16.31 *Courtesy of* MTA New York City Transit, from Sansone, G., *New York Subways: An Illustrated history of New York City's Transit Cars*, Johns Hopkins University Press, Baltimore MD, 2000, pg 163.
16.32 *Courtesy of* New York Transit Museum.
16.33 *Courtesy of* MTA New York City Transit, from Sansone, G., *New York Subways: An Illustrated history of New York City's Transit Cars*, Johns Hopkins University Press, Baltimore MD, 2000, pg 271, 272.
16.34 *Courtesy of* MTA New York City Transit.
16.35 *Courtesy of* MTA New York City Transit.

Chapter 17
The Best Laid Plans of Mice and Men

As magnificent as New York's transit system is, its history is full of plans and designs that, for better or worse, have never been implemented. Many of these have been noted in previous chapters. In this chapter, some of the more important, but unimplemented, projects are reviewed, together with some thoughts on how the city might have changed if they had been built. As New York looks towards the future, it is also appropriate to speculate on what role existing and perhaps expanded transit will play.

17.1 Introduction

The actual development of rapid transit in New York City occurred in four major phases:

- The development of the elevated railways, 1868–1890.
- The building of the IRT (Contracts #1 and #2), 1900–1904.
- The dual contracts (Contracts #3 and #4) expansion, 1913–1920.
- The building of the Independent Subway, 1932–1940.

In the era leading up to the age of the Els, a plethora of ideas and concepts had been developed and presented, including a variety of interesting and unique technologies (pneumatic tubes, trains propelled by sails, etc.). The ideas included more massive elevated structures (the Viaduct Plan) as well as subways (the Arcade Railway). When the IRT and the dual contract extensions were built, there were certainly alternative plans that were supported. The real era of significant expansions that never happened began with the Independent Subway. This chapter will discuss three major proposed expansions:

- The "second stage" and "third stage" of the Independent Subway
- The "Program for Action," or "New Routes" program
- The Second Avenue Subway

The latter, of course, was a part of the first two plans, as well as several other smaller schemes to be developed in between. The grandiose schemes of transit

R.P. Roess and G. Sansone: The Wheels That Drove New York, STTT 1, pp. 411–430.
springerlink.com

planners and advocates from the late 1920s through the present were "derailed" primarily by economic factors: the proposed expansions simply cost too much to build, and there was no realistic way to acquire the money to fund their construction, or their operation.

17.2 The Independent Subway: Stages 2 and 3

The idea for the Independent Subway was born in a report issued by the New York Public Service Commission (PSC) in 1920. Created by its chief engineer, Daniel L. Turner, it proposed a massive expansion of the existing system, which at that point, consisted of the IRT and the BRT (soon to reorganize as the BMT).

Turner created a plan that would connect Staten Island to the system, push rapid transit routes to the city line in all boroughs, and put virtually every New Yorker within a reasonable walk to a subway station. The plan clearly envisioned a two-stage process. The first stage would be to build a system around two new trunk line subways in Manhattan under 6th and 8th Avenue. The second stage would encompass a series of routes around a trunk subway under 2nd Avenue. Some of the extensions to the city line were part of a less detailed general plan which could have formed a third stage of construction.

While the Turner plan was the impetus for virtually all of the modified plans which came later, it was the driving force behind subway planning for 20 years. Mayor Hylan, whose agenda included recapture of the IRT and BMT, adopted many aspects of the plan as he pushed for development of an "Independent" system that would complete with the IRT and BMT.

When construction of the Independent (Stage 1) commenced in 1929, it was expected to be the first of three stages. In September of 1929, specific plans for the Stage 2 system were released. No firm plans for Stage 3 were ever created. The goal of the Stage 2 and Stage 3 system was to place a subway station within approximately one-half mile of all residents of New York. The planned expansion would have added 100 additional route-miles and 294 track-miles to the transit system.

The second system was aimed at making the outer boroughs more accessible, especially Queens, where the majority of residential growth had been expected. Unlike the original IND, the second-stage system served areas that had no competing subway or elevated service.

A map of the planned second stage of the IND is shown in Figure 17.1. It was created by the Board of Transportation and issued as part of its 1929 plan for subway/elevated extensions. Major components of the planned expansion are described below, and are illustrated on the map. Color-codes are used to help correlate the described extensions with their location on the map.

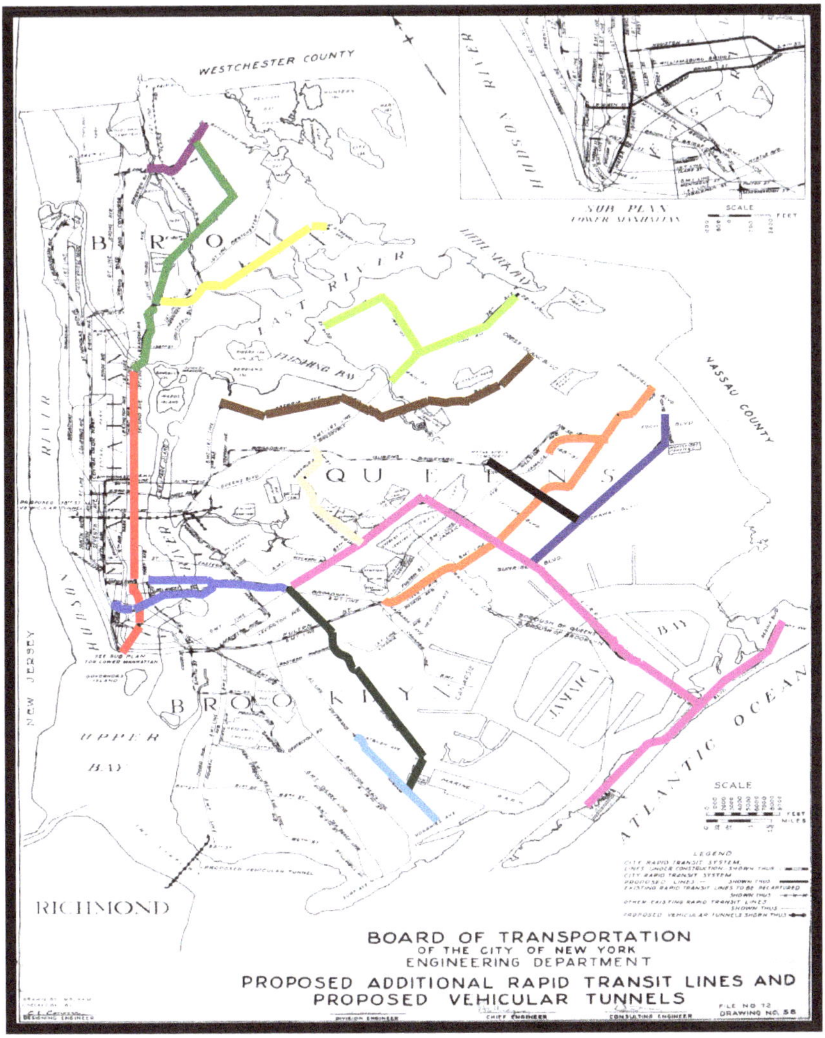

Fig. 17.1 Map of the1929 Board of Transportation Plan

Major components of the IND second system included:

- A Second Avenue Subway, created as a loop from the Chambers Street terminal of the 8 Avenue Subway. The loop would have run from Chambers Street to Water and Pearl St., then to the Bowery and Chrystie Street, then up 2nd Avenue to the Harlem River. The line was planned as a four-track trunk, with some sections having 2 or 6 tracks (Red).
- Two new river tunnels from downtown Manhattan to Brooklyn, both connecting to a proposed South 4th Street station in Brooklyn. The first tunnel was to

serve a new Worth Street line running under Church Street; the second to serve a new Houston Street line running from Essex Street to the East River (Blue).

- An extension of the 2nd Avenue Subway into the Bronx under Boston Road, with four tracks to 163rd Street, 2 tracks to 177th Street, and a short elevated connection to the White Plains Road line, which would have been converted to IND operation. (It remains an IRT line today) (Green).
- A two-track branch from the Boston Road line under 163rd Street to Hunts Point and Seneca Avenue, continuing under the Bronx River under Lafayette Avenue to East Tremont Avenue (Yellow).
- An extension of the Concourse Line from 205th Street to Baychester Avenue (Purple).
- An extension of the Fulton Street line (called the "Liberty Avenue Line") from Broadway Junction to Springfield Blvd. in southern Queens. The line would have used a portion of the Liberty Avenue El, and would have had a spur connecting to the Jamaica El at 168th Street (Orange).
- An elevated line along 120th Street and then Springfield Blvd. in Queens. One terminus would have been at the Rockaway Line at North Conduit Blvd./Hawtree Street. The other would have been at Foch and Springfield Blvds. (Dark Blue).
- A two-track Van Wyck Blvd. Line, running as a subway to 166th Street, then as an elevated to a junction with the 120th Street line at Rockaway Blvd. (Black).
- Two extensions of the Flushing Line from Main Street: (1) under Roosevelt Avenue to 155th Street, then as an elevated above the LIRR Port Washington ROW to 221st Street and 38th Avenue; (2) a branch from 147th Street and Roosevelt Avenue to 11th Avenue, then via 11th Avenue to 122nd Street in College Point (Lime Green).
- Extension of the Astoria Line along Ditmars Blvd., Astoria Blvd, 112th Street, then diagonally across local streets to Nassau Blvd (now the Long Island Expressway). The line would have been elevated, or in open cuts for some of its distance (Brown).
- A new line serving the Rockaways (called the Myrtle, Central Avenue, and Rockaway Line). It would have followed the LIRR Montauk Division and the LIRR Rockaway Branch, starting from Stuyvesant Avenue (where the Utica Avenue Line turns south). It would have provided direct service to Howard Beach, and would have crossed Jamaica Bay to a point on the Rockaway Peninsula near Hammels Avenue and Beach 83rd Street, then headed north to Mott Avenue (Violet).
- A South 4th Street-Utica Avenue Line in Brooklyn, connecting with the Houston Street Line at the East River. The line would wind its way through Brooklyn, serving areas in Williamsburg, eventually terminating at Voorhies Avenue (Dark Green).
- Extension of the IRT Nostrand Avenue Line as a two-track subway to Kings Highway, then elevated to meet the Utica Avenue Line at Avenue S (Light Blue).
- A connection between the IND Queens Blvd./Hillside Avenue line and the Myrtle/Central Avenue line (Tan/Flesh).

The 1929 plan was massive, adding 100 new route miles to the system, 52 of which would be in Queens. The Board of Transportation plan had already dropped a major project included in Turner's recommendations, that is, the connection to Staten Island. Other portions of the plan proved to be controversial.

Turner's original plan had focused on building trunk line capacity in Manhattan. When the Board of Transportation took control in 1924, with John H. Delaney as its chair, the focus shifted to building capacity in the outer boroughs, and to providing service around the rim of the city as well as into its core. The 1929 plan included only one trunk line in Manhattan, the 2nd Avenue Subway. Even Delaney, however, recognized the limitations of the Manhattan trunk system as additional feeder lines were built. In a general vision, perhaps, of the "third stage" program, Delaney indicated that after the 2nd Avenue was completed, another trunk line under Madison Avenue would be built, and that, eventually, every north-south artery in Manhattan would have a subway with the exception of 1st Avenue and 10th Avenue.

In Manhattan, there was broad support for the 2nd Avenue Subway. It was understood that the subway would replace the 2nd and 3rd Avenue Els. The commercial real estate market was expected to sky-rocket with the removal of the Els, and forecasts of a doubling of the population of the Lower East Side were expected to have a profound effect on residential real estate as well.

In Brooklyn, the Myrtle/Central Avenue line would have paralleled the Broadway El, and run directly under the Myrtle Avenue El. Expectations were that these elevateds, and perhaps even the Jamaica Avenue El, would be replaced by the new subways.

Despite the support of many public officials, opposition to the Brooklyn portion of the plan was not insignificant. Residents in areas not receiving new subway service were vocal in their opposition. Many Brooklynites expected the plan to include an extension of the 2nd Avenue Subway into Brooklyn, traveling along Atlantic Avenue all the way to Jamaica. Part of this expectation was fueled by the fact that the LIRR had already built a two-track tunnel along Atlantic Avenue and abandoned it 50 years previous. This tunnel is in current service, bringing LIRR passengers from Jamaica to the current terminal at Flatbush and Atlantic Avenues.

The biggest objection to the Brooklyn plan was that a number of the lines were to be built as elevateds, not subways. While the plan would have allowed some of the older elevateds to be removed, it was adding new els to the borough. One of the harshest critics of the plan to build new elevateds was Joseph V. McKee, an Alderman who would in a few short years become Mayor for a short time after the demise of Jimmy Walker.

No other borough would have benefited more from the 1929 plan than Queens, with more than half the route mileage occurring there. The Flushing line would have been extended to 221st Street, close to the City line. It would, however, have been built primarily as an elevated along Station Avenue, either right next to or above the Long Island Rail Road Manhasset Branch. A second extension would have been built along 149th Street and 11th Avenue into College Point, a particularly isolated part of Queens. This line too, however, would have been built partially as an elevated line.

The Astoria Elevated would have been significantly extended along Ditmars Blvd., 112th Street and Nassau Avenue (which would become Horace Harding Blvd., and is now the Long Island Expressway). This line could have provided a short link to LaGuardia Airport when commercial airline service commenced in 1939.

The largest single effort in Queens involved providing service to the Rockaways. Long-standing plans had existed to "take over" service to the Rockaways from the Long Island Rail Road. The 1929 plan, however, proposed to duplicate them, as negotiations with the LIRR had been stalled for some time. The take-over from the LIRR would, of course, happen decades later.

Despite the benefits that would accrue to Queens, there were a number of objections to the plan, primarily relating to the specifics of the planned routes. Residents in College Point wanted a more direct route to Flushing than the planned El provided, and they did not want an El. It had been expected by many that the plan would include an extension of the Queens/Hillside Blvd. line to the City line, and its absence caused some opposition.

In the Bronx, there was a great deal of support for the plan, especially for the extension of the Grand Concourse line into Baychester. This extension was colloquially referred to as the "Burke Avenue Line," as a portion of it would have travelled along Burke Avenue to Boston Post Road. Instead of this extension, the city took over an abandoned branch of the New York, Westchester, and Boston Railroad, running a shuttle between East 180th Street and Dyre Avenue. Sixteen years would pass before this shuttle was integrated into the IRT White Plains lines.

All of the discussion and controversy was for naught. The plan for the second system was released only months before the stock market collapse that spurred the Great Depression. All plans for any subway expansions were put off indefinitely. In fact, were it not for a unique federal grant under the Works Projects Administration, the first stage of the IND would not have been completed either.

Despite the fiscal travails of the Great Depression, the Board of Transportation continued to present capital plans for expansion of the subway system on a regular basis.

In 1932, the capital plan did not vary much from the 1929 blueprint. A 34th Street crosstown line was added, and the extension of the Queens Blvd./Hillside Avenue line was included.

In 1938, the capital plan featured the acquisition of the LIRR Rockaway line, and extensions to the Flushing, Queens Blvd./Hillside Avenue, and the Fulton Street lines. The Hillside Avenue extension was pushed further east, to Little Neck Parkway, and the Fulton line was pushed to 229th Street. The tunnel to Staten Island was put back into the plan, this time connecting to a newly-proposed IND route from the Church Avenue/Prospect Park line, running along 10th Avenue to 86th Street in Fort Hamilton.

In 1939 and 1940, the second system emerged once again as a revised proposal from the Board of Transportation. While it was, in essence, the second system, it incorporated a number of changes reflecting the fact that unification had occurred, and that all three divisions, the IRT, the BMT, and the IND, could be treated as a coordinated transit system.

The plan restored the subway tunnel to Staten Island, which would have connected directly to the Staten Island Rapid Transit System, and included an extension of the Hillside Avenue Line all the way to Little Neck Parkway. Figure 17.2 illustrates the plan.

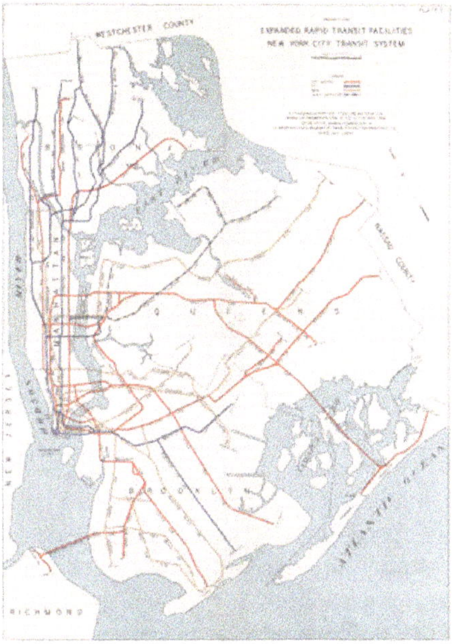

Fig. 17.2 1939/1940 Board of Transportation Subway Expansion Plan

The 1939/40 proposal suffered a fate similar to the 1929 plan: general economic conditions were not conducive to massive public investment in subways, and WWII was already visible on the horizon. One project, the extension of the Concourse Line beyond 205th Street, was funded. Bronx residents, however, argued for a rehabilitation of the Dyre Avenue line of the old NYW&B. As a result, money was diverted to its rehabilitation, and it began operation as a shuttle in 1941. It was eventually tied into the IRT division.

While it is impossible to ascertain the exact development outcomes that would have taken place if the 2nd Stage plan, or any significant part thereof, had been implemented, some reasonable estimates can be made. Certainly, Queens would have urbanized more rapidly, and the low-density residential neighborhoods of eastern Queens might have become more urban and less suburban. Connecting Staten Island to Brooklyn via rapid transit would have had the same effect there. Whether or not increased urbanization of Queens and Staten Island would have been a good thing is open for debate.

17.3 The 1968 MTA "Program for Action"

In 1967, with the full support of Governor Nelson Rockefeller and Mayor John Lindsay, the MTA announced an aggressive two-phase "Program for Action" which was also known as the "New Routes" program. In historical context, the plan was issued a little over a year after the transit strike of 1966, and somewhat before the city realized how bad its fiscal condition was becoming. Phase I included projects for "immediate" action and completion over a 10-year period. Phase II would begin at the completion of Phase I.

The Phase I plan included:

- Completion of the 63rd Street Tunnel, with connections to the 6th Avenue Line and the Broadway Line in Manhattan. (The 63rd Street Tunnel had already been planned, but construction had not yet started.)
- A Queens Super-Express, that is, a bypass running along the LIRR mainline, allowing trains from the IND Hillside Avenue/Queens Blvd Line to leave at 41st Avenue, run nonstop along the bypass, and return to the IND at Forest Hills. Originally conceived as a one-track, peak-direction-only service, it was expanded to a two-track bypass in 1973.
- A two-track subway diverting from the IND Queens Blvd. Line in Elmhurst, running along the Long Island Expressway to Fresh Meadows.
- A two-track subway, also diverting from the IND Queens Blvd. Line at Van Wyck Blvd., running along the LIRR Atlantic Branch to Springfield Blvd.
- The Second Avenue Subway, from 34th Street to the Bronx, with a connection to the 63rd Street Tunnel. It would be built as a two-track line, with provisions for expansion to a four-track line at a later date.
- Extension of the IRT Nostrand Avenue Line along Flatbush Avenue to Avenue U.
- Extension of the IRT New Lots Avenue Line to Flatlands Avenue and Linwood Street. The line was to have run at ground level in an abandoned LIRR ROW.

The Phase II Plan included:

- Completion of the 2nd Avenue Subway from 34th Street to Whitehall Street in the financial district.
- A midtown distribution system along 57th, 48th, 42nd, and 33rd Streets using automated technology to link terminals, offices, and other key points.
- Extension of the Jamaica Avenue El east, and razing and replacement of the El itself.
- Replacing the 3rd Avenue El in the Bronx with a new subway line running adjacent to the New Haven Line ROW.
- Extending the Pelham Bay Line to Co-op City in the Bronx.
- Extending the IND Concourse Line to White Plains Road.
- A new two-track subway under the Long Island Expressway to Kissena Blvd. and Queens College.

Figure 17.3 illustrates the plan.

Interestingly, the plan included a link from the LIRR directly into Kennedy Airport. While this aspect of the plan never moved forward, eventually a separate elevated light-rail line was built from the Jamaica LIRR terminal into Kennedy Airport, where it also provided inter-terminal service within the airport.

Unlike previous plans, construction of some elements of the New Routes program was actually started, and some small portions were actually completed, albeit in seriously diminished scope. On October 27, 1972, the Mayor and Governor broke ground at East 103rd Street and 2nd Avenue for the Second Avenue Subway. Construction of the 63rd Street Tunnel actually started on November 25 1969.

The anticipation of new routes also fueled the development of new high-speed, high-tech subway cars: the R-44s and R-46s. Both resulted in serious technological problems, including the infamous truck cracks experienced by the R-46s. The two companies that built them, that is, St. Louis Car, and Pullman, never built another subway car again. As none of the longer sections of the subway expansion were implemented, some of the technology and the high-speed capability incorporated into the cars could not be used anywhere in the system.

Once again, a disaster in city finances derailed the ambitious plans. Two bond issues needed to fund the improvements failed, one in 1971, and another in 1973. Finally, in 1975, with the city teetering on the brink of bankruptcy, all aspects of the plan were shelved. Construction of the 2nd Avenue Subway stopped, leaving 15 blocks of completed tunneling under 2nd Avenue.

The plan finally resulted in two additions to the transit system:

- The 63rd Street tunnel, completed in the early 1970s, opened for service in 1989, with a station on Roosevelt Island and a terminal station at Queensbridge/21st Street. In 2001, the "train to nowhere" was connected to the Queens IND. Today, the F-train uses the 63rd Street tunnel on its way to joining the 6th Avenue line in Manhattan.
- The Archer Avenue Subway opened in 1988. It was connected to the IND Queens Blvd. line at Van Wyck Blvd., and to the remaining portion of the Jamaica El. Originally intended to run along/above the LIRR mainline to the City Line, the subway is only two miles long and terminates at Archer Avenue, close to the LIRR Jamaica terminal.

In total, for all of the grandiose expansion plans, the New York City rapid transit system, the changes that have actually occurred have been far more modest. In addition to the two new subways noted above, the system has taken over former railroad services on Dyre Avenue in the Bronx, and to the Rockaways in Queens. The latter remains the most significant addition to the system since the completion of the IND Phase I plan in 1940. Along the way, miles of elevated routes have been lost on 3rd Avenue in Manhattan and the Bronx, and on Myrtle Avenue in Brooklyn.

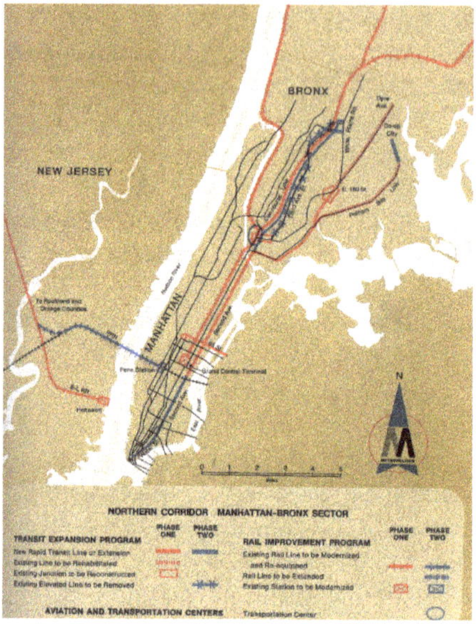

(a) Manhattan and the Bronx

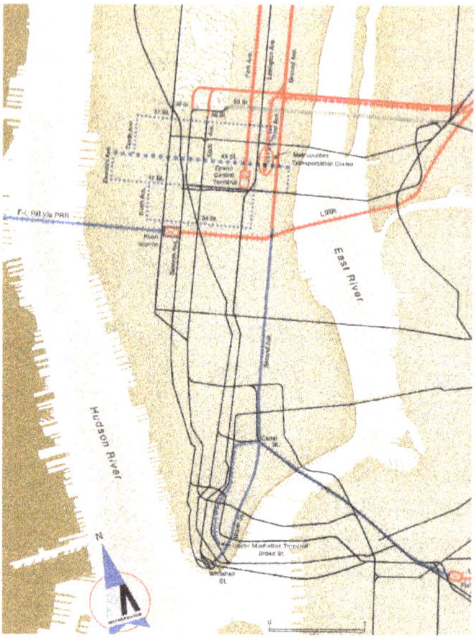

(b) Manhattan CBD Detail

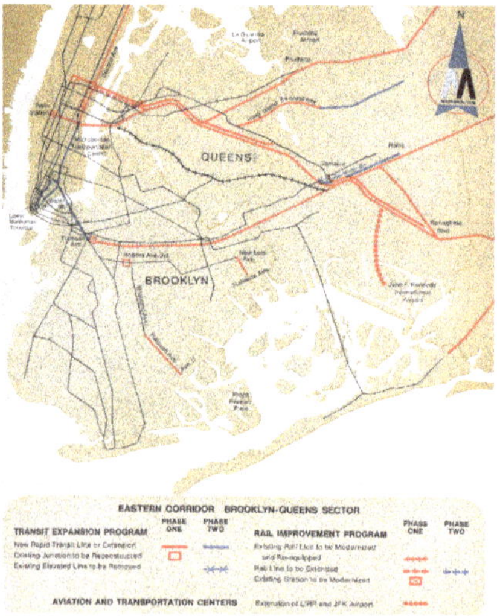

(c) Queens and Brooklyn

Fig. 17.3 The "Program for Action" Plan, 1968

17.4 The Second Avenue Subway

In the entire history of New York City transit, no line shares the long and tortured mythology of the Second Avenue Subway. So tangled is its history, that when ground was broken on April 12, 2007 for its latest incarnation, no one was sure whether this was the third or fourth ground-breaking for the subway.

The first formal mention of a Second Avenue Subway occurred in Daniel L. Turner's *Proposed Comprehensive Rapid Transit System*, in 1919. He wrote the paper as the formal recommendation of the NYS Public Service Commission, where he served as Chief Engineer. Turner's report recommended a massive new system that would dwarf subsequent plans. Turner believed that the only solution to keeping New York's economy growing was to build new subways virtually continuously to every corner of the five boroughs. His plan envisioned new routes under virtually every Manhattan Avenue, with numerous extensions into the other boroughs. His plan, naturally, included a 2nd Avenue Subway of six tracks running the length of the island.

In 1929, as part of the IND second system plan, the Board of Transportation recommended a 2nd Avenue Subway of four tracks, with connections to the Bronx. With the depression effectively derailing the second system project, the 2nd Avenue Subway did not reemerge until the revised plan of 1939. Like its predecessor, however, it too was derailed by the clouds of WWII.

In 1945, plans for the Second Avenue Subway were again revised, this time dropping the southern portion of the route. In 1947, plans reconnected the 2nd Avenue Subway to lower Manhattan, but this time via BMT trackage over the Manhattan and Williamsburg Bridges.

In 1947, the Board of Transportation accepted delivery of ten new prototype subway cars for the expected 2nd Avenue Subway. They were the first stainless steel cars ever ordered for the subway system, and became known as the R-11s. Only a few cars were ordered, but (of course) the cars never saw the 2nd Avenue Subway.

In the early 1950s, voters approved a $500,000,000 bond issue, intended to construct the 2nd Avenue Subway. In a somewhat bizarre financing gambit, the city proposed to include deep-rock stations on existing and planned subways as bomb shelters in the federal Civil Defense Bill of 1951 (1). It was proposed to use five existing deep-rock stations on the IRT and IND in the Washington Heights area of upper Manhattan as "bomb-proof" shelters. Several segments of the proposed Second Avenue Subway were also included. The attempt to combine bomb-shelter planning with subways eventually, however, failed.

The money from the $500,000,000 bond issue was quietly diverted to renewing and renovating parts of the existing system, including lengthening of IRT stations to accept longer trains. Voters were furious, but the 2nd Avenue Subway steadfastly remained an unfulfilled idea.

Part of the lower Manhattan plan was actually built in the early 1960s. The Chrystie Street Connection, which allowed for simplification of line routing through DeKalb Avenue, helped alleviate a major bottleneck in the system. The connection did increase the capacity of the BMT/IND system between Brooklyn and Manhattan.

In 1968, the 2nd Avenue Subway was back on the map, included in the previously discussed "New Routes" program. In 1972, a formal ground-breaking was attended by Governor Nelson Rockefeller, Mayor John Lindsay, MTA Chairman William Ronan, and Congressman Ed Koch. As reported in New York magazine:

> *"As reporters scribbled, Lindsay dryly noted that in the twenties, 'some people suggested a transit facility along Second Avenue. And it was such a good idea that I decided to follow up on it immediately.' The Pols took their swings with a pickax, but in an uncanny piece of symbolism, none could dent the pavement. A worker with a power rig was called in to break the concrete."*

As noted previously, some 15 blocks of tunneling were actually completed before work was halted as a result of the 1975 fiscal crisis. Later, Mayor Koch would seriously consider leasing out the tunnels for other appropriate uses. There was no real interest in alternative uses, however. This time, the pavement actually got dented, but the 2nd Avenue Subway curse ultimately prevailed.

17.5 Current Capital Plans and Projects

As of November 2011, the MTA Capital Construction Company was still pursuing four major capital projects that will have a major impact on the subway system:

The Second Avenue Subway
The LIRR East-Side Access Program
The Extension of the No. 7 Line
The Fulton Street Transportation Center

17.5.1 Second Avenue Subway (Again)

The problem of East Side subway access has remained as vexing today as in the past. With the removal of the 2nd Avenue El and the 3rd Avenue El in the 1940s and 1950s, the East Side of Manhattan is served by only one subway line, that is, the IRT Lexington Avenue Line. It is the single most overcrowded transit line in the country. On any given morning, over 180 passengers are crammed into each and every subway car on the line. The "Lex" carries 1.3 million daily riders, more than the entire ridership of the San Francisco, Chicago, and Boston rail transit systems. The West Side is served by four trunk lines: the IND 6th and 8th Avenue Subways, the IRT 7th Avenue Line, and the BMT Broadway Line.

After a comprehensive study of alternatives in 1995, a new plan for the Second Avenue Subway emerged. On November 8, 2005, New York State voters approved the Transportation Bond Act that will partially fund the state and city commitment to the project. Two years later, the Department of Transportation announced that the federal government would provide $1.3 billion for the project's first phase over a seven-year period. The Second Avenue Subway was once again on track.

The plan is to be implemented in four phases. In Phase 1, a connection to the Q train on the Broadway Line will be made at 63rd Street. Three stations will be built along 2nd Avenue at 72nd Street, 86th Street, and 96th Street. Phase 2 will extend the 2nd Avenue Subway uptown to a terminus at 125th Street. Phase 2 will make use of a tunnel completed in the 1970s, when work on the 2nd Avenue Subway began under the "New Routes" program. Phase 3 will extend the 2nd Avenue Subway downtown to Houston Street, and would introduce the T-train that will travel the length of the route. The 4th Phase will complete the plan by pushing further downtown to Hanover Square. Figure 17.4 illustrates the plan.

Fig. 17.4 Map of the Second Avenue Subway project

The construction of the first phase is underway. Phases 2 through 4 are in engineering, and there is not yet any funding in place at either the federal, state, or local governments for these phases. In addition to Phases 2, 3, and 4, there is some thought that the Hanover Square terminus can be connected to Brooklyn using existing facilities or a new tunnel.

On April 12, 2007, a ceremonial groundbreaking for Phase I took place at 99th Street. Actual construction work began on April 23, 2007 on the surface of 2nd Avenue between 91st and 95th Streets. On May 28, 2009, MTA awarded a $303.8 million contract to E.E. Cruz and Tully Construction Co. to construct the 96th Street station box. In June of 2009, the first of three contracts for the 86th Street station was awarded for advance utility work and construction of cut-and-cover shaft areas at 83rd and 86th Streets.

On May 14, 2010, the Tunnel Boring Machine was started at the 2nd Avenue launch box at 96th Street. The TBM was expected to dig approximately 50 ft of tunnel per day. When it reaches 63rd Street, it will be "backed out" and placed in the parallel tunnel to bore the second tube. The Boring of the east tunnel was completed in October 2011.

The 2nd Avenue Subway will be a two-track line along its entire length. Construction will be a mix of deep tunneling using a TBM, and cut-and-cover, which is needed to build stations and launch shafts.

The project has met with the usual problems of large-scale construction in a crowded environment. Relocation of utilities took longer than expected. In June of 2009, two buildings had to be evacuated as their foundations had been compromised, and they were in danger of collapse. This delayed blasting work in the area for some time. The first phase, originally scheduled for opening in 2014, has now been pushed back to December of 2016, a date which was holding as of March 2012.

The new subway will attempt to incorporate several interesting technological aspects not previously used in New York. The MTA is investigating the use of glass platform screens with doors. These would isolate station platforms from track fires, but would require that trains doors line up with screen doors, and could slow access and egress from subway cars. Stations will include lots of glass, to allow as much ambient light into stations as possible. Modern ventilation systems will be used to control air flow, and stations will be air-cooled.

Funding for the project is now in a critical phase. Due to the recession that began in 2008, it is expected that many aspects of federal funding will be cut. On November 7, 2011, however, the chairman of the House Transportation Committee, Representative John Mica, stated that funding for the Second Avenue Subway project would not be cut. This is critical, as the federal government is scheduled to provide $309 million in 2012 and 2013 to complete the federal commitment of $1.3 billion to the project. Even more encouraging was his statement that funding for Phase 2 would be included in the next six-year authorization for transportation infrastructure expenditures (2).

17.5.2 East Side Access Project

The plan to provide East Side access to the Long Island Rail Road has been in place for many years. Initially, a new East-Side terminal was to be built for the LIRR. Later plans placed the terminal beneath Grand Central. The 63rd Street Tunnel, completed in 1972, has a second level built to service the LIRR under the East River.

The current project involves connecting tunnels both in Queens and Manhattan to complete the route between Jamaica and Grand Central. In Queens, the connection will add a new station in Sunnyside at Queens Blvd. and Skillman Ave. The service will accommodate 24 trains per hour during peak hours, and is expected to serve an average of 162,000 passenger trips each weekday.

The station for the LIRR will be located far under the existing Grand Central Station. The LIRR will have eight tracks on two levels. High-rise escalators will be needed to transport passengers from the LIRR platforms to the rest of the terminal.

After approving a bond issue to provide state funding, the federal government committed $2.6 billion to the project in December on 2006. A construction contract for the one-mile tunnel in Manhattan was awarded on July 13, 2006 to Dragados–Judlau, a joint American–Spanish Co. One TBM was launched from the 63rd Street Tunnel in September of 2007 and reached Grand Central in July 2008. A second TBM began the parallel tunnel in July 2008. On the Queens side, construction involves soft-bore tunneling, track and switch facilities, and the new station. The Manhattan borings have been completed as of November 2011.

The original target date for initiation of service was 2013; a number of delays have now pushed it back to 2018, most involving parallel projects taking place in the LIRR East River tunnels. However, MTA recently (as of March 2012) accelerated construction, with an objective of returning to a 2013 completion. Direct East Side LIRR access is expected to relieve the E-train between Penn Station and Midtown East. On the other hand, it will throw additional passengers onto the Lexington Avenue line, already pushed to its limits. The Second Avenue Subway, expected to be completed in December 2016 (Phase I), becomes a critical link in serving the additional East Side subway demand created by LIRR service into Grand Central.

The East Side Access Project is expected to cost $7.3 billion. The Federal Transportation Authority (FTA) has pledged $2.6 billion for the project. In September 2011, MTA received $246 million from FTA for fiscal years 2011 and 2012, which will enable work to continue on the project (3).

17.5.3 Extension of the No. 7 Line

The plan to extend the No. 7 train (Flushing Line) from its present terminus at Times Square to a new station at 34th Street and Eleventh Avenue is intended to provide transit service to the expected redevelopment of the Hudson Yards. The Hudson Yards consist of storage tracks to the immediate west of Penn Station. It has long been recognized that the air rights over these yards comprises one of the most attractive sites for development in Manhattan. The development area above and around the Hudson Yards is bounded by West 28th Street on the south, West 43rd Street on the north, 8th Avenue on the east and Hudson River Park on the west.

Fig. 17.5 Aerial view of the West Side Yards

Initially, the redevelopment plan was centered on a West Side Stadium to house the New York Jets football team. Earlier plans had considered building a new Yankee Stadium on the site, before that project was shifted to the Bronx, right next to the "old" Yankee Stadium. The proposed stadium was also the center of New York City's attempts to be the site of the 2012 Summer Olympics.

With much public opposition, and the city's failed bid for the 2012 Olympics no longer an issue, the stadium plan was shelved for a more mixed approach to development of the valuable real estate. Because any development of the West Side Yards, which might now include expansion of the Javits Convention Center, would need improved public transit service, the No.7 extension project was kept active.

The $2 billon extension is to be financed using Tax Increment Financing (TIF). Bonds are sold to be repaid with increased property tax revenues from future developments in areas served by the extension.

In June of 2008, construction began along 11th Avenue in Manhattan. In February 2009, the first of two TBMs was lowered into its launch shaft at 25th Street and 11th Avenue; a second TBM soon followed. The boring machines dug two parallel 7,100-ft tunnels along 11th Avenue to the current Times Square terminus of the No. 7 train. The first broke through the Times Square station cavern wall on December 21, 2009. The second completed its work on July 15, 2010.

The original plan for the extension included a second station at Tenth Avenue and 41st Street. The fate of this station is still being discussed, although it was not funded in the initial construction contracts. It could be fully built, or a shell could be built for later completion.

The extension will have service tracks that extend to 23rd Street and 11th Avenue. This opens the possibility of a future extension further downtown, which would provide service to the Chelsea area and the Chelsea Piers, a popular recreation facility. In November of 2011, the thought of extending the No. 7 train across the Hudson River into New Jersey was also under discussion.

Figure 17.6 shows a map of the planned extension, which is currently scheduled for completion in 2013.

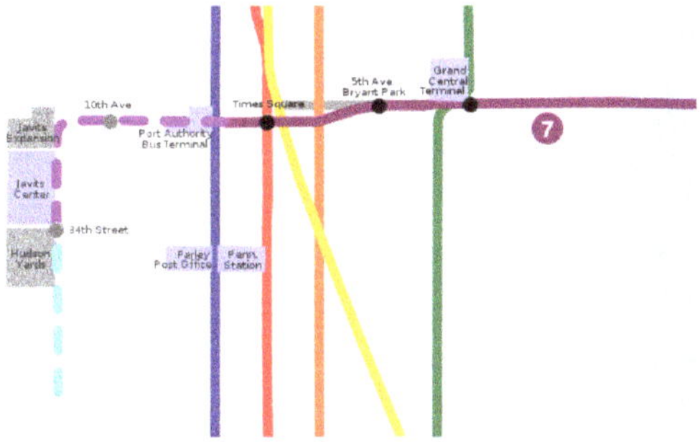

Fig. 17.6 Map of the No. 7 extension

17.5.4 Fulton Street Transit Center

The Fulton Street Transit Center is one of the more interesting current projects being undertaken. It involves no new subway services. It is essentially an effort to provide a series of new station platforms, connecting corridors, and new entryways for 10 subway lines in the vicinity of Fulton Street, PATH service, and the World Trade Center station in lower Manhattan.

The 10 subway lines involved are:

Division A: 2, 3, 4, and 5
Division B: A, C, E, J, Z, and R

The project was initiated in 2005, and some elements have been completed and are in current use. These include new station platforms for the Lexington Avenue and 7th Avenue lines, and several connecting corridors.

Funding was disrupted when the MTA found that it could not cover cost overruns, but the project was restarted in 2009 when it received stimulus funds from the American Recovery and Reinvestment Act of 2009. The project is now expected to be completed in 2014, although some elements have been simplified or eliminated.

The centerpiece of the original plan was a dramatic station entrance building to be located on the corner of Broadway and Fulton Street. The plans for this building have been revised almost continuously during the course of the project. On January 28, 2008, MTA announced that the revised plan would not likely include any substantial above-ground structure. In March of 2008, MTA reversed itself, and indicated that an above-ground structure would be built at the site, but held open that idea that it could lease the site to a private developer, with the station entrance being incorporated into a larger building.

When stimulus monies were received in January of 2009, the MTA released the following statement, which remains the current plan:

> *"Final details are being worked out for the above ground building. The 115-year old Corbin Building, at the corner of Broadway and John Street, will be restored and incorporated into the transit center entrance design The transit center will be a focal point with a vibrant design and a visible portal to downtown and the transit system below."* (The Future Beneath Us: Fulton Street Transit Center, http//:www.transitmuseumeducation.org 2009.)

17.6 Into the Future Again

As the old Who song goes, we all hope "we don't get fooled again." Construction is underway for Phase I of the new 2nd Avenue Subway. Phase II is viewed as a safe bet, as it will use previously-constructed tunnels for much of its length, but it is not yet funded. Phases III and IV are not at all certain.

There are real disagreements over what impact a completed 2nd Avenue Subway would have on New York. Another massive transit project, LIRR access to Grand Central, virtually depends on it. There is no more capacity on the Lexington Avenue Subway to handle the additional commuters who will arrive on the East Side and need to travel downtown. Some real estate experts expect the line to open up areas east of 2nd Avenue to luxury apartment development, and new commercial development all along 2nd Avenue.

It is doubtful that expansion of the subway system into regions of the outer boroughs will happen anytime soon. Residents of those areas have become very protective of the lower-density environments that currently flourish. On the other hand, Queens presents some very unique opportunities, where city transit lines could share, or operate above, existing LIRR rights-of-way.

The LIRR mainline out of Jamaica could accommodate an elevated above it serving parts of southern Queens, essentially completing the original plan for the Archer Street Subway.

The Port Washington branch could accommodate an extension of the Flushing Line, either by building an elevated over the existing right-of-way, or by taking over the branch and using the LIRR tracks directly. This is an attractive concept, as the Massapequa Branch has no at-grade crossings within the city. The issue of how to serve passengers from Nassau County now served by the line would have to be addressed. A shuttle could be used, or the subway system could provide direct service into Nassau.

The Second Avenue Subway will be built in small pieces. The MTA and the city realize that financing a more aggressive expansion of the subway system would be virtually impossible. Unless urban transit systems rise significantly as a national priority, the federal support for major expansions or new systems will not be easily attained. We have missed, perhaps, a great opportunity with federal stimulus funds. While some support has been obtained to save the Fulton Street Center project, there was no concerted effort to obtain larger amounts to accelerate the 2nd Avenue Subway or other major projects.

New York is a tough town to get something done in. Westway was continuously subjected to lawsuits until the deadlines on federal funding ran out. Despite numerous attempts to develop them, the air rights over the West Side Yards remain undeveloped, and the site itself is an eyesore. Yet, the area represents one of the most valuable undeveloped land parcels in the city, and indeed in the world.

We've come a long way since rail transit debuted in New York on Harvey's cranky cable-powered elevated on Ninth Avenue. In other cities, massive new systems have been built and/or expanded, while New Yorkers continue to use a system designed for their grandparents and great-grandparents. The fact that it still works so well is a tribute to the many far-sighted individuals who contributed to its birth and its development, and to those responsible for keeping it running today.

Unfortunately (or perhaps fortunately), we don't have the opportunity to go back and change history. Who knows what NYC would have looked like if the 2nd and 3rd stages of the IND were built? How would Staten Island have developed had it been connected to the rest of NYC through the subway system?

While NYC struggles to accommodate the needs of its nearly 8,000,000 residents, and the hundreds of thousands of workers who live outside the city, but work within it, other cities, unburdened with a city structure that literally cannot be changed, have moved forward with modern transit development.

Heavy rail systems have been built in San Francisco and Washington D.C. Both incorporate elevated sections, but over dedicated rights of way, never over city streets. Numerous cities have initiated light rail service, while Philadelphia and Boston continue to operate their more historic systems. Could light rail have solved the East Side transit problems years ago? Trolley buses are used in several major cities, providing the ultimate in clean-air bus systems.

The elevateds of Manhattan incorporated all of the worst elements of elevated systems that we can imagine: they blocked light, spewed metal dust, made noise, and rattled within a few feet of people's bedrooms. The current elevated portion of the Flushing Line over Queens Blvd., however, avoids most of these problems. Built on a solid reinforced concrete structure, these trains are quieter, and do not cause a significant amount of dust. Built with four traffic lanes on either side of the structure, light is not blocked from sidewalks and buildings, and no one is close enough to anyone's bedroom window to cause a problem. This concept could have been used elsewhere, particularly in the outer boroughs where many lines were built before any appreciable development existed.

History has left New York with perhaps the most extensive and efficient transit systems in the world. It is not perfect, but it does not have to be to serve its purpose. The City of New York, particularly Manhattan, could not exist in its current form without it. The first transit services moved the population out of lower Manhattan to the "outskirts" of the city beyond 14th Street. Today, workers commute from the far reaches of New Jersey, Long Island, and the outer boroughs with relative ease. Our transit system helps to define our city, and indeed, ourselves. In the end, we, and our city, are so much better for it.

General References

1. Ronan, T.: Subway Shelters to Cost $104,000,000 Proposed for City, December 29. New York Times, New York (1950)
2. Ferming, J.: 2nd Avenue Subway won't lose fed funding, November 8. New York Post, New York (2011)
3. Pentchoukov, I.: East Side Access Project Receives $246 Million, September 21. The Epoch Times, New York (2011) (available on-line)

References

1. Feinman, M.: History of the Independent Subway, online manuscript, `http://www.nycsubway.org`
2. Sargent, G.: The Line That Time Forgot, online at \ `http://www.nymag.com/nymetro/news/features/n_10109/`
3. MTA Signs Second Avenue Subway Contract. The New York Sun (March 14, 2007)

Figure Sources

17.1: Base map originally published by the New York City Board of Transportation, 1929. Highlights added by authors.
17.2: *Originally published by* New York City Board of Transportation, 1939.
17.3: *Courtesy of* Joe Korman, *Originally published in* "A Program for Action," Metropolitan Transportation Administration, New York, NY, 1968.
17.4: *Courtesy of* Wikimedia Commons.
17.5 *Courtesy of* Wikimedia Commons, photograph by Arturoramos, used under the Creative Commons Attribution – Share Alike 3.0 unported license..
17.6: *Courtesy of* Wikimedia Commons, photograph by Chris Ruvolo.

www.ingramcontent.com/pod-product-compliance
Lightning Source LLC
Chambersburg PA
CBHW061737180526
45293CB00003B/139
9783642304835